Pollution and Biomonitoring *of* Indian Rivers

Pollution and Biomonitoring *of* Indian Rivers

Dr. R.K. Trivedy

ABD PUBLISHERS
Jaipur ■ New Delhi

ABD PUBLISHERS

Regd. Off. "Bony Residency", Gate No. 2
Opp. Tilak Public School, Vishweswariya Nagar,
Gopalpura Road, Jaipur - 302018, India
Ph: 0141 - 2761280, Telefax: 2761381
Email - abdpublishersjpr@gmail.com

Delhi Office: 102, 1st Floor, Satyam House,
4327/3, Ansari Road, Darya Ganj,
New Delhi - 110002, India
Ph: 011-45652440

visit us: www.oxfordbookcompany.com

First published in India, 2019

ISBN: 9788183766104

Price : ₹ 1995/-

Typeset by Shivangi Computer, Jaipur

Printed and bound in India by Replika Press Pvt. Ltd.

Preface

There has been a steady deterioration in the quality of water of Indian Rivers over several decades.

India's fourteen major, 55 minor and several hundred small rivers receive millions of liters of sewage, industrial and agricultural wastes. Most of these rivers have been rendered to the level of sewage flowing drains. There are serious water quality problems in the towns and villages using this waters. Water borne diseases are rampant, fisheries is on decline, and even cattle are not spared from the onslaught of Pollution.

It is very sad and unfortunate that despite promulgation of water (Prevention and Control of Pollution Act) 1974 and a dozens of laws subsequent to that there has not been any tagible improvement in water quality of our rivers. In fact the situation has gone only from bad to worse. Even a highly cost intensive Ganga Water Purification Programme could not improve the River Ganga. Ambitions projects have been drawn for several other Rivers under National River Conservation Directorate, Government of India. It is hoped future River purification programme shall bring the desirved results.

Present volume is a collection of significant papers in this field which present a very grim picture of India's precious water resources. It is hoped that the information presented here shall awaken the authorities to rectify the situation as quickly as possible.

I am indebted to authors who have so kindly agreed to my invitation to contribute to this Volume.

R.K. Trivedy

Contents

List of Contributors

1. **A. K. Sinha,** Environmental Research Centre, Firoz Gandhi College, Rae Bareli - 229001 U.P
2. **V. P. Singh,** Environmental Research Centre, Firoz Gandhi College, Rai Bareli - 229001 U.P
3. **(Mrs.) Kiran Srivastava,** Environmental Research Centre, Firoz Gandhi College, Rai Bareli-229001 U.P
4. **S. A. Abbasi,** Centre for Pollution Control and Biowaste Energy, Pondicherry University, Pondicherry - 605014.
5. **Naseema Abbasi,** Centre for Pollution Control and Biowaste Energy, Pondicherry University, Pondicherry - 605014.
6. **R. Soni,** Centre for Pollution Control and Biowaste Energy, Pondicherry University, Pondicherry - 605014.
7. **K. Madhavan,** Centre for Water Resource Development and Management, Kozhikode - 673571.
8. **Ravi Kumar Gupta,** Centre for Mining Environment, Indian School of Mines, Dhanbad - 826004, Bihar (India).
9. **Gurudeep Singh,** Centre for Mining Environment, Indian School of Mines, Dhanbad - 826004, Bihar (India).
10. **Binay K. Gupta,** Bihar State Pollution Control Board, Dhanbad Regional Office, Dhanbad, Bihar (India).
11. **B. C. Rana,** Department of Biosciences, Sardar Patel University, Vallabh Vidyanagar - 388120, Gujrat (India).
12. **J. Jameson,** Department of Biosciences, Sardar Patel University, Vallabh Vidyanagar - 388120, Gujrat (India).
13. **M. K. Ghose,** Assistant Professor, CME, ISM, Dhanbad - 826004.
14. **S. Roy,** Bihar State Pollution Control Board, Dhanbad - 826004, Bihar (India).
15. **Ram Karan Singh,** Lecturer, Department of Civil Engineering, BITS, Pilani - 333031.
16. **Dr. Shiv Prasad,** Lecturer, Department of Mathematics, BITS, Pilani - 333031.
17. **N. S. J. Naidu,** Undergraduats Students, BITS, Pilani - 333031.
18. **Manoj E.,** Department of Zoology, M. S. University of Baroda, Vadodara, Gujarat (India).
19. **G. Ragothaman,** Department of Aquatic Biology, South Gujrat University, Surat - 395007, Gujarat (India).

20. **B. N. Pandey,** P. G. Department of Zoology, Purnea - 854301 (Bihar), India.

21. **A. K. Gupta,** P. G. Department of Zoology, Purnea - 854301 (Bihar), India.

22. **A. K. Mishra,** P. G. Department of Zoology, Purnea - 854301 (Bihar), India.

23. **P. K. L. Das,** P. G. Department of Zoology, Purnea - 854301 (Bihar), India.

24. **A. K. Jha,** P. G. Department of Zoology, Purnea - 854301 (Bihar), India.

25. **D. B. Ghatak,** Fisheries Laboratory, Department of Zoology, Kalyani University, Kalyani - 741235 (W. B.), India.

26. **S. K. Konar,** Fisheries Laboratory, Department of Zoology, Kalyani University, Kalyani - 741235 (W. B.), India.

27. **S. L. Garg,** Department of Chemistry, Government Girls Post-graduate College, Moti Tabela. Indore (M. P.) India.

28. **Miss Bum Preeti,** Department of Chemistry, Government Girls Post-graduate College, Moti Tabela, Indore (M. P.) India.

29. **Miss Bharadia Rama,** Department of Chemistry, Government Girls Post-graduate College, Moti Tabela. Indore (M. P.) India.

30. **Miss Rashmi Sharma,** Department of Chemistry, Government Girls Post-graduate College, Moti Tabela, Indore (M. P.) India.

31. **Miss Minakshi Jindal,** Department of Chemistry, Government Girls Post-graduate College, Moti Tabela, Indore (M. P.) India.

32. **Miss Kumudini Joshi,** Department of Chemistry, Government Girls Post-graduate College, Moti Tabela, Indore (M. P.) India.

33. **Miss Savita Vyas,** Department of Chemistry, Government Girls Post-graduate College, Moti Tabela, Indore (M. P.) India.

34. **Safal Kumar Mishra,** Department of Zoology, T. P. V. College, Narkatiaganj - 845455. W. Champaran, Bihar, India.

35. **B. K. Sahu,** School of Studies in Zoology, Jiwaji University, Gwalior - 474011, M. P. (India).

36. **R. J. Rao,** School of Studies in Zoology, Jiwaji University, Gwalior - 474011, M. P. (India).

37. **S. K. Behera,** School of Studies in Zoology, Jiwaji University, Gwalior - 474011, M. P. (India).

38. **R. K. Pandit,** School of Studies in Zoology, Jiwaji University, Gwalior - 474011, M. P. (India).

39. **Nitish Priyadarshi,** Department of Geology, Ranchi University, Ranchi - 834008, Bihar (India).

40. **H. Kaur,** Department of Zoology, Punjab University, Patiala - 147002, Punjab (India).

41. **S. S. Dhillon,** Department of Zoology, Punjab University, Patiala - 147002, Punjab (India).

42. **G. Mander,** Department of Zoology, Punjab University, Patiala - 147002, Punjab (India).

43. **K. S. Bath,** Department of Zoology, Punjab University, Patiala - 147002, Punjab (India).

44. **M. Gadhia,** Department of Aquatic Biology, South Gujarat University, Surat - 395007, Gujarat (India).

45. **P. B. Patel,** Department of Biosciences, South Gujarat University, Surat - 395007. Gujarat (India).

46. **S. Shaikh,** Department of Biosciences, South Gujarat University, Surat - 395007. Gujarat (India).

47. **P. K. Gadhia,** Department of Biosciences, South Gujarat University, Surat - 395007. Gujarat (India).

48. **G. Kamalak Kannan,** Institute of Environmental Science, Mahatma Gandhi Gramodaya Vishwavidyalaya, Chitrakut - 485331, District - Satna, M. P. (India).

49. **S. Kaushik,** School of Studies in Zoology, Jiwaji University, Gwalior - 474011, M. P. (India).

50. **D. N. Saxsena,** School of Studies in Zoology, Jiwaji University, Gwalior - 474011. M. P. (India).

51. **S. S. S. Verma,** School of Studies in Zoology, Jiwaji University, Gwalior - 474011, M. P. (India).

52. **V. Sankaran,** Department of Botany, N. G. M. College, Pollachi - 642001 T. N. (India).

53. **D. Tiwari,** Department of Botany, A. N. D. N. N. Mahila Mahavidyalaya, Kanpur, U. P. (India).

54. **P. Kaur,** Department of Botany, A. N. D. N. N. Mahila Mahavidyalaya, Kanpur, U. P. (India).

55. **S. Singh,** Department of Botany, A. N. D. N. N. Mahila Mahavidyalaya, Kanpur, U. P. (India).

56. **M. Chattopadhyay (nee Ray),** Department of Chemistry, Kalyani University, Kalyani - 741235, W. B.

57. **S. C. Lahiri,** Department of Chemistry, Kalyani University, Kalyani - 741235, W. B.

58. **B. B. Sandwar,** Samastipur College, Samastipur (Bihar), India.

59. **J. Prasad,** P. G. Department of Chemistry, Samastipur College, Samastipur (Bihar), India.

60. **Jagdish Kumar,** Shriram Institute for Industrial Research, 19, University Road, Delhi - 110007.

61. **D. P. Singh,** Shriram Institute for Industrial Research, 19, University Road, Delhi - 110007.

62. **R. K. Banerji,** Shriram Institute for Industrial Research, 19, University Road, Delhi - 110007.

63. **C. L. Jain,** Shriram Institute for Industrial Research, 19, University Road, Delhi - 110007.

64. **S. K. Raikar** (Ph. D.), Professor and Head, Department of Applied Chemistry, Jawaharlal Nehru National College of Engineering, Shimoga - 577201, Karnataka.

65. **Sri Ravindranath Sabhahita,** Department of Civil Engineering, Jawaharlal Nehru National College of Engineering, Shimoga - 577201, Karnataka.

66. **Arvind Kumar,** Environmental Biology Research Laboratory, Post-graduate, Department of Zoology, S. K. University, Dumka - 814101, Bihar (India).

67. **Rakesh Chandra,** Department of Chemistry, Bareilly College, Bareilly - 243001, U. P. (India).

Y. Bahadur, Department of Chemistry, Bareilly College, Bareilly - 243001, U. P. (India).

68. **B. K. Sharma,** Department of Botany, Barelly College, Bareilly - 243001, U. P. (India).

69. **R. K. Trivedy,** Department of Pollution Studies, Y. C. College of Science, Vidyanagar, Karad - 415124, (M. S.), India.

70. **S. S. Nakate,** Sabaero Organics Ltd., Sarigam, Gujarat.

71. **Rakesh Chandra,** Environmental Officer, Tata Chemicals Ltd., Babrala - 202521, District Badaun, (U. P.), India.

72. **K. Sankaran Unni,** Hydrobiology Laboratory, Botany Deptt. Govt. P. G. College, Chhindwara - 480001 (M. P.), India.

73. **Sanjay Pawar,** Hydrobiology Laboratory, Botany Deptt. Govt. P. G. College, Chhindwara - 480001 (M. P.), India.

74. **Surendra Bhade,** Hydrobiology Laboratory, Botany Deptt. Govt. P. G. College, Chhindwara - 480001 (M. P.), India.

75. **S. D. Khatavkar,** Deptt. of Pollution Studies, Y. C. College of Science, Karad - 415110 (India).

76. **B. L. Arjugada,** NFES Labs. Nehru Nagar, Karad - 415110 (India).

1

Physico-chemical studies on River Ganga and its tributaries in Uttar Pradesh - the present status

Uttar Pradesh with an area of 2,94, 413 Sq. Km has the proud privilege of being the most populous state of the country. Blessed with abundance of natural resources like the rivers and the forests mighty Himalayas, the state has been the abode of Saints and Rishis. The rivers Ganga and Yamuna forming the Ganga basin are regarded the life line for the people of not only this state but for whole of the country. The state can also boast of having the most fertile land in the indogangetic plains. The state is also famous for the 'TAJ MAHAL' and many pilgrim centres like Badrinath, Kedarnath, Gangotri, Varanasi, Mathura and Hardwar. Modern Allahabad, i.e.Tirthraj Prayag is situated on the confluence of Ganga, Yamuna and invisible Saraswati. These are some of the reasons for which Uttar Pradesh is one of the outstanding states of the country.

The origin of the rivers in the country, especially that of Ganga is associated with many fascinating stories. According to one such belief the celestial Ganga was brought to the earth due to the tireless efforts of the king Bhagirath for the sole purpose of salvation of his ancestors. One of the synonyms of Ganga is, therefor, Bhagirathi. Originating from gangotri, the river covers a distance of 2525 Km through the states of Uttar Pradesh, Bihar and West Bengal before its confluence with the Bay of Bengal. The river together, with its tributaries is the largest and a very important river basin of the country. According to Khoshoo (1986) the river system is interwoven with India's history, culture, religion and philosophy and some of the towns / cities on its bank date back to the dawn of the history. Ganga is regarded as a symbol of purity and sanctity and it is believed that only by reciting its name one gets salvation or 'MOKSHA'.

The important tributares of river Ganga i.e. Yamuna, Gomati and Sai have also been famous since ages, They have also played an important role in the development of human civilisation and settlements. Due to easy availability of water the industrial set ups sprang up on the banks of the rivers and thir tributaries. Stupendous increase in population, unplanned springing up of industries and urbanisation have put tremendous pressure on the rivers and tributaries, Unrestricted discharge of industrial effluents / wastes into the rivers coupled with dumping of half-burnt / dead human bodies and carcasses and the practice of cremation on the banks of the rivers have turned them into a polluted mass of water. The river water has become unfit for human consumption.

Scientific studies on river Ganga are of comparatively recent origin and were initiated under the "Man and Biosphere" Programme in the sixtees with emphasis on pollution and limnology. The literature on Physico -chemical characteristics of river Ganga were reviewed by Sinha et . al. (1992). The present paper deals with a short review of literature on Physico chemical studies of river Ganga and its tributaries in the state of Uttar Pardesh.

❑ RIVER GANGA

The river covers a distance of approximately 1550 Km in the state of Uttar Pradesh passing throught the districts of Garhwal, Shaharanpur, Bulandshahar, Aligarh, Farukhabad, Kanpur, Unnao, Fatehpur, Rae Bareli, Pratapgarh, Allahabad, Mirzapur, Varanasi, Ghazipur and Ballia. Incidentally it was in Uttar Pradesh at Varanasi and Kanpur - the former a famous pilgrim centre and the later being the industrial capital of the state of Uttar Pradesh that studies on physico-chemical characteristics of the river were initiated.

Lakshminarayana (1965) published a series of papers reporting the results of studies carried out at Varanasi during the period between March, 1957 and March, 1958. It was observed by him that the values of the most of the parameters decreased during rainy season while no marked variation was observed during winters and summers.

In the same year Chakraborty et al... (1965) from Kanpur reported the water quality of Ganga at J. K. Rayon's water intake point and at Golaghat and Bhairoghat pumping stations situated at the upstream of the river. It was concluded that the water quality gradually deteriorated as it passes from Bhairoghat pumping station to the J. K. Rayon water intake point in summers because in this stretch the river received waste waters from a number of sewage drains. A year later Saxena et al. (1966) who also studied the river at Kanpur concluded that the tanneries significantly increased the pollution load of river as they discharge huge amounts of effluents containing organic wastes and heavy metals. It was further reported that forty five tanneries, ten textile mills and several other industrial units discharged 37.15 million gallon per dayof waste water generating BOD load of approximately 61630 Kg./ day. In another publication in the same year Ray and David (1966) noted that the 35 Km. stretch of river Ganga at Kanpur was grossly polluted upto Jajmau and regeneration capacity of river water had failed to keep pace with the input of the pollutants.

After a decade of the above, another paper on the physico-chemical characteristics of river Ganga at bathing ghats and sewage outfalls in Varanasi city was published by Agarwal et. al. (1976 a.) They conducted studies on the distributon of total solids, Free and albuminoid ammonia, Nitrite and nitrate - Nitrogen, Chlorides, DO, and BOD of river water at bathing ghats and sewage discharge points and pointed out that inspite of the discharge of raw sewage the water quality at bathing ghats was fairly clean.

Subsequently Agarwal et. al. (1976 b.) studied the bacteriological population of the river water and concluded that addition of untreated waste and sewage was responsible for the presence of pathogenic organisms posing a threat to the residents of the city. Pandey and Pandey (1980) who analysed the river water at Kanpur for fifteen physico-chemical parameters at nine sampling sites concluded that river received huge discharge of effluents which had degraded the water quality severely during winter and summer seasons beyond the capacity of assimilation of the river.

The decade beginning with 1980 coincided with the spurt in studies on physico-chemical characteristics of the river at several places. The first such paper was authored by Dutt and Chaudhari (1980) who reported that the river water and sediments ware also shown to contain detectable amounts of mercury at Kanpur though the concentrations were not beyond the permissible limits. One year later Handa et al.. (1981) reported that the water of the right bank of Ganga at Kanpur near old bridge had high concentrations of Fe and Mn as compared to that of the left bank.

At the 1981 session of Indian Science Congress at Varanasi, scientists expressed concern at the growing pollution in the river Ganga in presence of the then Prime Minister revered Mrs. Indira Gandhi who inaugurated the session. At her instance, Dr. M. S. Swaminathan, the then member, Planning Commission asked the Central Board for Prevention and Control of Water Pollution, New Delhi to conduct studies on the state of the river Ganga. In collaboration with the State Pollution Control Boards of Uttar Pradesh, Bihar and Bengal and the centre for study of Man and Enviroment Calcutta, studies were conducted on the 'Sources' of pollution including all human activities, land use pattern and water quality of the river at selected sites during 1981-82 and a report entitled "Basin, sub-basin inventory of water pollution in the Ganga basin part-II" was published in 1984. According to this report sewage of 27 class I cities and towns and effluents from 137 major industries were the main source of the pollution of the river. In addition cremation of dead human bodies and dumping of carcasses aggrevated the pollution of the river. Later it was decided to entrust the universities and institutions situated on the course of Ganga to conduct integrated studies of the river. For this purpose, the upper Ganga, covering the river stretch from Gangotri to Narora, the middle Ganga, extending from Narora to Buxar, and the lower Ganga, from Buxar to Hooghaly. The newly created Ministry of Enviroment and Forests, Govt Of India was entrusted the task of funding the projects. Each university was asked to conduct the interdisciplinary studies on a definite stretch of the river. Eminent scientist Dr. C. R. Krishnamurthi was requested to head the scientific aspects as chairman. The universities of Uttar Pradesh took active part in this project and a list of institutions alongwith the names of Principal investigators with address have been summarised in Table - I

It was Chandra (1981) who conducted studies on the pollution status of river Ganga at Allahabad, pointed out that industries manufacturing nitrogenous fertilizers have significant role in polluting the river water. Bhargava (1981) in the same year noted that BOD and DO values in the river water were 6.0 - 6.8 mg./L and 6.5 mg / L respectively due to discharge of waste water in Ganga at Allahabad.

Table I

Name of UP Universities Sanctioned multi disciplinary research projects on River Ganga by Ministry of Environment and Forests, G.O.I.

Sr. No.	Title of the Project	Stretch of the river studied	Name of Principal Investigators with address
1.	Pollution study of the upper Ganga and its tributaries	Gangotri to Rishikesh	Prof. H.R. Singh Dept. of Zoology Garhwal University Srinagar, Garhwal.
2.	Integreated study of the river Ganga	Rishikesh to Garhmukteshwar	Dr. V. Shankar Dept of Botany Gurukul Kangri Vishwa Vidyalaya, Haridwar
3.	Pollution modelling of the upper Ganga basin	Garhmukteshwar to Narora	Dr. R. P. Mathur Dept. of Civil Engg. University of Roorkee, Roorkee.

4.	Integrated studies of the Ganga Ecosystem between Narora and Kannauj a. General Water qualtiy measurement b. Taxonomic studies of bacteria c. Man-river interaction from land use : Geographic points of view d. Pesticides and radio-activity in the river water	Narora to Kannauj	Prof. A. M. Siddiqui Dept. of Biochemistry, Aligarh Muslim University, Aligarh.
5.	Physio-chemical studies of the Ganga water from Kannauj to Shuklaganj	Kannauj to Shuklaganj	Dr. U.K. Pande Dept. of Zoology D.B.S.College, Kanpur (S.S.M.University, Kanpur)
6.	Biological studies of the Ganga Ecosystem between Kannauj and Shuklaganj	Kannauj to Shuklaganj	Dr. A.C. Shukla Dept. of Botany Christ Church College, Kanpur (S.S.M. University, Kanpur.)
7.	a. Comprehensive study of the Ganga and its dependents b. Environmental awareness Programme	Shuklaganj to Phaphamau	Dr. A. K. Sinha Dept. of Botany Feroze Gandhi College RaeBareli. (S.S.M. University, Kanpur)
8.	Plant Population Community Sturcture and function of riparian and aquatic macrophytes in relation to pollution of the river Ganga.	Mirzapur to Ballia	Dr. R. S. Ambasht CAS in Botany Banaras Hindu University, Varanasi
9.	Physico-chemical and biological characterization of the river Ganga in relation to pollution.	Mirzapur to Ballia	Dr. B. D. Tripathi CAS in Botany Banaras Hindu University, Varanasi

10.	Physico-chemical and Physi-ological assessment of heavy metal pollution in the river Ganga.	Mirzapur to Ballia	Dr. L. C. Rai CAS in Botany Banaras Hindu University, Varanasi
11.	Pollution of the river Ganga with heavy toxic metals	Mirzapur to Ballia	Dr. D. C. Rupainwar Instt. of Technology Banaras Hindu University, Varanasi
12.	Co-ordinated study of the river Ganga : Civil Engineering aspects	Mirzapur to Ballia	Dr. U. K. Chaudhary Dept. of Civil Engg. Banaras Hindu University, Varanasi
13.	Co-ordinated study of the river Ganga : Study of the river water pollution at Varanasi and Prevalence of gastrointestinal morbidities	Mirzapur to Ballia	Dr. T. C. Tiwari and Dr. P. C. Sen Dept. of Microbiology Banaras Hindu University, Varanasi
14.	Hydrogeological study regarding the pollution of the river Ganga (Hydrogeological aspects)	Mirzapur to Ballia	Dr. G. C. Chowdhari Dept. of Geology Banaras Hindu University, Varanasi
15.	A study of the submerged and aquatic plants of medicinal and antipollutant value in reference to the Ganga water pollution	Mirzapur to Ballia	Dr. R. H. Singh Dept of Pharmacology Banaras Hindu University, Varanasi
16.	Role of sediment in attenuation of trace metals in polluted aquatic environment of the river Ganga : Sedimentological aspects.	Mirzapur to Ballia	Dr. M. N. Mehrotra Dept. of Geology Banaras Hindu University, Varanasi
17.	Impact of heavy metal pollution in the river Ganga on fish	Mirzapur to Ballia	Dr. A. K. Mittal Dept. of Zoology Banaras Hindu University, Varanasi

18.	Water Pollution and endocrine imbalance with special reference to high thyroid physiology and reproduction in some important food fishes.	Mirzapur to Ballia	Dr. T. P. Singh Dept. of Zoology Banaras Hindu University, Varanasi
19.	A study of the river Ganga with reference to environmental and socio-economic awareness (extension activities)	Mirzapur to Ballia	Dr. Krishna Bahadur Dept. of Social Sciences Banaras Hindu University, Varanasi
20.	Co-ordinated study of the river Ganga ; (Agricultural aspects) Mirzapur to Ballia	Mirzapur to Ballia	Dr. R. C. Tiwari Dept. of Soil Chemistry Instt. of Ag.Sciences Banaras Hindu University, Varanasi.

Upadhyay et al.. (1982) studied the impact of mass bathing on the water quality of river Ganga at Sangam (Allahabad) on the occassion of Ardh Kumbha in 1982 and reported that most of the physico-chemical parameters exhibited high values. Total dissolved solids and total suspended solids were remarkably high and water was found to be unfit for human use. Their observations were in consonance with those of Sinha et al.. (1986), who assessed water quality of Ganga at Dalmau (Rae Bareli) on the occasssion of Kartik Purnima, an important mass bathing day of the district in 1985. Their findings also clearly revealed that the intensity of pollution was greater on the main bathing day though there was no discharge of public sewers or industrial effluents at and around the sampling sites. It has been proved beyond doubt that mass bathing on the occassion of Hindu festivals deteriorates the water quality of Ganga (Misra & Singh, 1987) and increases the bacterial population alongwith the organic pollutants in the river water significantly (Singh et al.. 1988)

Chandra and Mathur (1983) observed that change in colour and quality of Ganga river water was mainly due to discharge of raw sewage and industrial wastes at Kanpur. Similar conclusions were drawn by Rambilas and Kayastha (1983) during their course of studies of river water at Varanasi.

Impact of sewage disposal on physico-chemical parameters of Ganga water at Varanasi was studied by Sikandar & Tripathi (1984) at six sampling sites viz.

Assighat, Janakighat, Harishchandraghat, Dasashwamedhghat, Manikarnikaghat and one control at the upstream of the river where it receives least amount of sewage and other wastes. At these sites except the control one, diverse human activities viz, mass bathing and cremating dead bodies are common. It was reported that all the physico chemical parameters exhibited high values and water was found to be highly polluted at all the sites except at control one due to discharge of municipal and industrial waste water. It was further observed by Pandey & Tripathi (1984) that river water at almost all the bathing ghats was much deteriorated and contained several impurities.

Studies conducted on snowfed river Bhagirathi by Sharma (1984) revealed that river water was fairly clean and showed lower values of turbidity and pH and high concentrations of dissolved oxygen.

Regular water quality monitoring of river Ganga was initiated by Central Pollution Control Board (Das Gupta, 1984) since 1979 at sixteen stations between Rishikesh and Diamond harbour, out of which Rishikesh, Kanpur, Dalmau, Allahabad and Varanasi are in Uttar Pradesh. They observed highest BOD in the river between Kanpur and Allahabad indicating that the river is seemingly incapable of assimilating the organic load discharged into the river.

Another notable event occurred in the late 1984 when the central Ganga Authority was created under the chairmanship of the then Prime Minister Rajiv Gandhi after it was felt that the problem of Ganga pollution is of such multifaceted dimensions that it can be tackled only by an appropriate authority. The central Ganga authority formulated a Ganga action plan which envisages diversion and treatment of waste waters/effluents, use of bio-gas released from treatment of effluents as a principal source of energy, use of treated effluents as an irrigant and as a source of Pisciculture and Aquaculture and improving the water quality of the river.

Ajmal et al. (1984) reported bioaccumulation of metals in submerged plants and fishes of Ganga ecosystem. Agarwal and Srivastava (1984) at about the same time reported that direct discharge of waste water by various drains into river Ganga and Yamuna at Allahabad, polluted rivers badly at and around the discharge point. Other studies which indicate that sewage, industrial wastes and run off water from agricultural fields are main pollutants of river Ganga include those of Bechu Lal (1984), Bhargava (1984a), Bhargava (1985), Joshi (1986), Pandey (1985), Sahai (1985) Sharma and Goel (1985) and Upadhyay et al. (1985)

It was Singh (1985) who after an extensive survey of the Bhagirathi ecosystem studied the ecology from Gomukh to Deoprayag. He concluded that river water had high velocity, high dissolved oxygen and low temperature.

Chandra and Mathur (1986) conducted studies on the effect of tannery wastes discharge on water quality of Ganga at Bithoorghat, Gangaghat and Bengalighat in Kanpur and noted that huge discharge of waste water had significantly polluted the river water at Bengalighat ultimately causing public health hazards.

Improper disposal of domestic sewage was identified as a major water pollutant at Rajghat, Varanasi by Prasad et al. (1986). They further pointed out that bathing and washing by human beings, burning of dead bodies and throwing the carcasses intensify the problem of river pollution.

Tiwari et al. (1986) established a relationship between COD and BOD for the river Ganga at Kanpur and proposed a linear relationship between BOD and COD using the data of Chattopadhyay et al. (1984).

In their annual technical reports submitted to the Ministry of Environment and Forests, Govt. of India, the principal investigators reported the observations on physico-chemical characterstics of the river in the stretch alloted to them, Singh (1986) reported his observations on the river between Gangotri and Rishikesh, Shankar (1986) between Rishikesh and Garh mukteshwar, Mathur (1986) between Badrinath and Narora and Sinha (1986) between Shuklaganj and Kalakankar. The annual report of 1986 of ministry of environment and forests, Govt. of India also concludes that about 75% of the pollution of the river Ganga is caused by municipal wastes and the maximum BOD value was observed at Kanpur.

The river water of Alaknanda, Which joins the main stream at Devprayag was also reported to be contaminated by several pathogenic bacteria at Srinagar, Garhwal a river side town of himalayan region in Uttar Pradesh (Nautiyal et al. 1986). They concluded that water is so heavily polluted that further negligence may take a heavy toll of human life.

Sangu et al. (1987) during their course of studies of river Ganga at Garh Mukteshwar concluded that water quality of the river on the right bank was much below the prescribed limits recommended for bathing and drinking as compared to the right one. Goel (1987) reported that river water was much polluted at Kanpur as 186 polluting industries generate 40 MLD waste water which is discharged ultimately into river Ganga at different points. Simultaneously the water quality status of the river at Kanpur was discussed by Bhargava (1987)

Physico-chemical parameters of Ganga water in and around Varanasi were analysed between Jan. 1986 and June, 1986 by Mathur et al. (1987) with reference to heavy metal pollution within a stretch of 12 Km. They concluded that river was polluted with Cu, Cd, Cr, Co, Pb, Ni, Mn and Zn particularly at the points of sewage outfall.

Observations of physico-chemical and microbiological studies on river Ganga at Dalmau, Rae Bareli were reported by Sinha et al. (1988). According to them higher values of some of the parameters might be due to the practice of cremation of dead bodies and dumping of half burnt human bodies and carcasses. These contentions were confirmed by Srivastava et al. (1988) who showed that disposal of dead bodies and carcassess into the river Ganga was the main cause of the pollution of the river at Dalmau.

Singh and Srivastava (1988) noted high value of BOD and COD in Ganga water during summers due to high input of organic pollutants and reduced rate of flow between Ballia and Buxar. Saikia et al. (1988) at about the same time carried out studies on the impact of human activities and heavy metal pollution of Ganga from Badrinath to Narora within a stretch of 480 Km. A total of seven heavy metals viz. Fe, Co, Mn, Ni, Co, Cd and Zn were analysed and it was pointed out that the concentration of metals in river water was below the permissible limits probably due to the minimum human interference. Contrary to this Srivastava et al. (1988) reported alarmingly high concentrations of Cr., Cu, Co, Ni and Zn in river water and sediments of Ganga river.

Singh et al. (1988) studied the pollution status of the river at Allahabad before Kumbh Mela. These studies were continued by them (Singh et al. 1989 a) during Maha Kumbh Mela in 1989 at Allahabad for the months of January and February, 1989. They noted maximum values of BOD and bacterial counts on Mauni Amavashya which was the main bathing day of Maha Kumbh. It was concluded that drinking of such water may cause serious ailments to human beings. Gautam et al. (1989) conducted studies on river Bhagirathi for certain physico-chemical paramenters and their seasonal variation and concluded that the middle stretch of river was under pollution stress.

At about the same time Shankar (1989) studied river Ganga between Hardwar and Rishikesh to identify major sources of pollution. It was reported that upstream water was degraded in respect of BOD, COD and DO as number of drains viz. Sarswati Nalla and IDPL Nalla at Rishikesh and BHEL Nalla at Hardwar pour down huge quantities of industrial and domestic wastes alongwith number of small sewage drains. An average of bacterial counts exhibited an increasing trend down stream from Rishikesh as river received progressively huge amounts of domestic discharge.

Physico-chemical and biological characteristics of river Ganga water and sediments between Shuklaganj (Unnao) and Jajmau (Kanpur) was studied by Sinha et al. (1989) at four sampling sites. An alalysis of eighteen important physico-chemical parameters revealed that river water at Jajmau pumping station exhibited high values of almost all the parameters than those of remaining other three sites. Concentrations of BOD, COD, total nitrogen and conductivity were observed to be higher at all the sites while those of Ca, Mg, hardness, chlorides and sulphates were lower than the permissible limits. Physico-chemical characteristics of sediments varied with the site. Seaonal variations in the parameters studied were also observed.

Srivastava et al. (1989) carried out studies on physico-chemical properties of river Ganga water at Buxar (Unnao) during May, 1986 and April, 1987. An analysis of a total of twenty three parameters clearely revealed that extent of pollution varied in different seasons. Usually all the parameters showed high values in summer and lower during mansoons except turbidity which was high in rainy season. Values of BOD, COD, DO and H_2S were recorded high than the tolerance limits while concentrations of Ca, Mg, SO^-_4 and hardness were observed to be lower than the permissible limits throughout the course of study. They suspected that probably regeneration capacity of Ganga water could not keep pace with the input of water pollutants at Kanpur situated at 50 Km. upstream from the study point. At about the same year high concentrations of NO_3^- and PO_3^{---} at Jajmau pumping station, Dalmau, Gokana, Chandarpur and Buxar were reported by Srivastava et al. (1989).

Water quality of river Ganga at Kalakankar (Pratapgarh) was assessed by Sinha et al. (1989). It was concluded that even at such a remote and undisturbed place like Kalakankar the river water was not safe for drinking and bathing. It was also noted that the river showed an alkaline trend throughout the course of study.

The river Ganga between Karemanikpur and Phaphamau (Allahabad) was studied by Sinha et al. (1989) in the same year for certain important physicochemical parameters and their spatial and temporal variations. Water samples were collected from five sampling sites downstream from Karemanikpur viz. Kare Manikpur, Haudeshwar Nath ghat, Shringverpur, Phaphamau bridge eastern side and western side of the bridge during Sept. 1987 and August 1988. It was observed that the level of dissolved oxygen gradually decreased from Karemanikpur to Shringverapur due to input of rural wastes while total dissolved solids exhibited reverse trend. Turbidity was recorded maximum at Phaphamau and most of the parameter exhibited monthly variations at all the sites. It was concluded that despite appreciable improvement in the water quality as compared to Shuklaganj, river water was still not fit for human consumption. Water quality index (WQI) of river Ganga between Shuklaganj (Unnao) and Kalakankar (Pratapgarh) were calculated by Sinha et. al (1989) using the procedure of Tiwari et al. (1986). On the basis of water quality indices calculated, Shuklaganj, Jajmau pumping station and Dalmau were severely polluted while the rest of the sites were moderately to severely polluted.

Shukla et al. (1989) conducted studies on the physico-chemical and biological characteristics of river Ganga between Mirzapur and Ballia within a stretch of 225 Km. from Sept. 1984 to Aug. 1986. Their observations clearly revealed that river water was tolerably good at the points where it enters the city and was highly polluted at midstream due to large input of sewage and industrial wastes. They reported that down-stream water at Varanasi was highly polluted as compared to other sites because 63 MLD sewage and industrial effluents were discharged into the river through Rajghat Nalla while water quality at Ballia was fairly clean. Total coliform count was reported to be high in midstream samples of the river at all the sampling sites.

Ecological studies and physico-chemical characteristics in relation to phytoplankton productivity were conducted by Shukla et al. (1989) at Varanasi from March 1986 to Feb. 1987 at four sampling sites viz. Samneghat, Assighat, Dasashwamedhghat and Rajghat and noted that river was polluted at all the sites except Samneghat with respect to BOD and COD.

Physico-chemical properties of Ganga water at Farukhabad were studied by Kudesia (1989)who found that the main pollutants of Ganga at Farukhabad were domestic and industrial wastes. Singh and Misra (1989) evaluated the status of major rivers in Uttar Pradesh on the basis of data available with U.P. Pollution Control Board pertaining to the physico-chemical parameters of river Ganga and concluded that Ganga water from Farukhabad to Ghazipur within a stretch of 600 Kms. was highly polluted and was unfit for drinking and bathing without proper treatment. Water quality was, however, found to be satisfactory at Rishikesh while it was most polluted at Kanpur.

Concentration of heavy metals in river Ganga water between Karemanikpur (Pratapgarh) and Phaphamau (Allahabad) was studied by Sinha et al. (1990). It was observed by them that the values of Ni, Cd, Cr, Pb, Mn and Fe in river water were higher at all the selected sites indicating that the heavy metals have flowed down to the sites of study.

Sediment characteristics of river Ganga at Shuklaganj (Unnao), Gegasso (Rae Bareli) and Phaphamau (Allahabad) were reported by Sinha et al. (1991). It was concluded by them that mass bathing caused measurable change in the water quality of the river. The pollution load increased during mass bathing making the water unfit for even bathing.

Observations on a comprehensive study of river Ganga in the stretch between Kalaknkar (Pratapgarh) and Phaphamau (Allahabad) were reported by Sinha (1991). Based on a two year study of the river he concluded that the river was moderately to heavily polluted in this stretch. The river water in this stretch also showed high concentrations of certain heavy metals and presence of commonly used pesticides. The sediments of the river in this stretch was also observed to act as a reservoir of the pollutants that are thrown in the river.

Physico-chemical properties of city sewage discharged into river Ganga at Varanasi were investigated by Tripathi et al. (1991) at six sampling sites i.e. Assighat, Shiwalaghat, Harishchandraghat, Chaukighat, Rajendra Prasad ghat and Rajghat. It was reported that Rajghat sewage was highly concentrated while that of Assighat was less concentrated. There existed a monthly variation in pollution load of waste water which was highest in the month of June and the values reported for BOD and COD were 30428 Kg / day and 68,563 Kg / day respectively.

Sinha et al. (1992) conducted studies on the impact of mass bathing on diuramal variation of physico-chemical parameters on Makar Sankranti day at Kanpur and reported that most of the parameters were in lower range during night and higher during day time except Dissolved Oxygen and Temperature. Appreciable increase in bacterial count and Turbidity and lower values of D.O. were recorded when bathing activity was at its maximum. Shukla et al. (1992) assessed the water qulaity of river Ganga at Ghazipur in the same year and concluded that most of the important physico-chemical parameters exhibited higher values that the tolerance limt while Israili (1992) noted the presence of heavy metals in Ganga water and sediments throughout the course of study.

River Bhagirathi between Uttar Kashi and Tehri was surveyed for limnobiotic studies by Joshi et al. (1992) while Srivastava et al. (1993) conducted studies on the pollution status of river Ganga and its impact on sediments at Mirzapur. Water quality of river Ganga in District Rae Bareli was studied by Sinha et al. (1993) and that at Mirzapur and Ballia by Srivastava and Mehrotra (1993).

Srivastava et al. (1994) carried out investigations on pesticides residues in Ganga water between Kare Manikpur (Pratapgarh) and Phaphamau (Allahabad) during Jan. 1988 and June, 1988 and concluded that among organo-chlorine pesticide (OCP) the concentrations of DDT, BHC and Endosulfan were found to be higher when compared with the EPA limits for fresh water aquatic live stock at almost all the sampling sites. Organo-phosphorus pesticides (OPP) could not be detected. The entry of pesticides into the river system was attributed to agricultural run off. Physico-chemical characteristics of river Ganga water between Kannauj to Shuklaganj exhibited a variety of difference as the river in this stretch from Manimau to Bithoor traverses through agricultural belt receiving agricultural washouts and thereafter upto Shuklaganj (Unnao) through highly industrialised area. The industrial effluents enters into river water at Shuklaganj which contain major pollutants like Phosphates, Sulphates, Ca, Mg, Fe, Mn, Na,

Co, Ni, Cu, Pb, Zn, Cd, alongwith organic wastes viz. barks of tree, hair, fleshings and other tannery products. Thus river water quality progressively deteriorates from Manimau to Shuklaganj having maximum pollution load at Shuklaganj. Likewise the population of microbes showed increasing trend with increased level of pollution within the stretch (Shukla and Anjum, 1994)

Kumar et al. (1994) also studied the soil characteristics on the river banks and established a correlation between the soil characteristics and the growth of Bryophytes.

Singh and Sinha (1995) discussed the problems and prospects of water pollution especially those of river pollution and suggested certain remedial measures.

In an article entitled "Pollution status and Health Hazards of river Ganga", Sinha (1996) expressed his concern on the increasing pollution of the river inspite of Ganga Action Plan and feared that this may result in higher health risks.

❑ RIVER YAMUNA

Originating at Yamunotri and merging with river Ganga at Allahabad river Yamuna, though a major river of Uttar Pradesh, also passes through Delhi, the Capital of the Country. Chakraborty et al. (1959) were probably the first to study the physico-chemical condition of Yamuna at Allahabad during 1954-55. They were followed by Mohan and Sarkar (1961, 1965). and in their preliminary investigations on pollution status of river Yamuna at Delhi concluded that its water was degraded and was unfit for drinking and bathing. Mathur (1965) studied the river Yamuna at Delhi and reported deterioration in its water quality while Bhargava (1984) felt an immediate need to monitor the river Yamuna to prevent the pollution. Similar observations were made by Rai (1974a), Dakshini (1979), Madhwal (1983) and Kumar (1984).

Sangu and Sharma (1985) assessed the river quality at Agra and observed that it was significantly polluted and unfit for human use. Similar observations were made by Chaturvedi (1985) in the same year and it was concluded that the water quality of river Yamuna has significantly deteriorated in Uttar Pradesh.

Analysis of trace metals in river water and treated effluents of Delhi sewage disposal plant was carried out by Wassay and Jain (1988) alongwith the soil samples at a place where treated effluent is used for irrigation purposes. The presence of toxic metals were reported from the samples collected and remedial method was suggested for their removal by recycling of water with expanded bed of sand particles coated with naturally occurring polyelectrolytes extracted from the alluvial soil of the river.

Singh and Mishra (1989) assessed the water quality of river Yamuna at

Mathura, Agra, Etawah and Allahabad and observed that the river was badly polluted at Mathura, Agra and Allahabad as the BOD values within the stretch valued from 1.5 mg/lit. to 30.3 mg/lt. in different seasons while the total coliform count ranged between 1820 to 63,500. However, significant improvement in the water quality at Etawah was observed though the coliform (MPN) count was quite high, ranging between 7,000 to 210,000 probably due to sewage disposal. The river at Allahabad was found to be severely polluted with BOD values ranging between 7.7 mg/lit. and 30.3 mg/lit. in different seasons due to direct discharge of industrial effluents and municipal wastes. They found the water quality below the desired levels at all sites and suggested that the whole river needs proper treatments before any human usage.

Neeraj and Pankaj (1989) studied the effects of effluents on the sediments and mineralisation of river Yamuna in and around Delhi region and concluded that the river Yamuna water coming out of Delhi is heavily polluted with sewage effluents and high rate of sedimentation.

Studies on water quality and metal speciation in river Yamuna were conducted by Gadhi et al. (1991) between Dakpathar and Agra and progressively higher percentage of Cd, Cu and Zn were recorded in the downstream which clearly indicates that pollution load increases as the river receives industrial effluents along its course.

Singh et al. (1993) assessed the BOD load of river Yamuna in respect of potable water of Agra during Sept. 1991 to May, 1992 and reported that DO and BOD of river water were severely affected due to discharge of municipal sewage and industrial effluents of Mathura and Agra city. On the basis of the above studies it may be said that the water of river Yamuna is unfit for human consumption.

❑ RIVER GOMATI

River Gomati an important tributary of river Ganga and a perennial river of Awadh plains runs across the major part of Uttar Pradesh covering nine districts and a distance of approximately 940 Kms. Originating from Madhoganj Tanda Village in Pilibhit districts, it passes through the districts of Shahjahanpur, Kheri, Hardoi, Sitapur, Lucknow, Barbanki, Sultanpur, Jaunpur and ultimately merges in river Ganga near Saidpur town of district Ghazipur. During its course, it receives huge quantitities of untreated sewage and industrial wastes which alter the physico-chemical characteristics of river water significantly.

It was Bhaskaran et al. (1965) who initiated physico-chemical studies on the river at Lucknow. They concluded that the river water was significantly polluted and despite self purification capacity, it showed lower values of dissolved oxygen at certain points in summers.

Arora (1973) observed that river Gomati at Lucknow was severely polluted as 33 waste outfalls were discharging approximately 21.26 MGD untreated waste water into it. It was also noted that the river water quality did not improve upto the desired level even after it had travelled 65 Km. down stream from Lucknow.

Hydrobiological studies on the river Gomati with reference to eutrophication were carried out by Prasad and Saxena (1980) within a stretch of 11 Km. in Lucknow city between Feb. 1975 and Jan 1976. They also carried out investigations of physico-chemical parameters in relation to occurrence of blue green algae. From the observations they concluded that there was clear relationship in total solids, pH, rainfall and atmospheric temperature. Direct relationship between total solids and rain fall was reported during the course of study while the same was not well defined with temperature. An inverse relationship was noted between pH and total solids at all the sampling sites. It was further reported that there exhibited a direct interdependance between pH and Carbonates which confirmed that river was rich in Carbonates. According to them blue green algae showed more or less definite zonation in relation to physico-chemical parameters and exhibited a gradient through progressive degree of pollution.

Prasad and Singh (1982) carried out studies on physico-chemical characteristics and species divesity of diatoms which are reliable bio-indicators of pollution. River water was analysed for fifteen physico-chemical parameters at four sampling stations includine one at a point where river enters in Lucknow city. They concluded that addition of sewage in the river has caused a reduction in mean values of DO and transparency and an increase in bicarbonates, total alkalinity, chlorides, total hardness, calcium, magnesium and free and saline ammonia. Floral spectrum of the dominats taxa at the various stations in decreasing order alongwith monthly mean percentage importance values clearly indicated that *Navicula heufleri* var. *leptocephala* dominated at the point where river receives least sewage wastes while *Nitzschia palea* and *Melosira grenulata* dominated at the stations where input of municipal wastes was maximum.

Kumar and Kant (1987) analysed the water of river Gomati at Lucknw for certain heavy metals and observed that it was polluted with Cu, Zn and Cr. The Concentration of metals was found much higher the permissible limits and it was suggested that the river water was not safe for human usage without proper treatment.

Impact of industrial waste water of distilleries and sugar mills on the water quality of river Gomati and its tributaries was studied by Singh an Misra (1989) who concluded that BOD values in river water ranged between 5.2 mg/L (Gaughat upsteram) and 59.5 mg/Lit. (Wazirganj Nala) in summer, 3.4 mg/Lit

(Naimisharanya) and 210 mg/lit (Mohan Meakin Nala D/S) in post mansoon and 5.1 mg/lit. (Gaughat U/S) and 850 mg/lit (Mohan Meaking Nala U/S) in winter seasons. The total Coliform counts (MPN) were observed to range from 20 (Mohan Meakin Nala D/S) to 2,40,000 (Wasiranj Nala D/S and Jaunpur D/S) in summers, 780 (Namisaranya, Sitapur) to 2500 (Mohan Meaking Nala D/S and Wazirganj Nala D/S) in post mansoon and 2200 (Gaughat U/S) to 24,000 (Jaunpur D/S) in winter seasons at different places. Physico-chemical parameters and microbial counts (MPN) clearly revealed that river was grossly polluted at Lucknow and Jaunpur due to discharge of large quantities of raw sewage and industrial wastes.

Studies conducted on the tributaries of river Gomati viz. Sarayan at Sitapur and Gone at Kamalapur in the same year by Singh and Misra (1989) indicated that both of them were significantly polluted due to discharge of sugar mill effluent and distillery waste water in their catchment areas.

Bhatt and Pathak (1992) conducted studies to assess the pollution status of river Gomati within a stretch of river draining the Kumaon region and observed that several physico-chemical parameters exhibited seasonal variations. They further reported that the natural water of river was BOD free and possessed a low ionic strength with a dominance patterna Ca^{++} Na^{+} Mg^{++} K^{+} HCO_3^{-} Cl^{-} SO_4^{--}. The river water was found to be tolerably clean and contained biologically important nutrients while it gets highly polluted down stream due to human interference and input of municipal and industrial wastes water.

Misra et al. (1994) studied river Gomati and its tributaries viz. Kathna, Sarayan and Gone at Shahjahanpur, Kheri, Sitapur, Hardoi, Lucknow, Barabanki, Sultanpur, Jaunpur and Varanasi before its confluence with river Ganga at Ghazipur. The river Gomati water was monitored at nineteen sampling sites between Nov. 1987 and Oct. 1990. From the observations they concluded that the river was grossly polluted at Shahjahanpur, Kheri and Lucknow as BOD and COD values were quite high at these sites in winter and summer seasons. Comparatively lower values of BOD and COD at Sultanpur and Jaunpur in the same seasons during course of study clearly indicated the self purifying capacity of river water. Dissolved oxygen and pH values of river water ranged between 4.8 mg/Lit. to 9.4 mg/Lit. and 7.1 to 8.8 respectively at all the sampling stations during the study period which indicated that river showed alkaline trend throughout. Mean values of twelve selected parameters of Sarayan river indicated that it was polluted at downstream Sitapur as BOD and COD values were quite high particularly in winter and summer seasons. Lower values of BOD and COD in the river water during mansoon were attributed to the dilution of the wastes discharged into it while turbidity was observed to be maximum at all the sampling stations in same season. River Kathna which joins river Gomati near Qutabnagar (Sitapur) was surveyed at upstream Kheri and upstream Qutabnagar and was found grossly polluted at both the sites. Similarly river Gone which is a tributary of river Sarayan receives considerable domestic and industrial effluents near Hargaon town (Sitapur) and was observed to be most polluted throughout the year at Kamalpur and Hargaon town.

❑ RIVER SAI

River Sai a major tributary of river Gomati originates from Uchaulia Nalla situated in the northern side of Pihani in Hardoi district. It traverses through the districts of Unnao, Rae Bareli, Pratapgarh and Jaunpur covering a distance of approximately 450 Km. In the recent past these districts have witnessed rapid industrial growth and urbanisation. Ultimately river Sai merges with river Gomati at Rajapur Trimuhani near Jalalpur in Jaunpur district. The river has been used as sink for dumping off domestic sewage since long which has increased enormously in last few years due to increased urbanisation and population growth. Moreover rapid industrialisation in Unnao and Rae Bareli districts has increased the effluent quantity considerably which is discharged into the river Sai.

Studies on the water quality of river Sai were initiated by Sinha et al. (1985). It was concluded from their preliminary studies on the physico-chemical properties of the river at Rae Bareli that pH, DO, BOD, hardness, conductivity, total solids, and total dissolved solids were under tolerance limits while concentrations of Cr, Cd, Pb, Fe, Co, Ni and Mn were more than the permissible limits of WHO standards. It was confirmed in later studies by Vajpayee et al. (1988).

Kumar and Kant (1987) analysed river water for certain trace elements and noted that it contained detactable amount of Zn and Mo which was in confirmity with the earlier observation that river was polluted with toxic metals.

Sai was reported fairly clean at Pihani as it exhibited lower values of BOD (1.2 mg/lit.) and bacterial counts which ranged between 240-480 (Singh and Misra, 1989)

Detailed studies on the hydro-chemical characteristics of Sai water were conducted by Srivastava (1991) from Gausganj (Hardoi) to Rajapur Trimuhani (Jaunpur) between Nov. 1990 and Dec. 1991 in Mansoon, winter and summer seasons. Analysis of water samples collected from eleven sampling sites clearly indicated that river water at Rae Bareli and Pratapgarh was highly polluted in winters and summers. This was probably due to untreated discharge of sewage and industrial effluents into the river. It was also reported that river water was fairly clean at Gausganj and Rajapur Trimuhani as compared to Rae Bareli, Pratapgarh and upstream of Jaunpur city indicating clearly the regeneration capacity of river. The river water was observed to be belowthe desired levels for drinking and bathing except as Gausganj when compared with ISI and WHO standards.

Comparison of water quality of river Sai at the point of origin and confluence was studied by Srivastava et al. (1992).

It was concluded from the studies that the quality of water of the river at the origin was better than that at its confluence presumably because the river gets polluted due to discharge of urban sewage, industrial effluents and agricultural run off.

In the same year Srivastava et al. (1992) monitored the quality of the river water at Bela (Pratapgarh) as affected by mass bathing during Navratri festivals. It was concluded from the results that mass bathing caused a significant deterioration of the river water quality which might pose a health hazard to the users of the river water.

Sinha et al. (1994) conducted studies of the physico-chemical properties of industrial waste waters of the paper mill at Rae Bareli which is discharged into river Sai and noted that it showed high values of COD, BOD and Total suspended solids. The concentrations of dissolved oxygen was much lower than the permissible limits through out the study period and the effluent was suspected to damage the local flora and fauna.

Srivastava et al. (1994) conducted studies on the impact of mass bathing on the physico-chemical characteristics of river Sai water on an important Hindu festival 'Mahashivratri' near Rampur Sudauli village at Rae Bareli where an age old temple of Lord Shiva 'Bhaureshwar Nath' is situated on the left bank of the river. The place has a marked geographical importance as the boundaries of the districts of Lucknow, Rae Bareli and Unnao meet here and because of its mythological and historical importance, mass bathing and a fair is organised every year. The river water was analysed for a total of fiften important physico-chemical parameters and was reported that most of them exhibited higher values. They noted an increase in the values of pH, Turbidity, DO, BOD, Hardness, total solids, total dissolved solids, total suspended solids and chlorides due to mass bathing while alkalinity of river water surprisingly decreased with an increase in the bathing activity. Moreover Ca, showed higher values and Mg was found in lower concentrations when compared with the recommended limits of ISI (1983). They also reported that phytoplankton and zooplankton density and populations of benthic flora and fauna were also remarkably affected by an increase in bathing activity and exhibited a gradual decreasing trend.

On the basis of data obtained by analysing eight physico-chemical parameters of river Sai water during 1988 to 1990 at six sampling sites viz. Lohanipur, Rajghat, Munsiganj, Dariyapur, Jagdishpur and Behtra Khurd, water quality index (WQI) of the river was calculated by Sinha et al. (1994). It was concluded that water at Rajghat, Dariyapur, Jagdishpur and Behtra Khurd was severely polluted while at remaining sites moderately to excessively polluted. It was observed that WQI usually decreased during August and increased in May due to variation in the volume of water in mansoon and summer seasons.

Impact of sugar factory effluents on physico-chemical characteristics of river Sai at Rae Bareli was studied and reported by Srivastava et al. (1995). It was concluded by them that the physico-chemical characteristics of the river water changed due to the addition of sugar factory effluents into the river. The river water got deteriorated and became unfit for human consumption during the season of the operation of the sugar factory.

CONCLUSIONS

It would be clear from the account of the available literature on river Ganga and its major tributaries (Yamuma, Gomati and Sai) presented above that the water of these major rivers of the state of Uttar Pradesh are unfit for human use. In addition a perusal of the literature published on other minor tributaries and sub-tributaries Gupta and Gupta (1980), Kudesia and Verma (1986), Bisht et al. (1989), Singh et al. (1991), Tiwari et al. (1991), Mohan et al. (1991), Chauhan (1991), Mazhar and Kapoor (1992), Bisht and Chopra (1992), and Pandey (1994) indicated that the water of these are also not suitable for human consumption. It is, thus, inferred that the major factor of the deterioration of water quality of the rivers in Uttar Pradesh is human interference coupled with population explosion. Therefore, while planning stretegies for cleaning the major rivers like Ganga and Yamuna, the tributaries and subtributaries of these should also be paid attention. All the efforts to control pollution of the major river and neglecting tributaries would only result in aggrevating the problem of the pollution of these rivers.

❑ REFERENCES

Agarwal D.K., Gaur S.D., Tiwari T.C., Narayanswami N. and Marwah S.M. 1976 a. Physico-chemical characteristics of Ganges water at Varanasi. *Indian J. Environ. Hlth 18* (3) : 201 - 206.

Ajmal M., Nomani A. A. and Khan M. A. 1984. Pollution in the Ganges river India. *Water Sci. Technol. 6 (5-7)* : 347-358.

Arora H.C.; Chattopadhya S.N., Sharma V.P., Routh T. and Elyas S.I. 1973. A short term study on the pollution of Gomati river in Lucknow region. *Proc. symp. on*

Environmental pollution. 44-58.

Bechu Lal, Tripathi B.D., and Sikandar M. 1984. River pollution Perception and Awareness. *Proc. Natl. Conf. River Poll Human Hlth. Varanasi, Feb. 20-21 1983* : 89-93

Bhargava D.S.1981. Pollution Control strategy for Ganga. *In International Śymposium on water resources conservation Pollution and Abatment, Roorkee,* Dec. 11-13.

Bhargava D.S. 1984 a. Water Quality Survey of the Ganges. *Asian Environ. 1* : 4-6

Bhargava D.S. 1985. Variation in quality in the Ganges. *Effluent water treatment J. Instn.,Engrs. India. 25 (2)* : 61-65.

Bhargava D.S. 1987. Nature and the Ganga. *Environ. Conserv 11(1)* : 307-318.

Bhaskaran T.R., Chakraborty R.N. and Trivedi R.C. 1965. Studies on the river pollution I. Pollution on purification of Gomati river near Lucknow. *J. Instn Engrs India 45(6)* : 39-50.

Bhatt S.D. and Pathak J. K. 1992. *J. Environ. Biol., 13(2)* : 113-126.

Chandra K., 1981. Pollution from wastes of Industries manufacturing nitrogenous fertilizer. A case study from river Ganga near Allahabad. In *Proc. Symp. W.R.C.P.A. Roorkee* 11-23 Dec. : 141-151.

Chandra S. and Mathur S.S., 1983. Pollution of Ganga water due to tanneries effluents in Kanpur. *Souv. and Abst. Nat. Conf. on river pollution and Human Health.* P-5.

Chandra S and Mathur J.S. 1986. Impact of discharge from Tanneries on water quality of river Ganga in Kanpur. *Indian J. Public Hlth. 30(2)* : 97-100.

Chattopadhya S.N., Routh T., Sharma V.P., Arora H.C. and Gupta R.K. 1984. A short term study on the pollution status of river Ganga in Kanpur region. *Indian J. Environ. Hlth. 26(3)* : 224-257.

Chakraborty R.D., Ray P and Singh S.B. 1959. A quantitative study of the plankton and the physico-chemical condition of river Yamuna at Allahabad, 1954-1955. *Indian Jr. Fish 6(1)* : 186-203.

Chakraborty R.N. Saxena K.L. and Khan A.Q. 1965. Stream pollution and its effect on water supply. *A report of survey, Proc. Symp. Problems in Water treatment.* Oct. 29-30, 1964 (CHERI, Nagpur) : 211-219.

Dakshini K.M.M. and Soni J.K. 1979. Water quality of sewage drains entering Yamuna in Delhi. *Indian J. Environ. Health. 21* : 354-360.

Das Gupta S.P. (eds) 1984). *Basin-Sub-basin inventory of water Pollution of Ganga Basin, Part - II.* Central Board for Prevention and Control of Water Pollution, New Delhi.

Dutt A. and Chaudhuri M. 1980. Mercury levels in the river Ganga water and the sediments near Kanpur. *J. Inst. Engrs. India. 61 (1)* : 48.

Gautam A., Singh H.R. and Sati O.P. 1989. Seasonal variation of certain oxidation reduction characteristics of the river Bhagirathi (India). *Proc. Indian Natn. Sci. Acad. 55(2)* : 111-114.

Goel S.L. 1987. Water Pollution status at Kanpur, *Civic Affairs. 34 (9)* : 9 - 16.

Godh, Ranu, Singh O.V., Tandon S.N. and Mathur R.P. 1991. A study of water quality and metal speciation of Yamuna river. *Asian Env. 13(2)* : 3-10.

Handa B.K., Kumar A. and Goel D.K. 1981. Trace Elements of surface in Uttar Pradesh. *TAWPC TECM ANN* 8 : 11-17.

Israili A.W.1992. Occurrence of heavy metals in Ganga river water and sediments. *Indian J. Environ. Hlth. 34(1)* : 63-66.

I.S.I. 1983. *Indian Standard Tolerance Limits for Industrial Effluents.* IS : 2490, Part I. General Limits (II revision). Indian Standard Institution, New Delhi.

Jhingran A.G. 1988. Aquatic pollution with special reference to the Ganga River System. In : *Souvenir National Symposium on Environmental impacts on Animals and Aquaculture.* Kalyani, May 14-16, 1988.

Joshi H.C. 1985. Pesticide Residue Monitoring in the Bhagirathi Hooghly stretch of the Ganga river system. *Proc. Natl. Semin. Polln. Centr. Environ. Manag.*, Nagpur, 17-19 March : 80-85.

Khoshoo T.N. (eds) 1986. *Environmental concerns and strategies.* APH, New Delhi.

Joshi B.D., Pathak J.K., Singh Y.N., Bisht R.C.S. 1992. A study of minor limnobiotic components of river Bhagirathi from Garhwal Himalaya. *Himalayan J. Env. Zoo., 6(2)* : 152-157.

Kudesia V.P. 1989. Physico-chemical properties of Ganga river at Farrukhabad. *Indian J. Environ. Agric. 3.* (1 and 2) : 81-82.

Kumar, Suresh and Kant, Sunil. 1987. *Indian J.Env. Agric. 2 (1 & 2)* : 22-28.

Kumar, Adarsh, Sinha A.K. and Pandey D.C. 1994. Ecological studies of *Riccia frostii* growing on the banks of River Ganga between Shuklaganj and Kalakanakar. In : "*Environment Management in Developing Countries. Vol. II Water and its Manage-*

ment". Edited by Sinha et al., Published from Gyanodaya Prakashan, Nainital, pp. 309-312.

Kumar V. 1984. A Ph. D. thesis of Meerut University, Meerut.

Lakshminarayana J.S.S. 1965. Studies of the phytoplankton of the river Ganges, Varanasi, India, Part -I, Physico chemical characteristics of River Ganga. *Hydrobiologia. 25* : 119-175.

Mohan C. and Sarkar H.L., 1961. Preliminary studies on pollution of the river Jamuna, *Bull. CPHERI, 1* : 76-82.

Mohan C. Balani and Sarkar H.L. 1965. Some observation on the pollution of Yamuna river at Okhla water works intake, Delhi. *Environ. Hlth. 7(2)* : 84-86.

Mathur R.P. 1965. Pollution study of river Jamuna at Delhi. In Proc. of Symposium on water pollution control, *CPHERI*, Nagpur, 28-30.

Madhwal B.P., Chopra A.K. and Singh H.R. 1983. *J. Zool. 3* : 157-158.

Mathur R.P. 1986. Pollution Modelling of upper Ganga Basin. A report. Department of Civil Engineering, Roorkee University, Roorkee.

Mathur A., Sharma V.C., Rupainwar D.C., Marthi R.C. and Chandra S . 1987. A study of river Ganga at Varanasi with special emphasis on Heavy metal Pollution. *Polln. Res. 6(1)* : 37-44.

Misra G.N., Bhatt C.S. and Sikandar Mohd. 1994. Physico-chemical characteristics of River Gomati and its tributaries. In : "*Environment Management in Developing Countries. Vol. II - Water and its Management*", Edited by Sinha et. al. Published from Gyanodaya Prakashan, Nainital, pp. 321-332.

Nautiyal P., Dobriyal A.K., Chaukiyal D.G. and Singh H.R. 1986. Quality and Potability of the Alaknanda water at Shrinagar-Garhwal : A note on Bacterial density as an indicator of Pollution. *Polln. Res. 5(3 and 4)* : 143-145.

Neeraj and Pankaj 1989. aquageo-chemistry of ground water and effect of effluents on the sediments and mineralisation of river Yamuna in and around Delhi region. In "*Water Pollution (Conservation and Management)*" edited by Sinha et al.. Published from Gyanodaya Prakashan, Nainital, pp. 43-50.

Prasad B.N. and Saxena, Manjula 1980. Ecological study of Blue Green Algae in river Gomati. *Indian J. Environ. Hlth. 22(2)* : 151-168.

Prasad B.N. and Singh Y. 1982. Species diversity of benthic diatoms in lotic Environment. *Indian J. Environ. Hlth. 24(4)* : 308-315.

Prasad B.N. and Singh Y. 1982. On diatoms as indicators of water pollution. *J. Indian Bot. Soc. 61* : 326-336.

Pandey P.K. and Pandey G.N. 1980. Physico-chemical characteristics of river Ganga at Kanpur. *J. Inst. of Engrs. 61(1)* : 28-30.

Pandey D.N. and Tripathi B.D. 1984. Contamination of Ganga water and its impact on Mammals. *Proc. Natl. Conf. river Polln. human Hlth.* Varanasi, Feb. 1983. 94-99.

Pandey N.C. 1985. Pollution of Rivers in U.P. with specific reference to Varanasi. *Civic Affairs. 32(11)* : 53-69.

Prasad G.V., Vlabhaje V. and Mehrotra M.N. 1986. Some Aspects of Pollution of River Ganga near Rajghat, Varanasi (U.P.). *Journal of the IPHE, India, 6(1)* : 15-27.

Rambilas and Kayastha S.L. 1983. Public awareness towards the river pollution along the ghats of River Ganga at Varanasi. A study of Environmental Assessment and perception. Souv. and Abst. *National Conf. on River Pollution and Human Hlth.* p. 4.

Ray P. and David A. 1966. Effects of Industrial wastes and sewage upon the Chemical composition and fisheries of river Ganga at Kanpur (U.P.) *Environ. Hlth. 8* : 307-339.

Rai H. 1974 a. Limnological studies on the river Yamuna at Delhi, India Part I : Relation between the chemistry and state of pollution in the river Yamuna. *Arch. Hydrobiol. 73(3)* : 369-393.

Sangu R.P.S., Shankar V. and Sharma S.K. 1987. An Assessment of water quality of River Ganga at Garhmukteshwar, Ghaziabad (U.P.). *Indian. J. Ecol. 14(2)* : 278-282.

Sangu R.P.S. and Sharma K.D. 1985. Studies on water pollution of Yamuna river at Agra. *Indian J. Environ. Hlth. 27* : 257-261.

Sahai I.M. 1985. Control of Ganga Pollution in U.P. *Civic Affairs, 32(11)* : 5-11.

Saikia D.K., Mathur R.P. and Srivastava S.K. 1988. Heavy metals in water and sediments of upper Ganga. *Indian J. Environ. Hlth. 31(1)* : 11-17.

Saxena K.L.Chakraborty R.N., Khan A.R., Chattopadhya S.N. & Chandra H. 1966.Pollution studies of the river Ganga near Kanpur. *Environ.Hlth.8* : 270-85.

Sharma R.C. 1984. Potamological studies on the lotic Environment of the upland river Bhagirathi of Garhwal Himalaya. *Environ. Eco. 2(4)* : 239-242.

Sharma S.M. and Goyal P.L. 1985. Sisamau Nala Polluting the Ganga at Kanpur. *Civic Affairs, 32(11)* : 49-52.

Shankar V. 1986. Integrated study of river Ganga. A brief Report (1984-86). Submitted to Dept. of Environ. Ministry of Environ. and Forests, Govt. of India. Gurukul Kangri University, Hardwar.

Shankar V. 1989. An Integrated study of Ganga at Rishikesh and Hardwar. In "*Water Pollution (Conservation and Management)*". edited by Sinha et al.. Published from Gyanoday a Prakashan, Nainital, pp. 277-284.

Shukla S.C., Kant R. and Tripathi B.D. 1989 a. Ecological Investigation on Physico-chemical characteristics and Phytoplankton productivity of river Ganga at Varanasi. *Geobios. 16(1)* : 20-27

Shukla S.C., Tripathi B.D., Kant R. Kumari D. and Pandey V.S. 1989 b. Physico-chemical and Biological characteristics of River Ganga from Mirzapur to Ballia. *Indian J. environ. hlth. 31(3)* : 218-227.

Shukla S.C., Tripathi B.D., Mishra B.P. and Chaturvedi S.S. 1992. Physico-chemical and bacteriological properties of the water of river Ganga at Ghazipur *Comp. Physio Eco. 17(3)* : 92-96

Shukla A.C. and Anjum 1994. Biological Fabric of Ganga Ecosystem : A Glimpse : In "*Environment Management in Developing Countries Vol. II Water and its Management*" Edited by Sinha et al.. Published from Gyanodaya Prakashan, Nainital, pp. 206-209.

Sikandar Mohd. and Tripathi B.D. 1984. The physico-chemical characteristics of Gangawater and Varanasi. *Proc. Natl. Cohf. River Polln. Human. Hlth.* Varanasi, Feb. 1983 : 53-62.

Singh H.R. 1985. The Bhagirathi Ecosystem. *Environ. Natural Resources* : 151 - 155.

Singh H.R. 1986. Pollution study of the upper Ganga and its Tributaries. Progress report submitted to Dept. of Environ., Govt. of India, Garhwal University, Srinagar.

Singh J.R., Yadav P.K. and Singh L. 1988 a. Pollution status on Sangam and its Adjoining river before the Kumbha Mela at Allahabad. *IJEP 8(11)* : 839-842.

Singh S. and Srivastava V.K. 1988. Variations in water quality of Ganga River between Buxar and Ballia *Polln. Res. 7(3-4)* : 85-92.

Singh S.K., Bhatt C.S. and Pandey K.K. 1988 b. Impact of Mass bathing on water quality at Sangam and adjoining Rivers during Magh Mela at Allahabad. *IJPE, 8(4)* : 275-278

Singh J.P., Yadav P.K. and Singh L. 1989 a. Mass Bathing effect on water quality of Sangam during Maha Kumbha Mela at Allahabad. *IJEP, 9(3)* : 189-193.

Singh J.P., Yadav P.K. and Singh L. 1989 b. The assessment of water quality of Sangam and its adjoining rivers Ganga and Yamuna after Maha Kumbh Mela at Allahbad. *Indian J. Environ. Prot., 9(5)* : 372-375.

Singh S. and Misra G.N. 1989. Status of major rivers in Uttar Pradesh. In "*Water Pollution (Conservation and Management)*" edited by Sinha et al.. Published from Gyanodaya Prakashan, Nainital, pp. 251-276.

Singh J.P., Singh S.K., Singh, Sakun, Yadav P.K. and Joshi Madhu 1993. Assessment of BOD load in river Yamuna in respect of potable water of Agra. *Indian J. Environ. Prot. 13(2)* : 119-123.

Singh V.P. and Sinha A.K. 1995. Problems and Prospects of Water Pollution. In "Assessment of Water Pollution" Edited S.R. Mishra, APH, New Delhi.

Sinha A.K., Vajpayee P., Kumar A., Singh R.P. and Pande D.P. 1985. Water quality of river *Sai* at Rae Bareli-Preliminary report. *Res. J. Pl. Environ. 2(2)* : 87-90.

Sinha A.K., Srivastava D.K., Pande D.P., Singh R.P., Srivastava S., Pandey D.C., Santhi Bhusan K., Srivastava K.N., Kumar A. and Tayal S. 1986. Water quality assessment of Ganga Water at Dalmau (Rae Bareli) India on Kartik Purnima 1985. *Bull. Environ. Sec. 3* : 3-6.

Sinha A.K., Srivastava Seema, Vajpayee Poornima, Dwivede Anjali, Srivastava K.N. and Pande D.P. 1988. River Ganga at Dalmau U.P. : Physico-chemical and Microbiological characteristics. In "Threatened Habitat" Edited, Agarwal et. al. Published by Society of Biosciences, Muzaffarnagar, India, pp. 433-444.

Sinha A.K., Pande D.P., Srivastava Seema, Vajpayee Poornima, Dwivedi Anjali and Srivastava K.N. 1989 b. Physico-chemical and Biological characteristics of River Ganga between Shuklaganj (Unnao) and Jajmau (Kanpur), India. In "*Water Pollution (Conservation and Management)*". Edited by Sinha et al.. Published from Gyanodaya Prakashan Nainital, pp. 345-371.

Sinha A.K. 1986. A Comprehensive study of Ganga and its Dependents. First Technical Report submitted to Deptt. of Environ., Govt. of India. Environment Research Centre, Feroze Gandhi College, Rae Bareli.

Sinha A.K., Srivastava Seema and Srivastava K.N. 1989 a. Physicochemical studies of river Ganga water at Kalaknkar (Pratapgarh). *IJEP. 9(3)* : 194-197.

Sinha A.K., Srivastava Seema, Srivastava K.N. and Pande D.P. 1989 c. Water quality index for River Ganga water between Shuklaganj (Unnao) and Kalakankar (Pratapgarh). In "*Management of Aquatic Ecosystems*" by Agarwal et al.. (eds.) Published by Society of Biosciences, Muzaffarnagar, India. pp. 233-242.

Sinha A.K., Srivastava R.K., Pande D.P. and Modak D.P. 1989 d. Water quality characteristics of Ganga River from Kare-Manikpur to Phaphamau - A case study *IJEP, 9(11)* : 845-852.

Sinha A.K., Srivastava R.K. and Pande D.P. 1990. Concentration of heavy metals in river Ganga water between Kare-Manikpur (Pratapgarh) and Phaphamau (Allahabad) - A Preliminary report. *Sci. and Cult. 56(11)* : 451-452.

Sinha A.K., Pande D.P., Srivastava R.K., Srivastava P., Srivastava K.N., Kumar A. and Tripathi A. 1991. Impact of Mass Bathing on the water quality of the Ganga River at Haudeshwarnath (Pratapgarh) India - A Case Study. *The Science of the total Environment* 101 : 275 - 280. (Elsevier Science Publishers B.V. Amsterdam.).

Sinha A.K. 1991. A comprehensive study of River Ganga in the stretch between Kalakankar and Phaphamau. Final Technical Report submitted to Ganga Project Directorate, G.O.I., New Delhi.

Sinha A.K., Tayal S., Srivastava D.K., Pandey D.C., Singh R., Vajpayee Poornima and Srivastava Seema. 1992. Diurnal Variations of Physicochemical and biological characteristics of Ganga water at Kanpur. U.P. on Makar Sankranti Day, 1987. *Comp. Physiol. Ecol. 17(1)* : 31-40.

Sinha A.K. Srivastava Seema and Srivastava K.N. 1992. Physicochemical characteristics of River Ganga State of Art. In "*Aquatic Ecology*". Edited by S.R. Mishra and D.N. Saksena. pp. 131-152, Ashish Publishing House, New Delhi, India.

Sinha A.K., Srivastava Seema and Srivastava K.N. 1993. Quality of River Ganga Water in Rae Bareli district. In "*Proceedings of Seminar on Ganga in the service of the nation.*" Univ. of Roorkee, III-9-III-17.

Sinha A.K., Srivastava Kiran and Srivastava K.N. 1994. Water quality Index of River *Sai* at Rae Bareli. *I.J.E.P., 14(12)* : 888-890.

Sinha Kanchan, Bisht Y.P.S., Singh R.P. and Srivastava Geeta. 1994. Physico-chemical characteristics of Industrial Effluents of Rae Bareli Industrial Area-I : A Paper Mill. In "Environment Management in Developing Countries. Vol. II - Water and its Management" Edited by Sinha et al.. Published from Gyanodaya Prakashan Nainital, pp. 157-161.

Srivastava K.N., Vajpayee Poornima, Srivastava Seema and Sinha A.K. 1988. Pollution of river Ganga in Rae Bareli district due to disposal of dead bodies and carcasses. In *Proceeding of the Symposium on "Environmental hazards in Rae Bareli and Amethi."* held at F.G. College, Rae Bareli. 14th August, 1986. pp. 49-52.

Srivastava D.K., Awasthi B.B. and Srivastava R.C. 1988. Analysis of Metallic elements in sediments and water of the river Ganga with reference to Effluents. A Preliminary study. Natl. Acad. Sci. Letter., 11(7) 209-210.

Srivastava K.N., Srivastava Seema and Sinha A.K. 1989 a. Hydrochemical Assessment of River Ganga water at Buxar (Unnao). *Res. J. Pl. Environ. 5(1)* : 33-40.

Srivastava Seema, Srivastava K.N. and Sinha A.K. 1989 b. NO_3^- (Nitrate) and PO_4^{---} (Phospate) concentration in the River Ganga water. In *Proc. Indian Sci. Cong. 76th Session, 1989* No. 310, pp. 175-176.

Srivastava K.N. 1991. Studies on River *Sai* with special reference to its fauna. First Technical report, submitted to Department of Environment, Govt. of Uttar Pradesh. Environment Research Centre, Feroze Gandhi College, Rae Bareli, U.P.

srivastava K.N., Srivastava Kiran and Sinha A.K., 1992. Comparative water quality of River *Sai* at the point of origin and confluence. Proc. National Symp. on Env., BARC, Bombay : 213.

Srivastava K.N., Srivastava Kiran and Sinha A.K. 1992. Monitoring of water quality of River *Sai* at Bela Pratapgarh as affected by Mass Bathing. In "*Environment and Biodegradation*" Ed. V.P. Agrawal, Published by Society of Biosciences, Muzaffarnagar, U.P., pp. 179-184.

Srivastava Ajai, Mehrotra M.N. and Tiwari R.N., 1993. Study of pollution of the river Ganga in thr Mirzapur region (India) and its impact on sediments. *Int. J. Environ. Stud. 43(2/3)* : 201-208.

Srivastava A. and Mehrotra M.N. 1993. Some aspects of pollution of River Ganga in the Mirzapur region and its impact on sediments. In Proceedings of seminar on *Ganga in the service of the Nation.* Univ. of Roorkee, III-36-III-43.

Srivastava R.K., Sinha A.K. and Pande D.P. 1994. Pesticide residues in water of River Ganga between Kare-Manikpur (Pratapgarh) and Phaphamau (allahabad). In "*Environment Management in Developing Countries. Vol. II. Water and its Management*". Edited by Sinha et al.. Published from Gyanodaya Prakashan, Nainital, pp. 395-399.

Srivastava Kiran, Singh V.P., Srivastava Ajit K., Srivastava K.N., Reuben D.E. and Sinha A.K. 1994. Impact of Mass Bathing on the water quality of River *Sai* at

Bhaureshwarnath (Rae Bareli), India on "Maha Shivratri" In : "*Environment Management in Developing Countries - Water and its Management*". Edited by Sinha, et al., published from Gyanodaya Prakashan, Nainital, pp. 150-156.

Srivastava Kiran, Srivastava K.N. and Sinha A.K. 1995. Impact or Sugar Factory Effluents on physico-chemical and Biological characteristics of River *Sai* at Rae Bareli - A case study. *Proc. of the IV National Symposium of Environment.* Feb. 7-10. 1995 : 39-40.

Tiwari T.N., Das S.C. and Bose P.K. 1986. A relation between COD and BOD for the Ganga at Kanpur, *IJEP, 6(3)* : 183-184.

Tripathi B.D., Sikandar M. and Shukla Suresh C. 1991. Physico-chemical characterization of city sewage discharge into river Ganga at Varanasi, India. *Env. Int. 17(5)* : 469-478.

Upadhyay R. Dubey A.P. and Pandey G.N. 1982. Monitoring of Pollution at Sangam during Ardha-Kumbha. *Poll. Res. 1(1-2)* : 11-20.

Upadhyay R., Diney A.P. and Pandey G.N. 1985. Stretegy for Environmental Pollution Control at Kanpur. *J. Inst. Engrs Pt. Env. 65(2)* : 41-45.

Vajpayee Pornima and Sinha A.K. 1988. Heavy metal pollution on River *Sai* at Rae Bareli. In "*Proceedings of the symposium on "Environmental hazards in Rae Bareli and Amethi*". held at F.G. College, Rae Bareli; 14th August, 1986. pp. 29-34.

Wasdy S.A. and Jain P.K. 1988. Heavy metals in raw sewage entering river Yamuna at Delhi and their treatment by natural Polyelectrolyte. *J. Instn. Engrs. (India) - Env. Engng.; 68(2)* : 81-85.

Authors :
A.K.Sinha, V.P.Singh and (Mrs) Kiran Srivastava Environmental Research Centre Firoz Gandhi College, Rae Bareli 229 001. U.P.

2

Studies on the water quality of Kuttiadi river (Near Kozhikode), Kerala

❑ INTRODUCTION : LOCATION AND SALIENT FEATURES OF KUTTIADI RIVER

Kuttiadi river (Figure 1) originates from the Narikota Ranges on the Western slopes of the Wynad hills, a part of Western Ghats, at an elevation of 1220 metres above the mean sea level. The river flows through Badagara, Quilandy and Kozhikode Taluks in Kerala State before it falls into the Arabian sea at Kottakal 7 km South of Badagara. The river has a length of 74 km and along with its tributaries it drains an area of 583 Km^2.

Tha major tributaries of the river are Kadiyangadpuzha, Muttatupuzha, Onpuzha, Alampuzha, Kadantaropuzha and Madapallipuzha, The river passes through Oorakuzhi, Kuttiadi, Tirvalur, Myyipot, Maniyur, and Karuvancheri villages. The historical Kottakal Fort is situated at the mouth of the river. The river is used mainly for collection of sand from the bed, traditional fishing, navigation, irrigation and power generation.

Physiography

The river basin can be broadly classified into three distinct physiographic zones, namely (i) the highland on the Eastern side lying between 1220 m and 90 m above mean sea level, (ii) the midland lying between 90 m and 45 m above mean sea level, and (iii) the low land on the western side lying between zero to 45 m above mean sea level. On the side of physical geologic formings, the basin can be classified into three parallel zones :

(i) the estern zone consisting of crystalline rocks of Archaean group, ii) the central zone consisting of mainly residual laterites, and iii) the Western coastal zone consisting of recent sands and silt.

Climate

In the coastal towns of Badagara and Quilandy, the average annual rainfall is estimated to be 3070 mm and 3660 mm, respectively. In the upper reaches in the Kuttiadi and Kinalur estate the records show an average rainfall of 4522 m and 4150 mm respectively. The South-West monsoon (June-August) and the North-East monsoon (September-November) cause heavy rains in the basin. The average ambiant temperature in the basin is close to 30^0 C and rarely exceeds 33.5^0 C.

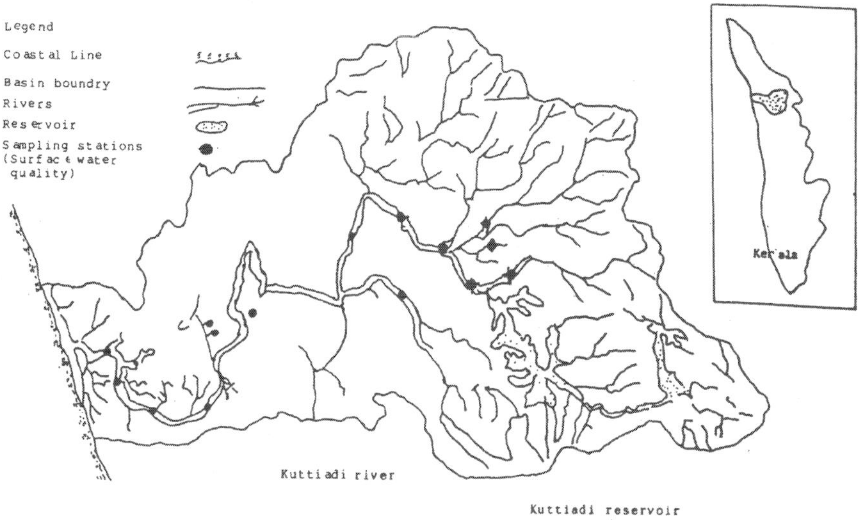

Figure 1 : Kuttiadi River basin showing the reservoir and the sampling stations.

Catchment area and land use

The catchment area of the basin consists of about 10 km^2 water spread reservoir area, 16 km^2 gardenlands, other plantations and rural settlements, 8 km^2 rubber plantations and 34 km^2 forested area.

Water resources projects

A hydroelectric project has been commissioned, at Kakkayam (11° 33' N latitude and 75° 56' E longitude) with an installed capacity of 3 x 25 MW. The existing major irrigation project (Kuttiadi Irrigation Project) in the basin is situated at Peruvannamuzhi (11 ° 35' N latitude and 79° 49' E longitude) (Figure 1); its distributaries command a gross area of 36414 hectares. In the coastal belt the river is used for navigation and fishing. The salient features of Kuttiadi irrigation project are presented in Table 1.

❑ EXPERIMENTAL

Sampling stations

Thirteen sampling stations were set after a reconnissance survey at Thurasserykadar, Idinjakadar, Payyoli, Payyochirp, Valiyakumbalam, Cheriyakumbalam, Paleri, Avala, Kadiyangad, Kadantarapuzha, Onipuzha and Muthhatupuzha (Figure 1).

Van dorn type samplers were employed for the collection of water samples; the samples were of 3 litre capacity of rigid polyvinyl chloride (PVC) with plated brass out fittings.

Analysis

The in situ analysis was done onsite for the changeable parameters like temperature, pH, dissolved oxygen, and electrical conductivity. The Sechi disc transparency was also measured. Parts of th samples were filtered and preserved on site for the analysis of nutrients and heavy metals.

The preserved and unpreserved samples were transported to the laboratory and stored carefully. The analysis of various parameters was carried out as per internationally accepted analytical methods (Rand et al 1987; Abbasi 1989, 1996).

The samples of aquatic organisms, were also collected and analyzed for the presence of zoolplankton, phytoplankton, amphibians, and fish as per the internationally accepted methods of collection, preservation, storage, enumeration, and identification (Rand et al 1975, Abbasi 1989, Beavan 1982, Jhingaran 1982).

Table 1 .

Salient features of Kuttiadi river basin

Basin area, Km^2	:	583
Basin area in Kerala State, Km^2	:	583
District of Kerala in which basin is located	:	Kozhikode
Origin of river	:	Narikota Elevation, m : 1220
Length of main stream, km	:	74
Main tributaries	:	Onipuzha, Thottilpalampuzha, Kodiyangadupuzha, Mannathilpuzha, Madappalipuzha
Water requirement for wetland for three crops, Mm^3	:	240
Water requirement for greenland, Mm^3	:	112

Projected water requirement for domestic use (2021 AD), Mm^3	:	58.8
Projected water requirement for industrial use (2021 AD), Mm^3	:	45
Existing major/medium irrigation project (Commissioned/partially commissioned)	:	Kuttiadi
Existing hydroelectric project	:	Kuttiadi
Hydroelectric projects investigated	:	Kuttiadi augmentation
Navigable length of river, km	:	9.6

amphibians, and fish as per the internationally accepted methods of collection, preservation, storage, numeration, and identification (Rand et al 1975, Abbasi 1989, Beavan 1982, Jhingaran 1982).

Water quality standards

The water quality standards cited in the discussion are

ISI - 1981	:	Indian standards Institution (now BIS), New Delhi (1981) for effluent discharge on Inland surface waters.
ICMR - 1975	:	Indian Council of Medical Research (1975), New Delhi for domestic use.
ISI - 1979	:	Indian Standards Institution (now BIS), New Delhi (1979) for short-term and long-term use in irrigation.
ISI - 1981	:	Indian Standards Institution (now BIS), New Delhi (1981) for effluent discharge on Inland Surface Waters.
ISI - 1982	:	Indian Standards Institution (now BIS), New Delhi (1982) draft standards for drinking waters.
NAS - 1972	:	National Academy of Sciences (1972), Washington D.C., for short-term and long-term use in irrigation.
MWH - 1976	:	Ministry of Works and Housing (1976) Government of India, New Delhi, for drinking water supply.
USDI - 1968	:	United States Department of Interior (1968), Washington D.C., for short-term and long -term irrigation waters.
USEPA - 1979	:	United States Environmental Protection Agency (1979), Washington D.C., for Protection of fish and aquatic life and for prevention of nuisance algal growth.
WHO - 1971	:	World Health Organisation (1971), Geneva, for drinking waters.

❑ RESULTS AND DISCUSSIONS

Status of water quality in Kuttiadi River

From the 13 survillance stations mentioned above samples were collected from surface, middle and bottom. The discussion presented below focusses on the river water quality from the reservoir bund size to the confluence of the river with sea.

Temperature (T) : The temperature in Kuttiadi river varies in the range 24°C to 35.5°C. The maximum temperature was recorded during pre-monsoon and the lowest during monsoon. No significant difference in temperature values was found in the surface and the bottom layers.

Dissolved oxygen (DO) : The pattern of seasonal variation of DO across the river length is presented in Figure 2.

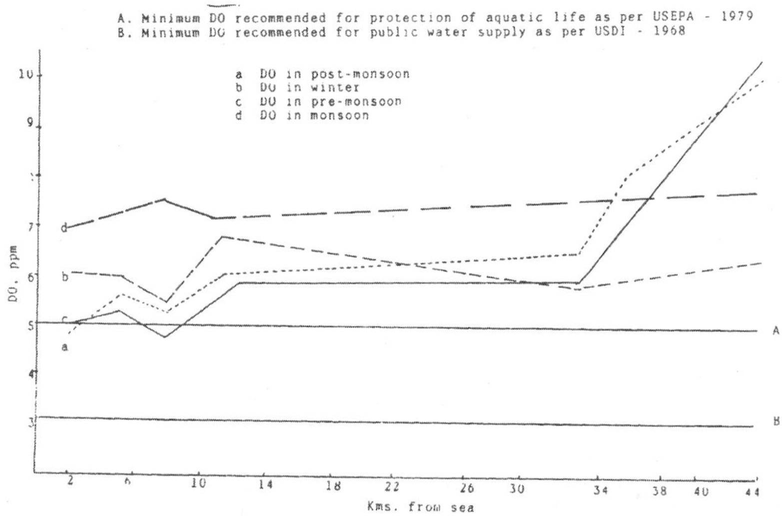

Figure 2 : Seasonal variation of dissolved oxygen in Kuttiadi river.

The dissolved oxygen levels are found to vary in the wide range of 2.7 ppm to 18.6 ppm in the river. In general, dissolved oxygen levels were found to decrease with increaseing depths. In a few cases the oxygen levels were seen to increase with increase in distance form the sea. For example the oxygen concentration in surface samples during winter at 2 km from the sea was 5 ppm whereas the corresponding level at 44 km was 18.5 ppm. At intermediate sampling points this trend was consistent showing almost gradual increase in DO with increasing distance from the sea. The possible reason for this observation is that coastal region is densely populated and there is stress on the river DO due to inflow of sewage and such activities as batching and laundry.

The fluctations in dissolved oxygen values during various seasons and at various distances from the sea (upstream) are represented in Figure 2. It can be seen that dissolved oxygen levels vary widely from season to season, without showing any uniform pattern beyond 22 km in the direction away from the sea. Before this distance the highest levels are evident during monsoon followed by winter and post-monsoon periods.

The lowest levels are found during pre-monsoon period. At the 8 km point, a slight decrease in oxygen level was seen during post-monsoon, winter, and pre-monsoon. This point is near the Payyoli Village, from where a small canal called 'Payyoli Canal' joins the river. The canal carried various organic wastes, arising from the market of the mixing of such wastes carried by Payyoli canal into the river.

pH

In Kuttiadi river the pH varies within a range of 6.1 to 7.8 units. In exceptional cases pH as high as 8.4 and as low as 5.7 was recoreded during pre-monsoon period. The lowest pH (3.4) recorded in bottom samples collected from Kurad (2 km from the sea) and the highest pH (5.7) was recorded from a small tributory Onipuzha (42 km from the sea).

Electrical Conductivity (EC)

The electrical conductivity in the Kuttiadi river varies in the range of 32 microsiemens/cm to 48500 microsiemens/cm. The electrical conductivity was found to decrease gradually from sea upto a distance of 33 km (upstream). Beyond this no significant changes in electrical conductivity were observed. The highest EC (48500 microsiemens/cm) was recorded during pre-monsoon period (March) from bottom layer at Mura (2 km upstream) and the lowest (30.6 microsiemens/cm) was recorded from Chariyakumbalam (~ 34 km upstream) during monsoon (Julyu). In many cases the electrical conductivity was found to be higher in bottom layers than surface layers specially during pre-monsoon, later winter, and monsoon. During late monsoon (end of July) and post-monsoon (October) the electrical conductivity levels in surface were decreased considerably throughout the course of river compared to winter and pre-monsoon.

Total Dissolved Solids (TDS) or Salinity

A representative set of observations pertaining to the seasonal variation of TDS across the river length is presented in Figure 3.

The patterns of variation of TDS in the river ran parallel to the patterns observed with EC. The TDS levels gradually decreased upto a distance of 33 km from the bar month (Figure 3) and thereafter did not vary with the season. The highest levels of TDS were recorded during summer (pre-monsoon; Figure 4) period and the lowest during monsoon (Figure 5) and post-monsoon periods. For example in surface layers at Murad (2 km upstream) the salinity levels during winter and pre-monsoon were found within the range 28062 ppm to 43875 ppm and 31875 ppm to 43812 ppm respectively, while during monsoon and post-monsoon the corresponding salinity levels were 139 ppm to 2956 ppm and 152 ppm respectively.

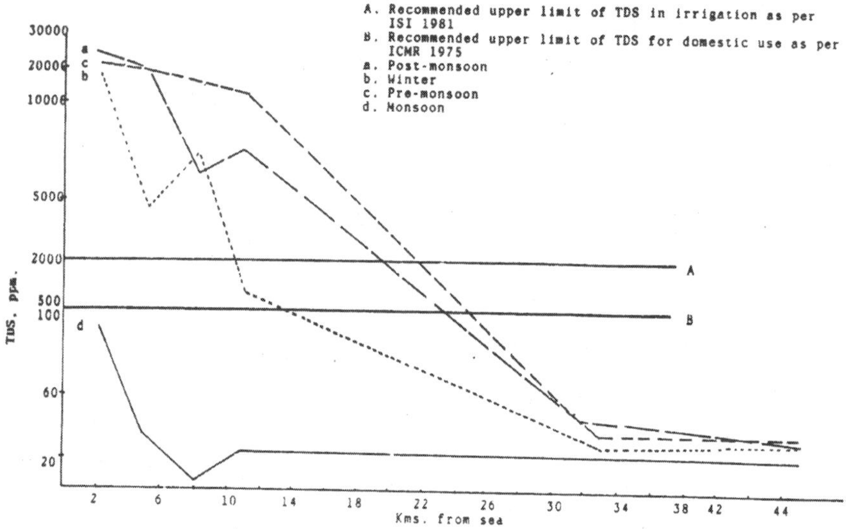

Figure 3 : Seasonal variation of total dissolved solids (TDS) in Kutliadi river

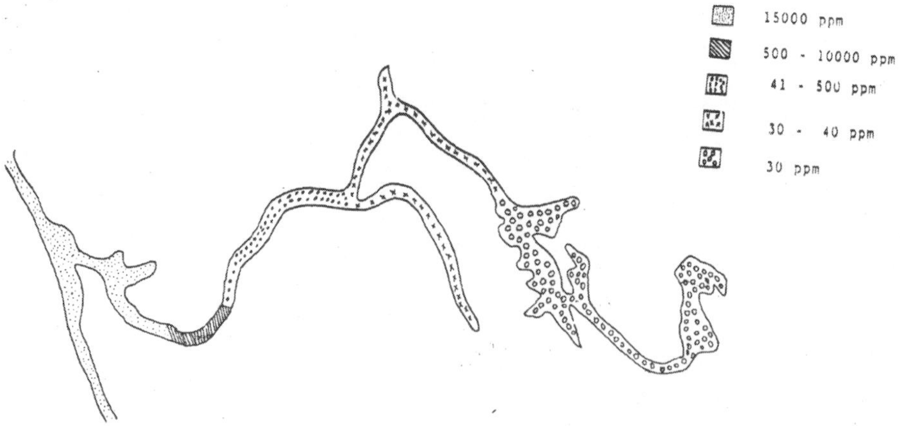

Figure 4 : Illustrative figure showing salinity in Kuttiadi river & reservoirs (summer)

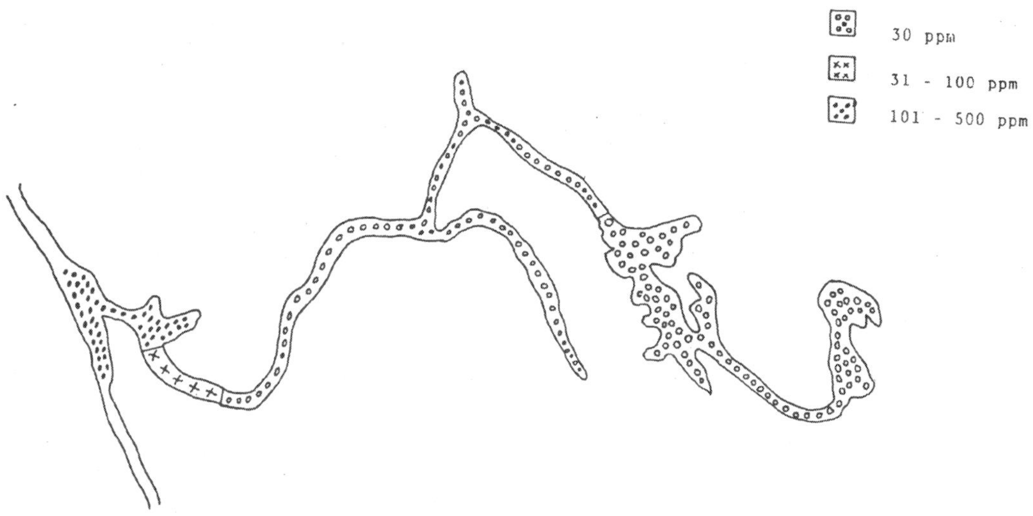

Figure 5 : Illustrative figure showing salinity in Kuttiadi river and reservoirs (monsoon)

A comparison of the findings with the recommonded levels of salinity for irrigation (ISI-1979), and domestic use (ICMR-1975), reveal that in many cases salinity levels are not acceptable for these uses specially upto a distance of 10 and 24 km from the sea during winter, pre-monsoon respectively (Figure 3). However in monsoon salinity levels were generally within acceptable levels - specially in surface layers.

Total hardness

The seasonal variations in total hardness more or less followed the pattern of salinity. The total hardness levels in Kuttiadi river varied widely from 6 ppm to 8000 ppm depending upon the season and place of sampling. The higher levels found upto a distance of 8 km upstream (from the sea); lower levels were recorded beyond this distance.

In terms of spatial variations, the total hardness levels in the river gradually decrease with increasing distance from the sea. In many cases the total hardness levels also increase with increasing depth. For example at Murad (2 km upstream) the total hardness in surface layers during December was 4300 ppm and at Valiyakumbalam to Muthatupuzha (10-44 km upstream) the hardness was ≤10 ppm respectively. During December, the total hardness levels in surface and bottom layers at Murad and Turasserykadav (4 km upstream) were 4300 and 6100 ppm, and 500 and 3500 ppm respectively.

In terms of temporal variation, the total hardness level decreased significantly during monsoon and post-monsoon and increased during the onset of winter and pre-monsoon. For example, total hardness levels in Murud (surface samples) were 3900 to 8000 ppm in winter, 3199 to 7100 in pre-monsoon, 14 to 230 ppm in monsoon and 27 ppm in post-monsoon months.

The highest levels of total hardness (7000-8000 ppm) in Murad were recorded during February and March and lowest (14-27 ppm) during July and October.

Exceptionally high levels of total hardness levels (80-4000 ppm) were recorded in water samples collected at the confluence of Kuttiadi river and Poyyoli canal also known as 'Payyoli Chirp', which is 8 km upstream.

A comparison of total hardness in the Kuttiadi river with prevailing drinking water standards set by MWH 1976 (200 - 600 ppm) reveal that the hardness levels in river beyond 8 km upstream only are within the acceptable levels for drinking purposes.

Chlorides

The chloride levels were found to vary in the range 5.5 ppm to 21539 ppm. In case of chlorides also the pattern of gradual decrease with increasing distance (upto ~ 8 kms upstream) was observed. The chloride levels upto 8 km upstream were within a range of 2200 ppm to 21539 ppm, whereas beyond this distance the chloride levels were ranging from 5.3 to 15.2 ppm.

In terms of temporal variations the highest chloride levels were recoreded in samples collected during winter and pre-monsoon and lowest during monsoon. For example in surface samples from Murad (2 km upstream) during winter and pre-monsoon, chloride levels varied from 12895 to 21539 ppm and 11548 to 20719 ppm respectively and during monsoon within a range of 37.5 to 772 ppm.

A significant variation in chlorides was recorded in samples collected from surface, middle and bottom layers of Kuttiadi river. For example, during February the chloride levels in surface, middle (3 meters), and bottom (6 meters) layers were 17469, 17875 and 19500 ppm respectively.

Exceptionally high levels of chloride (upto 200 ppm) were recorded in samples collected from the confluence of Kuttiadi river and Payyoli canal (8 km upstream), even during monsoon when the chloride levels in other areas nearer the sea (~ 4 km upstream) got diluted to ~ 125 ppm.

A comparison of allowable chloride levels for discharge on inland surface water (600 ppm; ISI 1974) and irrigation (600 ppm; ISI 1979) and potable use (< 25 ppm; ISI 1982; IMCR 1975) with the levels in Kuttiadi river reveals that the levels are not acceptable for these uses in many cases, specially upto a distance of ~8 km upstream.

Alkalinity
The alkalinity in Kuttiadi river varies from 6 ppm to 3600 ppm. In general, alkalinity levels were found to decrease following an increase in distance from the sea (upstream).

In terms of spatial variation, the alkalinity levels in winter vary from 88 to 400 ppm at Murad (2 km upstream) to 13-79.7 ppm at Muthattupuzha (~44 kms upstream). Similarly it has also been recorded that alkalinity varies significantly without showing any regular pattern with depth. For example, during March the samples collected from surface, middle (3 meters), and bottom (6 meters) showed alkalinity levels as 89.9, 96.4 and 109.2 ppm respectively, and during May the corresponding levels of alkalinity at these points were 130, 90 and 80 ppm respectively.

The patterns of seasonal variation in alkalinity in the river indicate that generally during winter (11-400 ppm) and pre-monsoon (11-1100 ppm) the levels are higher than monsoon (6-150 ppm).

Calcium
The calcium levels were found to vary in the range 1.6 ppm to 3000 ppm depending upon the variation in distance from the sea and the season of collection of the samples.

The calcium levels were found to decrease with the increase in distance from the sea. For example during February, the calcium level at Murad (2 km upstream) was 3000 ppm in surface samples, whereas at Muthatupuzha (44 kms upstream) the calcium concentration was only 3.2 ppm.

The calcium levels also vary within a wide range from season to season. In samples collected from Murad the levels varied from 200-3000 ppm in winter, 26-360 ppm in pre-monsoon and 0.8 - 20 ppm in post-monsoon periods.

As with other parameters, in case of calcium also a pattern of change across the depth of the river, was recorded.

For example, during December the calcium levels in surface, middle, and bottom samples at Murad were 200, 320 and 360 ppm and during May the correspoding levels were 320, 360 and 400 ppm.

Magnisium
The magnesium levels fell in the range 0.24 to 1464 ppm. The concentrations in the river were found to be highest at Murad and gradually decreased with increase in distance (upstream) from the sea upto 8 km Thereafter no significant

variation in magnesium concentration was noticed. For example, during February the magnesium level upto 8 km upstream were found to vary in the range 317 to 1464 ppm but in the 8 - 44 km stretch subsequently, the levels varied only narrowly in the range 0.5 - 0.98 ppm.

In terms of seasonal variations it has been found that magnesium levels were highest during winter and pre-monsoon and lowest during monsoon. For example, in samples collected from surface at Murad the magnesium levels in winter and pre-monsoon were 768-076 ppm and 88 - 1488 ppm, while during monsoon the levels varied in the narrower range : 1.46 - 51.24 ppm.

In many cases magnesium varied significantly with increasing depth. For example, in surface, middle and bottom layers at Murad the magnesium levels during January and June were 768, 812 and 976 ppm and 51.24, 75.6 and 917.4 ppm respectively.

Total phosphorus

The phosphorus concentrations ranged from zero (not dectectable) to 0.8 ppm.

In all samples collected during pre-monsoon, total phosphorus was below detection limit, while during winter and monsoon total phosphorus in the river samples were found to vary in the range below detection limit to 0.53 ppm, and below detection limit to 0.8 ppm respectively.

Phosphorus concentrations in the samples collected from 2 km upstream to 8 km upstream during winter and monsoon varied from below detectable levels to 0.53 ppm and 0.006 ppm to 0.8 ppm; while beyond 8 km upstream to 44 km upstream the sea, the corresponding levels varied in the range below detection limit to 0.1 ppm (winter) and below detection limit to 0.6 ppm respectively. All-in-all phosphorus levels were sometimes high enough to indicate eutrophic tendencies (Dhillon and Rigler, 1975).

Zinc

A representative set of observations pertaining to the seasonal variation of zinc in Kuttiadi river is presented in Figure 6.

The concentrations of zinc in Kuttiadi river fall in the range below detectable limit to 0.38 ppm (Figure 6). The highest zinc concentrations were found at a distance of 5 to 8 km from the sea (post and pre-monsoon) and lowest values were seen during monsoon. Exceptionally high levels of zinc were recorded in samples collected near the confluence of Payyoli canal and Kuttiadi river (8 km upstream), specially during winter and pre-monsoon. The levels of zinc in Kuttiadi river are found within acceptable limits (2-10 ppm) for drinking water (ISI 1982) and irrigation water (NAS 1972), but could yet be higher-than-safe (Abbasi and Soni 1986a, b, 1987; Abbasi et al 1985).

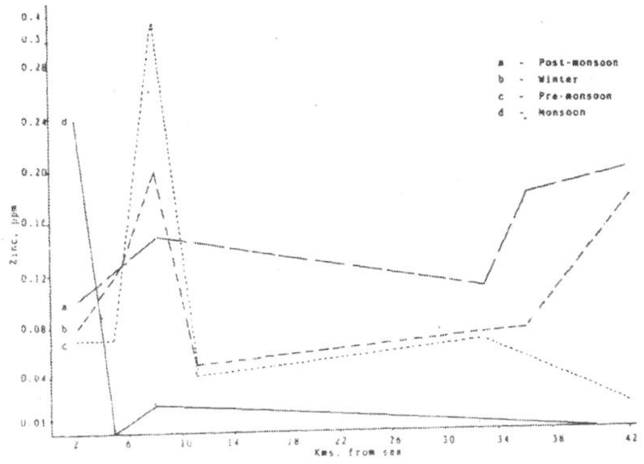

Figure 6 : Seasonal variation of zinc in Kuttiadi river.

Manganese
The pattern of seasonal variation of manganese across the length of kuttiadi river is illistrated in Figure 7.

The manganese levels in Kuttiadi river vary from below dection limit (0.005 ppm) to 0.15 ppm, in various seasons (Figure 7). The highest levels were seen at a distance of 8 km from sea during winter. The gradual decrease of manganese levels after 8 kms were observed with increasing distance from sea upto 44 kms. A comparison of drinking water standard of manganese (WHO 1971) with existing manganese levels of river indicate that in many cases manganese levels were objectionable, especially in the stretch between 2 to 11 km upstream.

Copper
The presence of Copper in the water of Kuttiadi river varied as functions of season and the length of the river, as illustrated in Figure 8.

The copper levels were found in the range 0.01 - 0.17 ppm. It can be seen from the figure that copper levels were higher in pre-monsoon, post-monsoon and winter specially upto 11 km upstream. Beyond 11 km a gradual decrease was observed with increasing distance. In post-monsoon and winter a slight increase in copper was observed beyond 34 km A comparison between the water quality standards of copper for domestic use (IMCR 1975) with the existing levels in Kuttiadi river indicates that the copper levels are objectionable for domestic use upto 35 km upstream (during post-monsoon), 19 kms upstream (during pre-monsoon), but they were found to be within acceptable limits during monsoon), but they were found to be within acceptable limits during monsoon period. The copper concentrations were found to be acceptable for irrigation as they were well below the NAS (1972) standards prescribed for irrigation water use.

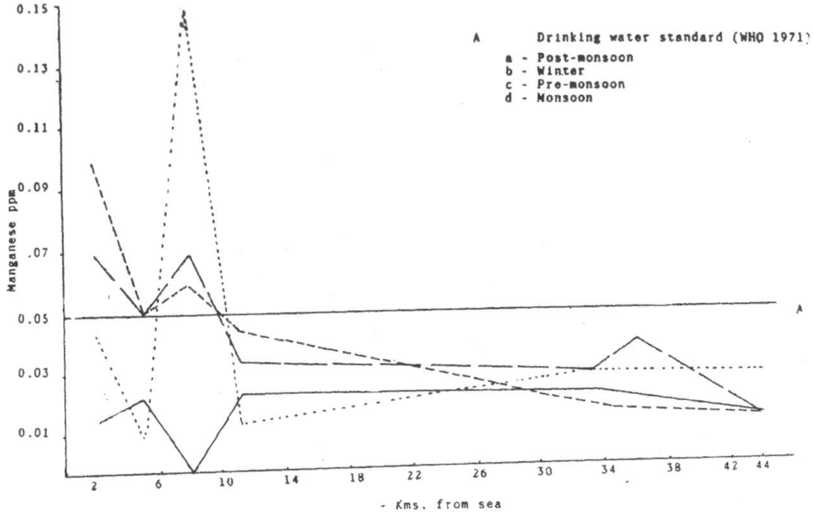

Figure 7 : Seasonal variation of Manganese in Kuttiadi river.

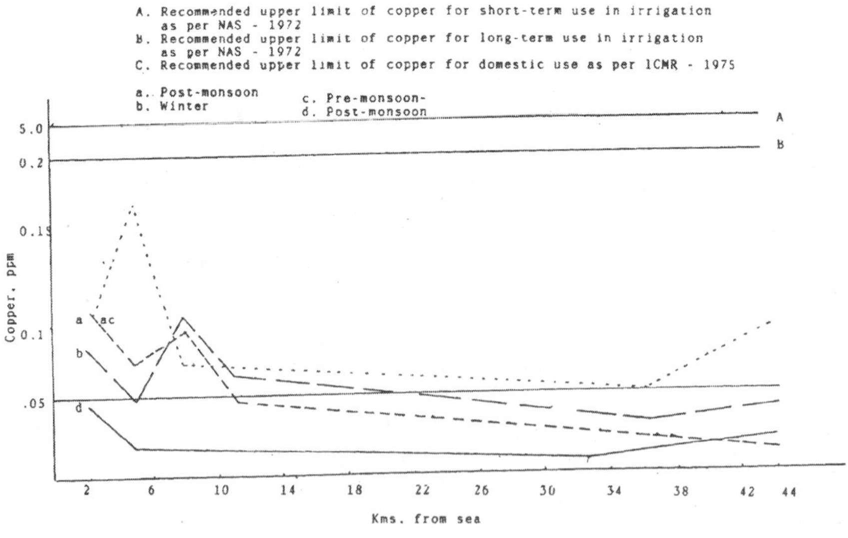

Figure 8 : Seasonal variation of copper in Kuttiadi river.

Summary of the status of water quality in Kuttiadi River.

i. In general the water quality of Kuttiadi River depends significantly on the following factors; seaons, saline water intrusion, demographic pressure around the river, and topography.

ii. The dissolved oxygen in water followed a consistent trend of increase with increasing distance from sea beyond 2 km upstream. The river showed exceptionally low levels of DO at 8 km upstream, (near Payyoli canal) where a stream carrying domestic and industrial wastes joins the river.

iii. The river salinity decreased gradually from the bar mouth upto 33 km upstream. A comparison of the existing levels of salinity for the river with the permissible levels of salinity of domestic use, drinking water and irrigation water reveal that except during monsoon the salinity levels in the stretch between bar mouth to 10 km upstream are not generally acceptable.

iv. Exceptionally high levels of tatal hardness were recoreded in waters collected near Payyoli canal. Beyond 8 km upstream the hardness levels were found to be adequate vis a vis drinking water quality.

v. Chloride showed a pattern of variation similar to the one found with salinity. The chloride concentrations were largely higher than acceptable for irrigation and domestic water use.

vi. Exceptionally high levels of zinc were recorded in samples collected near Payyoli canal specially during winter and pre-monsoon.

vii. In case of manganese a gradual decrease in its concentration with increasing distance from the sea was observed. The manganese levels between 2 to 11 km (upstream) were unsafe for drinking.

viii. The concentration of copper increased gradually as the river a approached the sea. A comparison of existing copper levels in Kuttiadi river with the levels of the metal allowed for domestic and irrigation waters indicate that the metal levels are objectionably high for domestic use upto 11 km (except during monsoon), but they are generally acceptable for irrigation.

Status of aquatic organisms in Kuttiadi river basin

We encountered 87 genera of aquatic organisms representating various taxonomic groups :

A. **Green algae :**
Ankistrodesmus, Arthrodesmus, Actinastram, Scendesmus, Tetraedon, Sphaerocystis, Pediastrum, Hephrocytim, Treubaria, Zygnema, Pachycladon, Staurastrum, Xanthidium, Navicula.

B. **Yellow green algae :** Dinobryon, Sonylux

C. **Yellow brown algae :** Peridinium

D. **Blue green algae :** Chroococcus

E. **Euglenoides :** Phacus, Euglena

F. **Diatoms :** Diatoma, Fragillaria, Synedra, Asterionella, Frustulia

G. **Rotifers :** Brachionus, Keratella, Euchlanis, Asplanchnopus

H. **Protozoans :** Diffugia, Verticella, Paramoecium

I. **Cladocerans :** Moina, Daphnia, Calanus, Sida, Bosmina

J. **Copepods :** Nauplius, Cyclops

K. **Fishes :** Brachidanio, Wallago, C.mrigila, Tilappia, Etroplus, Ophiocephalus, Arius, Barbus, Anguilla, Lutjanus, Belon, Puntius, Aemecheilus

L. **Macrophytes :** Nymphaea

A representative set of data pertaining to planktonic organisms is presented in Table 2. Comparison of the number of aquatic organisms found per unit volume of water at surface and at sechi disk depth reveals that, as expected, the abundance is significantly greater at surface than at the sechi disk depth. This is particularly tree for the organisms which contain chlorophyll (such as Ankistrodesmus, Treubaria, Tetraedon and Ulothrix). The concentration of organisms is higher in summer (June) than during monsoon (August) evidently due to the dilution effect of rains.

Table 2.

Seasonal Variations in the abundance of planktonic organisms in Kuttiadi river.

Organism	Number per litre			
	June		August	
	Surface	Secchi depth (1.3 m)	Surface	Secchi depth (1.3 m)
1. Ankistrodesmus	710	110	-	110
2. Asterionella	350	320	220	120
3. Arachnoidea	-	-	-	-
4. Cosmerium	80	70	-	-
5. Cyclotella	10	-	370	160
6. Cyclops	70	-	-	-
7. Daphnia	20	20	-	-
8. Dinobryon	-	-	-	-
9. Gonyaulax	50	80	-	70

10.	Malosira	180	370	250	160
11.	Nauplius	-	-	-	-
12.	Ostracoda	-	-	-	-
13.	Pediastrum	10	40	130	-
14.	Peridinium	600	160	110	-
15.	Rhizosolenia	40	150	-	-
16.	Staurastrum	380	520	270	-
17.	Surirella	10	-	-	-
18.	Treuberia	140	120	130	-
19.	Tetraedron	200	8	-	-
20.	Ulothrix	1020	180	-	-
21.	Xanthicium	10	30	130	-

❑ REFERENCES

Abbasi S. A., Kunhahamed T., and Soni R. 1985 "Relative toxicity of seven heavy metals and their safe concentrations". Technical Report, Centre for Water Resources Development and Management, Calicut, 32 p.

Abbasi S. A., and Soni R. 1986 a An examination of environmentally safe *levels* of zinc (II), Cadmium (II) and Lead (II) with reference to impact on channel *fish Nuria denricus. Environmental Pollution* (Series a) 40, 37-52.

Abbasi S. A., and Soni R. 1986 b Evaluation of permissible levels of Cobalt, Nickel and Copper in irrigation of waters and industrial effluents following impact studying on fresh water Teleost Nuria Denricus (Hamalton) In : Proceeding of International Conference on Chemical in the Environment, Lisbon.

Abbasi S. A., and Soni R. 1987 'Studies on Environmental Management of Mercury (II), Chromum (II) and Zinc (II) with respect to the impact of on some arthropods and protozoans - toxicuty of zinc (II). *Internation Journal of Environmental Studies* (Under publications).

Abbasi S. A. (1989) *Experimental aquatic ecology and water pollution,* Pondicherry University Press; 293 pages.

Abbasi S. A. (1996) : Water sampling for pollution analysis, *J. Inst. Public Health Engineers,* In press.

Beavan R. (1982) : *Handbook of the Fresh water Fisheries of India* - 245.

Chapra S. C., and Reckhow K. H. (1979) : 'Expressing the phosphrous loading concept in probabilitic terms, *Journal of Fisheries Research Board,* Canada, 36 (2) 22-

229.

Dillon P. J., and Rigler F. H. (1975) : A simple method for predicting the capacity of lakes for development based on lake tropic status, *Journal of firsheris research board*, Canada, 1519-1531.

ICMR (Indian Council of Medical Research; (1975) '*Manual of Standards of Quality for drinking water supplies*', New Delhi.

ISI (Industrial Standards Institution; 1981). Tolerance Limits for Industrial Effluents', Part - II, General Limits IS 2480(I) New Delhi.

ISI (Indian Standards Institution; 1982) : *Draft Indian Standard specialisation for drinking water*. DOC; CDC, 26, 8686, New Delhi.

Jhinghran V. G. (1982) : *Fish and Fisheries in India*. 2nd edition. Hindustan Publishing Corporation.

NAS (National Academy of Sciences; 1972) '*Water quality for agriculture*', Washington DC.

USDI (United States Dept of Interioir; 1968) 'Report of the Committee on Water Quality Criteria'. Federal Water pollution Control administration US Government printing office Washington.

USEPA (United State Environment Protection Agency; 1979) : Quality Criteria for water. R. E. Train Ed., Caste House Publication Limited, Great Britain.

WHO (World Earth Organisation; 1971) : *Internation Standards Drinking water. Geneva.*

Authors :

1. **S. A. Abbasi, Naseema Abbasi and R. Soni** Centre for Pollution Control & Biowaste Energy, Pondicherry University PONDICHERRY - 605 014.
2. **K. Madhavan** Centre for water Resource Development & Management KOZHIKODE - 673 571.

■ ■

3

Damodar River Pollution Due to Coal Washeries

❑ INTRODUCTION

Damodar river originating from Khamarpet near Chuthanpani with total catchment area extended over 18,100 Sq. km. flows from west to east. The river has been regarded as the lifeline for the Jharia coalfield because it is the only source of all major water supplies including drinking water. Jharia coalfield is the sole repository of prime coking coal in the country. There has been a number of coal based industries grown up in the area which forms an important industrial part of India. The inevitable consequences of this industrialisation and urbanisation has been increasing the pollution load on the river.

There has been widespread concern of Damodar river pollution in the highly industrialised and densely populated Jharia coalfield. The river water has been indiscriminately polluted due to continuous, uncontrolled and untreated discharge of industrial and domestic effluents (Gurdeep Singh, 1989, Prasad and Singh, 1990, Bandopadhyay, 1994). Coal washeries have been implicated as one of the major sources of Damodar river pollution (Gurdeep Singh, 1986,1990, Bandopadhyay, 1987, Gupta & Gurdeep Singh, 1993). There is an urgent need to assess the pollution of the river during effluents discharge from the coal washeries. Though a number of earlier studies have been made on Damodar river (Gopalakrishnan et.al. 1966, Dhaneheshwar et.al. 1972, Bandopadhyay, 1987, Gurdeep Singh, 1986, 1988, Gurdeep Singh et.al. 1989, 1990, Banerjee and Gurdeep Singh, 1993), no systematic study has been conducted so far on pollution of Damodar river due to coal washeries. The present study attempts to highlight the pollution of Damodar River due to coal washeries.

Background of Coal / Beneficiation

With the continuously increasing demand of cooking coal in conjuction with the exhaustion of good quality coking coal (XII to XVIII upper seams) in Jharia coalfield, it has been necessitated to mine lower seams coking coal. However, coal as mined usually contains useless rock, dirt, maceral and mineral matter from the strata above and below the coal from pocket of rock within the same coal seam itself. Run-of-mine (ROM) coal must be cleaned and dried before supply to the downstream industries. The whole cleaning process evolved various physical and chemical methods known as coal washing/beneficiation. Coal beneficiation is a generic term that is used to designate various operations performed on ROM coal to meet the size requirement, reduce the amount of ash and increase the heating value, prepare it for specific use. The range of options for physical coal cleaning is given in Figure 1. The washeries are at present receiving inferior grade lower seams coal both from underground as well as opencast mines. In earlier days the crushing sections of washeries were producing 3 inch coal but due to exploitation of lower seams of coal having indifferent washability characteristics the crushing units are supposed to crush the coal upto -1/2 inch. During size reduction huge amount of coal fines are generated due to fragile nature of coal. These coal fines mixed with process water and constitute the effluents. In washery operation about 25% of ROM coal reported as rejects and after processing it is simply dumped all around the washery. The major potential sources of contamination from coal preparation include storage and handling, coal transportation, coal refuse disposal, operating machines and liquid effluents streams.

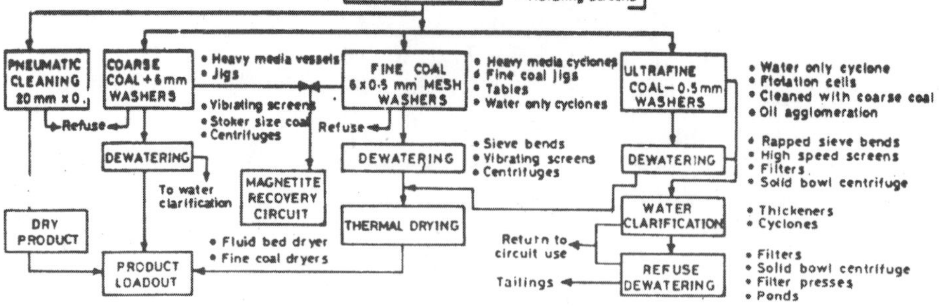

Figure 1 : Physical Coal Cleaning - Range of Options.

Characterisation of Coal Washery Effluents

Presently there are 25 washeries out of which 18 are located in the vicinity of the Damodar river. These 18 coal washeries utilise Damodar river water for different washing processes and subsequently huge amount of coal fines is generated which is mixed with process of wet washing and constitute the effluents. The amount of process water used in each washery varies with the plant capacity. On the basis of our survey and estimates, average water demand at present in the washeries reaches 60-170 gallons/tonne of feed coal and the average effluents discharge various from 150 m^3 to 750 m^3 back into the Damodar river.

To know the pollution profile from coal washeries effluents discharge from the final outlet of washeries were sampled and characterisation of coal washery effluents were carried out (Gupta and Gurdeep Singh, 1995). The minimum, maximum and average values are reported in data form in Table 1. Suspended solids, oil & grease and COD are identified as the three main pollutants. Direct discharge of coal washery effluents give rise to visual and aesthetic pollution of Damodar river which appears blackish to brownish black in colour. Suspended solids contents in the washery effluents exceed 20-200 times the permissible limits of 100 mg/l as per IS:2400. There is considerable load of suspended solids predominantly in the form of coal fines. COD value exceeded the permissible limit of 10 mg/l as per IS:2490. A thin film of oil and grease exists over the surface of Damodar River which seriosuly impede the self purification capacity.

While all the washeries in India have been carefully to operate on close circuit conception, the washeries often dishcarge considerable quantity of coal fines into the Damodar river system due to increase in amount of slurry and poor performance of recovery system. This apart from posing serious water pollution problem also results economic losses in the form of coal fines of highly combustiable value (Ravi Gupta & Gurdeep Singh, 1994).

Reasons for Unavoidable Discharge of Effluents from Washeries

All the washeries operating at present fail to keep the exact record of process water and chemicals to the process plant, virtually not maintaining the close water circuit conception. The solid particles generated during washing process settled down in thickener and clarified water is collected in a sump and then pumped back to the system. However, under unavoidable conditions tailings are allowed to be bleeded out due to the following reasons (Gupta & Singh, 1995) :

There occurs wide variation in characteristics of coal due to the supply of heterogeneous coal even from the same coal seam creates imbalance in performance of equipment and necessitates bleeding of effluents.

Thickness blockage causes overflow with subsequent loss of coal fines along

Table 1

Characterisation of Coal Washery Effluents

Parameters in mg/l except colour pH, temp. & cond.(Gmhos)	Sudamdih Mi - Mx	Ag	Dugda Mi - Mx	Ag	Chasnalla Mi - Mx	Ag	Jamadoba Mi - Mx	Ag	Patherdih Mi - Mx	Ag	Bhojudih Mi - Mx	Ag	Permissible Discharge Limit IS : 2490
Colour	Black		Black		Black		Black		Black		Black		Colourless
pH	7.2-7.9	7.5	6.8-7.2	7.0	7.6-8.0	7.8	6.8-7.0	6.9	7.0-7.2	7.0	7.9-8.2	7.5	5.5-9.0
Temp (°C)	26-28	27	30-32	29	30-36	30	31-38	32	29-35	30	24-30	28	<40-40°C
TSS	2000-300	2150	1811-2100	1850	2172-2826	2405	782-1798	1225	982-1419	1080	320-700	450	100 mg/l
TDS	120-480	275	860-1100	840	235-510	255	76-312	175	86-206	139	356-375	359	2100 mg/l
Conductivity	280-320	260	560-720	575	250-336	268	200-258	185	228-226	225	219-250	228	-
Oil & grease	9.2-16.8	10	12.2-18	12.5	7.8-19	11.2	12.0-14.0	11.0	9.8-18.0	12.0	4.8-12.8	7.2	10.0 mg/l
DO	2.2-3.4	2.4	2.8-4.2	2.9	2.8-3.0	2.9	2.5-3.8	3.0	2.6-3.4	2.9	3.8-4.7	4.0	-
COD	340-660	468	283-796	425	432-960	545	498-814	525	376-896	585	230-790	445	250 mg/l
Hardness	140-152	132	120-235	120	135-142	138	98-112	101	146-180	160	104-157	110	-
Calcium	32-38.5	29	29-110	55	29-36	32	26-58	38	38-42	38	64-126	75	-
Magnesium	20-34	22	17-120	61	19-31	24	22-46	30	28-36	30	28-81	45	-
Sulphate	25-35	27	26-78	49	20-42	29	30-48	34	35-68	48	44-56	49	100 mg/l
Chloride	60-82	62	29-68	39	60-90	65	40-80	59	56-78.8	65	26-40	29	100 mg/l
Fe	BDL		0.0-1.04	0.9	0.0-0.16	0.16	BDL		0.0-0.2	0.2	0.0-0.4	0.4	-
Cr	BDL		BDL		0.0-0.4	0.4	BDL		0.0-0.8	0.8	0.0-1.2	0.8	2mg/l
Cu	BDL		0.2-1.4	1.0	1.0-1.08	0.8	BDL		0.4-1.2	1.0	0.0-0.6	0.6	3 mg/l
Mn	BDL		0.0-0.6	0.4	BDL		0.0-1.06	0.9	BDL		0.0-2.8	2.8	-
Pb	BDL		BDL		BDL		BDL		BDL		BDL		0.1 mg/l
Zn	BDL		BDL		2.0-2.08	1.2	BDL		BDL		BDL		15 mg/l
Cd	BDL		BDL		BDL		BDL		BDL		BDL		2 mg/l
Hg	BDL		BDL		BDL		BDL		BDL		BDL		0.01mg/l
As	BDL		BDL		0.0-0.01	0.01	0.0-0.01	.01	BDL		BDL		0.2 mg/l

BDL - Below detection limit, i.e. 0.001 mg/l, Mi - Minimum, Mx - Maximum, Ag - Average.

with washery effluents, finally contaminating the surface water/Damodar river.

Uncontrolled consumption of water for clarification, desliming, dewatering and other processes cause overloading of various sections and consequent overflow from these sections from the washery effluents.

Though the power failure is not regular, shut down of any unit causes unwanted overflow/bleeding of the effluents.

If vaccum filter is not operating efficiently, there occurs unwanted internal flow of slurry containing high concentration of coal fines, which forms the washery effluents.

All the washeries are planned with a maintenance shift. However, it has been observed in most of the washeries that there is extra operation during maintenance period which interferes with the normal functioning of the system and lead to form the washery effluents.

During physical operation, lots of coal fines are generated and spread over a large adjoining areas. Floor washing, spraying, spill water of maintenance period, rain water etc. are a few other factors contribute to form the washery effluents through mixing with the coal fines, which are rushed to the Damodar river system causing environmental pollution.

Some attempts are made to recover the coal fines escaping from the coal washery effluents in nearby settling ponds, but in most of the cases these discharged washery effluents are recovered by private parties by settling the fines in ditches. The overflow of these ditches go to the Damodar river. During rainy season the settled solids are frequently flushed to the river system thus polluting the Damodar river.

Most of the washeries do not have well built settling ponds. These are uneven in size and do not support the plain sedimentation process. Timely excavation of deposited solids is essential for proper functioning of the process, but effluents occasionally flow over the filled ponds and subsequently get into the Damodar river.

Though recirculation pumps are installed in the washeries for recycling of the clarified water and even effluents, these do not work efficiently due to improper siting of settling ponds in a proper series and also not excavating the settled material in a timely and periodic basis. This increase the water demand resulting continuous discharge of effluents ot the Damodar river.

Monitoring Stations at Damodar River

Three monitoring points viz. upstream, confluence point and downstream of Damodar river around the six coal washeries at Bhojudih, Patherdih, Sudamdih, Jamadoba, Chasnalla and Dugda have been selected for this study. The study was conducted over one year period. (Feb. 1993 to Feb. 1994). Data ware recoreded from each sampling point at all the selected washeries except in Patherdih where the data recorded only from confluence point. The sampling points and approximate location of coal washeries are illustrated in Figure 2.

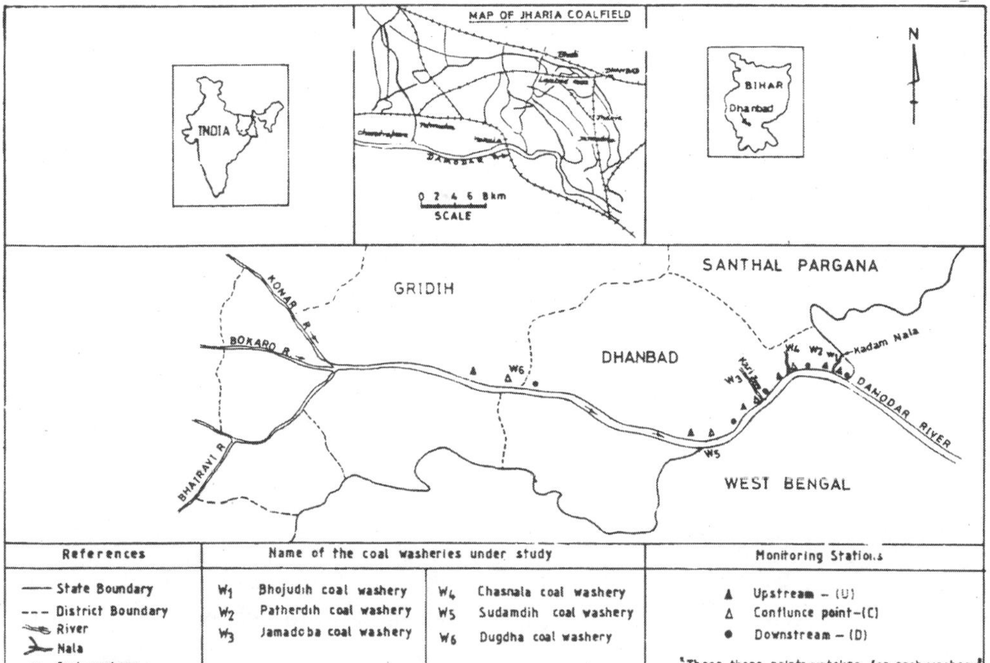

Figure 2 : Coal Washeries Under Study and Monitoring Stations on River Damodar

Experimental Methodology

Water samples were collected as per standard sampling techniques at monthly intervals from selected sampling points between 8 a.m. to 10 a.m. on the day of sampling. In some cases sampling time may differ due to non operational shift of coal washeries. Several grab samples collected from each point were finally made into one composite sample. All these samples were analysed as per standard methods (IS : 2488, APHA 1985, NEERI Manual 1986). Some of the sensitive parameters such as pH, temperature, TDS, conductivity and dissolved oxygen were directly monitored in the field itself under in-situ conditions using Century portable kit CK - 710. Heavy metals analysis were carried out using GBC - 902, double beam Automic Absorption Spectrophotometer.

Result and discussion

The entire Damodar river water appears greyish black or brownish black in colour in the entire stretch of the Jharia coalfield particularly around coal washeries. The visual pollution of Damodar river is mainly caused due to the uncontrolled and untreated discharge of coal washery liquid effluents into the river. Suspended solids, COD and oil & grease are identified as major water pollutants of the coal washery effluents. Mean values of major pollutants recorded at selected sites are depicted in Figure 3 and various other water quality parameters ar presented in Table 2.

It is revealed from the study that water quality of Damodar river at all the selected monitoring stations was neutral to slightly alkaline in nature with pH ranging between 7.2 to 8.1. This indicates that decomposition of carbonaceous organic matter contributed by washeries effluents supressed the pH all along the study area.

The suspended solids concentration level which are mainly in the form of coal fines exceeded the parmissible limit of IS:2490 at all the monitoring stations except the upstream and downstream of Bhojudih washery. As the size of suspended solids were found to be less than that of 300 mm which behaves like colloidal particles, these do not settle down easily. The coal fines of colloidal nature tend to remain suspended in river water for long period of time, thus reflecting river colour black. Due to continuous and uncontrolled discharge of these effluents, the suspended solids hinder the movement and breathing of the aquatic life and make the water unsuitable for their breeding. The asphyxiation of benthic life form has been observed due to choking of their pores with coal fines.

COD in general exceeded the permissible limits of 250 mg/l as per IS:2490 at the confluence point of the washeries at Damodar river, which are mainly due to the high concentration of suspended solids and their reducing character in the form of coal fines. This seriously impedes the self purification capacity of Damodar river which is further corroborated with considerably low dissolved oxygen value observed at confluence point of all washeries.

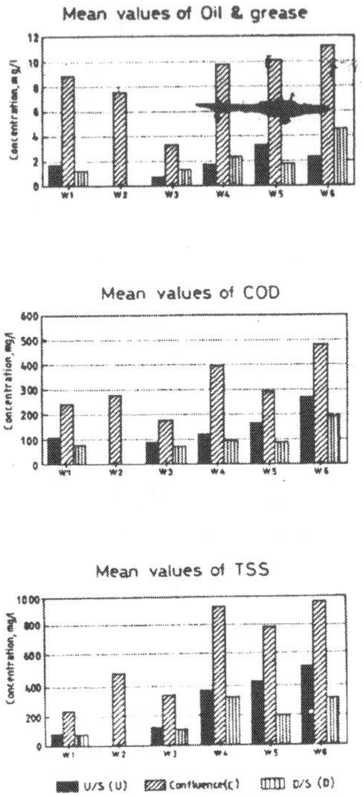

Figure 3 : Major water pollution indicators of coal washery effluents.

Discharge of coal washery effluents containing oil & grease at significant concentration has severely implicated the Damodar river water quality. Oil and grease concentration levels exceeded the permissible limit of 10 mg/l as per IS:2490 at the confluence point of the Dugda, Sudamdih and Chasnalla. Because of its poor solubility and low specific gravity characteristics in water, a flating layer of oil & grease can easily be observed over the surface of river water around the study area. This seriously affects the biological life, photosynthesis and self purification processes and overall carrying capacity of the Damodar river. This also impede the contact of dissolved oxygen with the aquatic life and there has been occurrences of massive fish kills in certain pockets of Damodar river along the washery area. Discharge of coal washery effluents containing high oil & grease contents has also been adversely affecting the water treatment plants situated at the bank of Damodar River.

Table 2

Mean values of Physico-Chemical Parameter recorded at different sites during effluents discharge from Coal washery in Damodar river (during 1993).

Parameters in mg/l except colour pH, temp. & cond.(Gmhos)	W1	W2	W3	W4	W5	W6	Permissible Discharge Limit
	A - B - C	A - B - C	A - B - C	A - B - C	A - B - C	A - B - C	IS : 2490
Colour	Gb Bb Gb	Bb	Gb Bb Gb	Gb Bb Gb	Gb Bb Gb	Gb Bb Gb	Colourless
pH	7.5 - 8.1 - 7.6	NA - 7.9 - NA	7.5 - 7.8 - 7.2	7.8 - 8.0 - 7.6	7.4 - 7.9 - 7.5	7.6 - 6.8 - 7.2	5.5 - 9.0
Temp (°C)	31 - 28 - 30	NA - 32 - NA	31 - 29 - 30	29 - 33 - 30	28.4 - 31 - 30	30 - 33 - 29	<40-40°c
TSS	86 - 238 - 74	NA - 482 - NA	124 - 342 - 110	367 - 928 - 328	427 - 782 - 201	517 - 962 - 312	100mg/1
TDS	108 - 291 - 170	NA - 325 - NA	210 - 282 - 254	198 - 504 - 169	228 - 349 - 136	276 - 434 - 168	2100mg/1
Conductivity	128 - 335 - 215	NA - 368 - NA	224 - 312 - 276	212 - 532 - 175	261 - 380 - 181	216 - 412 - 189	-
Oil & grease	1.8 - 8.9 - 1.2	NA - 7.6 - NA	0.0 - 3.4 - 1.2	1.8 - 9.8 - 2.4	3.4 - 10.1 - 1.8	2.4 - 11.3 - 4.6	10.0mg/1
DO	7.1 - 6.0 - 6.9	NA - 5.2 - NA	6.2 - 6.8 - 7.4	6.2 - 5.8 - 6.8	6.5 - 5.4 - 7.2	6.7 - 5.9 - 7.0	-
COD	110 - 247 - 79	NA - 280 - NA	89 - 180 - 75	124 - 398 - 97	165 - 296 - 89	272 - 486 - 198	250mg/1
BOD	1.8 - 2.0 - 2.4	NA - 3.1 - NA	1.2 - 2.9 - 2.6	1.5 - 3.8 - 1.4	1.5 - 2.8 - 1.2	2.4 - 4.1 - 2.0	30mg/1
Hardness	90 - 118 - 102	NA - 136 - NA	87 - 112 - 102	112 - 143 - 90	112 - 133 - 90	106 - 197 - 132	-
Calcium	75 - 114 - 98	NA - 76 - NA	24 - 46 - 19	70.8 - 82 - 62	70.5 - 82 - 64	60 - 102 - 55	-
Magnesium	20 - 42 - 24	NA - 39 - NA	29 - 32 - 27	23 - 46 - 19	26 - 39 - 31	32 - 54 - 26	-
Sulphate	25 - 36 - 21	NA - 37 - NA	29 - 48 - 32	30 - 22 - 13	31 - 38 - 24	23 - 46 - 39	1000mg/1
Chloride	29 - 26 - 33	NA - 19 - NA	59 - 64 - 41	14 - 54 - 34	14 - 54 - 38	10 - 39 - 13	-
Fe	0.8 - 1.4 - 1.2	NA - 2.12 - NA	BDL-1.02-BDL	0.34 - 0.46 - 0.14	0.34 - 2.26 - 0.19	2.4 - 1.03 - .05	-
Cr	BDL - 0.8 - BDL	NA - BDL - NA	BDL - 2 - 08	BDL-BDL-BDL	BDL-BDL-BDL	BDL-BDL-BDL	2mg/1
Cu	BDL - 0.6 - 0.2	NA - 1.0 - NA	BDL - 1.04 - 24	BDL-BDL-BDL	BDL-BDL-BDL	1.19 - 1.8 .4	3mg/1
Mn	BDL - 1.2 - 0.8	NA - 0.6 - NA	BDL - 0.6 - 10	BDL-BDL-BDL	BDL-BDL-BDL	BDL-BDL-BDL	-
Pb	BDL-BDL-BDL	NA - BDL - NA	BDL-BDL-BDL	BDL-BDL-BDL	BDL-BDL-BDL	BDL-BDL-BDL	0.1mg/1
Zn	BDL - 0.4 - 0.4	NA - 2.0 - NA	BDL- 1.06 BDL	BDL- 1.02 -BDL	BDL- 2.03 -BDL	BDL- 1.83 -BDL	15mg/1
Cd	BDL-BDL-BDL	NA - BDL - NA	BDL-BDL-BDL	BDL-BDL-BDL	BDL-BDL-BDL	BDL-BDL-BDL	2mg/1
Hg	BDL-BDL-BDL	NA - BDL - NA	BDL-BDL-BDL	BDL-BDL-BDL	BDL-BDL-BDL	BDL-BDL-BDL	0.01mg/1
As	BDL-BDL-BDL	NA - BDL - NA	BDL-BDL-BDL	BDL-BDL-BDL	BDL-BDL-BDL	BDL-BDL-BDL	0.2mg/1

NA - Not assesed. **BDL** - Below detection limit, i.e.-.001 mg/1. **A,B,C** - Respectively stands for Upstream, Confluence point and Downstream

Trace elements such as Fe, Cr, Cu, Mn, Zn etc. are present at low concentration levels which are below the permissible limits as per IS:2490 and as such do not seem to pose any serious pollution problem. Other parameters in general are either below or within permissible limits.

Overall the Damodar river has been seriously polluted over the entire study area due to continuous discharge of uncontrolled and untreated effluents from coal washeries.

CONCLUSION

This study highlight the pollution of Damodar river due to suspended solids, oil and grease and COD contents from coal washery effluents. There occurs visual and asesthetic pollution of Damodar river. Discharge of coal washery effluents impart blackish and brownish black colour and a film of oil & grease over the surface of the Damodar river impeding seriously its carrying capacity. The tailings laden settling ponds of the washeries situated in the close range of the river, always overflow and pollute the river. Recently coal washeries attempt to maximise water recycle in order to reduce the quantity of effluents discharge outside and it is expected that coal washeries will tend towards close water circuit as originally designed for. However, strict environmental compliance is required to check the pollution load contributed by the Coal washeries into the Damodar river.

❑ REFERENCES

Gopalakrishnan V. 1966. Observati[illegible] the breeding of major crops in the Tilaiya and Panchet reservoirs, *Bull. Cent. Inland Fish. Res. Int.*, Barrackpore No. 9, 9p (Mimeo).

APHA. 1976. *Standard methods for examination of water and waste water 14th ed.*, American, Pub. Health Association, Inc., New York.

Bandopadhyay P. 1987. Emission of water pollutants from coal washeries and water pollution control technology, Seminar on Water Pollution and Land Reclamation in mining areas with Special Reference to Jharia Coalfield, ISM, Dhanbad. June 5, 1987.

Gurdeep Singh. 1986. Impact of Coal mining on water quality; National seminar on Environment pollution control & monitoring, CSIO, Chandigarh, Oct. 21-23, 1986.

Gurdeep Singh 1986, Status of Water Quality in a mining area. A case study of Jharia coalfield International symposium on Environmental Management, Istanbul Turkey, June 5-9, 1986.

Gurdeep Singh. 1988. Impact of Coal Mining and Mine Water Quality, *Int. J. Mine*

Water (3) 49-59, 1988.

Gurdeep Singh and Ram. 1989. Status of Water Quality in Mining area - A case study in Jharia coalfield, *Minetech 10 no. (l)*:73-78.

Prasad, Bably and Gurdeep Singh. 1990. Environment impact due to coke-plant effluents - An overview, Proceeding of the colloquium on World Env. day celebration, 5 June, 1990, p 49-54, held at ISM, Dhanbad.

Banerjee, S. P. and Gurdeep Singh. 1993. Problems of water pollution in mining areas - Issues related to its protection and control. Proceedings of the First National Seminar on Environment Policy Issues in Mineral Industry, Organised by The Mining Geological and Mettallurgical Institute if India, Calcutta, Jun. 11-12, 1993.

Banerjee, N. N. Investigation into physico-chemical characteristics of coal washery effluents and their utilisation in optimisation of the effluents treatment practice. *Ph.D. thesis, Indian School of Mines,* Dhanbad.

Gupta, Ravi and Gurdeep Singh. 1993-94. Characterisation of Coal washery liquid effluents (unpublished).

Gupta, Ravi and Gurdeep Singh. 1994. Pollution potential of Coal Washeries in Jharia Coal field and its Impact on Damodar river, International symposium on Mineral Beneficiation-Recent trends & Beyond - 2000 A.D. Nagpur, India Oct. 3.5, 1994.

Gupta, Ravi and Gurdeep Singh. 1995. Water Pollution profile of coal washeries, *Poll. Research 14(2)* 203-213 (1995).

Wills, B.A. 1987. Mineral Processing Technology, 2nd Ed., Pergamen Press Oxford.

Leonard, Joseph. P.1979 Coal repartation 4th Ed. *The American Institute of Mining Metallurgical and Petroleum Engg. Inc. New York.*

Chandra, D. 1992. Jharia Coalfield, published by - Geological Society of India Banagalore, 1992.

Bandopadhyaya, P., Processing and Utilisation of Minerals and Pollution of Waer Resources, Second National Seminar on Minerals and Ecology. ISM, Dhanbad, Jan. 1994.

Authors :

1. **Ravi Kumar Gupta and Gurudeep Singh,** Centre for Mining Environment Indian School of Mines, DHANBAD - 826 004. Bihar (India)

■■

4

Damodar River Water Quality Status along Dugda-Sindri Industrial, Belt of Jharia Coalfield

❑ INTRODUCTION

Damodar river is the lifeline of thickly populated and highly industrialised areas of Jharia coalfield and all the major water supplies including drinking are met from it. The river Damodar once known as the 'River of Sorrow' for its seasonal ravages, has now become a constant agony from pollution point of view. This is highly polluted due to continuous and uncontrolled discharge of toxic and hazardous effluents into it by over 50 industries located along the river. With the objective to abate the pollution of water in the Damodar river caused by continued and uncontrolled discharge of toxic and hazardous effluents from industries located on its banks or in its vicinity the Central Pollution Control Board decided to launch a time-bound action oriented pollution control programme on the Damodar river with top priority (CBPCWP, 1985). However, no major success has been observed regarding abatement of pollution and rather more pollution generating industries have been sited in the recent past around the banks of Damodar river. Recently Damodar river has been declared as one of the 17 most polluted rivers of India. The Ministry of Environment and Forest, Government of India under its second phase of Ganga Action Plan envisages to launch Operation Damodar River Clean.

That Damodar river, the prime source of drinking water for about 1.2 million of People of Jharia coalfield, is being polluted by cluster of industries like, coal washeries, thermal power, - fertiliser, - steel, - coke oven, - cement plants lead- zinc smelter and several refractories. Pollution of Damodar river in the Bokaro-Sindri industrial belt of Jharia coalfield was reported on water pollution due to coal mines (Rawat et. al. 1981, Gurdeep Singh 1990, and coal washeries (Gupta and Singh, 1995, Bandopadhyaya, 1994). Status of water quality of Damodar river in the Jharia coalfield revealed that the river is grossly polluted in the entire coal mining area due to continuous discharge of uncontrolled and untreated effluents from various industries situated around Damodar river (Gurdeep Singh, 1990, Gurdeep Singh and Bigan Ram, 1989). Major sources which result water quality deterioration in the region, identified as coal washeries, thermal power plants, coke ovens and domestic wastes.

Siting of Sampling Points

A stretch of 50 km from Dugda to Sindri, industrial stretch of coalfield was selected for the detailed analysis of the Damodar river. Sampling points and approximate locations of each industry in and around the river are shown in the plan given in Fig. 1.

Five sampling points were selected from Dugda to Sindri. Industrial effluents of various industries situated in between these points were also collected. River sampling points were selected in such a way so as to include the effluents discharged from all major polluting industries in between to provide the overall pollution profile.

Description of sampling points of Damodar River :

1. U/S of Damodar river near Dugda Coal Washery
   This sampling point reflects the quality of river water just before Dugda Coal Washery and it also gives baseline data for the study. River water here contains effluents of Chandrapura Thermal Power Plant.
2. D/S Damodar river near Dugda Coal Washery
   This sampling point gives the combined effects of pollution caused by Chandrapura and Dugda Coal Washery.
3. Near Telmucho bridge (D/S of Dugda Coal Washery)
   This sampling point is selected to know the status of river water after mixing and dilution at a distance of 2 km from Dugda Coal Washery sampling point at Damodar river.
4. River Damodar near Jamadoba Water Works
   This sampling point gives the extent of pollution caused by Loyabad Coke Oven Plants, Loyabad Power House, Moonidih and Jamadoba Coal Washeries. This sampling point is very vital as CMADA's Water Works are situated here and drinking water supply is being made after treatment of river water at this point to a sizeable section of mining population.
5. Damgarh Water Works
   This sampling point reflects the combined effects of various industrial and domestic effluents on Damodar river water quality. Additionally effluents from Sindri Township complex and also the domestic wastewater are being discharged here.

Sampling Procedure

Water and effluents were sampled as per IS:2488, 1966 Part 1, NEERI Manual and APHA. Samples were collected in cleaned polyethylene bottles, which were rinsed initially with a portion of sample. Monitoring was done on monthly basis, starting from May to October, 1992.

For oil and grease sampling, a pre-cleaned small wide mouth glass container was used because this particular parameter requires whole sample for the analysis. For Bacteriological analysis pre-sterilised (at 121^0 C for 15 minutes and paper wrapped) BOD bottles were used. Collection, preservation and analysis of samples was carried out as per APHA (1980)

Damodar River water quality analysis results

There is a wide variation in the water quality parameters as analysed during May to October 1992 of Damodar river water of all five different monitoring stations along Dugda - Sindri industrial belt in Jharia coalfield. The water quality parameters as analysed each station wise are given in Table 1-5. Water quality varies from season to season and is very much relevant to discharge valocity and self-purification capacity of the river. All the five sampling stations were selected so as to include the effluents discharges from all major sources of pollution from various industries into the river as already described and also account for the variation of river water quality at these stations. Some of the significant water quality parameters are presented below.

Table 1

Analysis of U/S Damodar river before Dugda Coal Washery.

Parameters	1992					
(mg/*l*)	May	June	July	August	September	October
pH	7.2	7.5	7.8	8.0	7.6	7.5
Temperature A/W (^{0} C)	38/35	35/33	31/32	31/30	32/31	30/29
Turbidity (NTU)	768	744	785	991	811	696
Suspended Solids	1242	1220	1268	1652	1305	1042
Dissolved Solids	208	184	182	186	184	183
Alkalinity	92	108	128	143	102	90
Hardness	105	125	117	186	139	144
Calcium	19.0	19.5	28.4	28.2	27.8	24.0
Magnesium	38.8	28.2	39.4	50.8	46.8	35.8
Chloride	18.8	18.2	19.4	17.8	16.8	15.8
Sulphate	38.2	67.5	36.6	76.8	48.2	69.7
DO	5.2	5.8	7.0	7.2	6.8	5.8
BOD	2.3	1.8	1.2	1.0	1.2	1.2

COD	468	405	454	1620	484	360
Oil and Grease	0.8	0.72	1.0	2.4	1.1	0.6
Lead	BDL	BDL	BDL	BDL	BDL	BDL
Total Chromium	BDL	BDL	BDL	BDL	BDL	BDL
Copper	BDL	BDL	BDL	BDL	BDL	BDL
Zinc	0.5	0.5	0.5	0.5	0.5	0.5
Iron	0.001	0.001	0.002	BDL	BDL	0.001
Manganese	BDL	BDL	BDL	BDL	BDL	BDL
Selenium	BDL	BDL	BDL	BDL	BDL	BDL
Nickel	BDL	BDL	BDL	BDL	BDL	BDL
Ammoniacal nitrogen	Nil	Nil	Nil	Nil	Nil	Nil
Nitrate	Nil	Nil	0.1	0.2	0.11	Nil
Total Coliform (MPN/100 ml)	1200	1400	1800	1800	1500	1200
Faecal Coliform (MPN/100 ml)	400	500	600	600	500	480

Table 2

Analysis of D/S Damodar river after Dugda Coal Washery.

Parameters	1992					
(mg/*l*)	May	June	July	August	September	October
pH	7.0	7.2	7.8	7.8	7.2	7.5
Temperature A/W (°C)	38/30	36/30	31/30	30/29	31/30	30/29
Turbidity (NTU)	984	801	1042	1682	900	704
Suspended Solids	1642	1308	1968	4382	1598	1143
Dissolved Solids	248	268	242	128	198	180
Alkalinity	90	96	100	110	129	130
Hardness	119	108	110	190	148	146
Calcium	30	32	36	48	49	58
Magnesium	13	17	19	22	24	28
Chloride	27.8	32.2	28.1	25.8	49.6	38.8
Sulphate	51	26	27	66	25	21
DO	6.2	6.4	7.0	6.4	6.0	6.0
BOD	2.3	2.2	1.8	2.8	1.0	2.9
COD	1120	410	1208	1760	1018	328
Oil and Grease	1.2	1.0	1.8	9.4	1.1	0.8
Lead	BDL	BDL	BDL	BDL	BDL	BDL
Total Chromium	BDL	BDL	BDL	BDL	BDL	BDL

Copper	BDL	BDL	BDL	BDL	BDL	BDL
Zinc	0.5	0.5	0.5	0.5	0.5	0.5
Iron	0.005	0.004	0.003	0.002	0.003	BDL
Manganese	BDL	BDL	BDL	BDL	BDL	BDL
Selenium	BDL	BDL	BDL	BDL	BDL	BDL
Nickel	BDL	BDL	BDL	BDL	BDL	BDL
Ammoniacal nitrogen	Nil	Nil	Nil	Nil	Nil	Nil
Nitrate	0.18	0.14	0.25	0.3	0.16	Nil
Total Coliform (MPN/100 ml)	1440	1600	1800	2100	1800	1800
Faecal Coliform (MPN/100 ml)	450	430	600	650	600	600

Table 3

Analysis of River water at Telmucho bridge

Parameters	1992					
(mg/*l*)	May	June	July	August	September	October
pH	7.2	7.0	7.4	7.5	7.4	7.5
Temperature A/W (^{0} C)	38/32	36/30	32/30	30/29	31/30	30/29
Turbidity (NTU)	550	460	179	110	240	282
Suspended Solids	698	500	213	136	380	316
Dissolved Solids	109	100	198	128	200	202
Alkalinity	78	80	98	86.8	92.5	98
Hardness	142	96.8	170.0	128	110	135
Calcium	30.8	30.2	20.6	19.8	18.4	19.2
Magnesium	15.2	14.8	7.2	8.4	9.2	9.4
Chloride	36	25	28	32	34	37
Sulphate	56	21	86	49	34	61
DO	5.2	5.8	6.8	6.8	5.6	5.8
BOD	2.8	2.6	2.9	2.3	3.0	3.2
COD	48	52	46	48	58	60
Oil and Grease	0.45	0.3	0.3	0.4	0.6	0.9
Lead	BDL	BDL	BDL	BDL	BDL	BDL
Total Chromium	BDL	BDL	BDL	BDL	BDL	BDL
Copper	BDL	BDL	BDL	BDL	BDL	BDL
Zinc	0.5	0.5	0.4	0.5	0.5	0.5
Iron	BDL	BDL	BDL	BDL	BDL	BDL
Manganese	BDL	BDL	BDL	BDL	BDL	BDL

Selenium	BDL	BDL	BDL	BDL	BDL	BDL
Nickel	BDL	BDL	BDL	BDL	BDL	BDL
Ammoniacal Nitrogen	Nil	Nil	Nil	Nil	Nil	Nil
Nitrate	0.11	0.12	Nil	0.1	0.13	0.15
Total Coliform (MPN/100 ml)	3280	3220	3180	3180	3280	3620
Faecal Coliform (MPN/100 ml)	2600	2400	2280	2000	2280	2284

Table 4
Analysis of River water at Jamadoba Water Works

Parameters	1992					
(mg/*l*)	May	June	July	August	September	October
pH	7.5	7.8	8.0	7.8	7.5	7.5
Temperature A/W (^{0} C)	37/32	36/31	32/30	31/30	31/30	30/29
Turbidity (NTU)	598	582	514	318	316	450
Suspended Solids	784	742	696	512	500	600
Dissolved Solids	168	180	215	284	280	168
Alkalinity	90	92	94	98	94.8	92.4
Hardness	125	118	136	102	142	122
Calcium	19.0	18.0	28.4	28.2	27.0	24.8
Magnesium	6.8	5.2	8.4	8.2	8.0	8.4
Chloride	25	32	26	26	32	28
Sulphate	62	48	66	32	38	57
DO	5.8	5.8	6.4	6.0	5.6	5.2
BOD	3.4	3.2	3.0	2.8	2.2	3.2
COD	108	104	94	82	100	110
Oil and Grease	1.0	0.8	0.5	0.6	0.8	0.9
Lead	BDL	BDL	BDL	BDL	BDL	BDL
Total Chromium	BDL	BDL	BDL	BDL	BDL	BDL
Copper	BDL	BDL	BDL	BDL	BDL	BDL
Zinc	0.1	0.1	0.1	0.1	0.1	0.1
Iron	BDL	BDL	BDL	BDL	BDL	BDL
Manganese	BDL	BDL	BDL	BDL	BDL	BDL
Selenium	BDL	BDL	BDL	BDL	BDL	BDL
Nickel	BDL	BDL	BDL	BDL	BDL	BDL
Ammoniacal Nitrogen	Nil	Nil	Nil	Nil	Nil	Nil
Nitrate	Nil	Nil	0.10	0.11	0.15	Nil

Total Coliform (MPN/100 ml)	4086	3984	6800	9240	8528	6758
Faecal Coliform (MPN/100 ml)	2890	2830	4250	6200	4280	3400

Table 5

Analysis of River water at Damgarh Water Works.

Parameters	1992					
(mg/l)	May	June	July	August	September	October
pH	7.8	7.8	8.0	7.8	7.8	7.8
Temperature A/W (^{0}C)	38/32	36/31	31/29	30/29	31/29	30/29
Turbidity (NTU)	708	646	600	312	304	572
Suspended Solids	943	690	630	480	450	600
Dissolved Solids	168	152	144	208	200	198
Alkalinity	98	102	98	88	89	90
Hardness	92	98	105	110	98	94
Calcium	28	26	22	23	22	21
Magnesium	6	7	14	13	13	12
Chloride	26	24	22	22	22	21
Sulphate	25	36	31	43	34	30
DO	5.2	5.8	5.8	6.0	6.2	5.8
BOD	4.1	3.9	3.4	3.2	2.4	2.8
COD	440	434	422	286	158	184
Oil and Grease	1.2	1.0	0.8	0.6	0.5	0.6
Lead	BDL	BDL	BDL	BDL	BDL	BDL
Total Chromium	BDL	BDL	BDL	BDL	BDL	BDL
Copper	BDL	BDL	BDL	BDL	BDL	BDL
Zinc	0.4	0.5	0.5	0.5	0.4	0.4
Iron	BDL	BDL	BDL	BDL	BDL	BDL
Manganese	BDL	BDL	BDL	BDL	BDL	BDL
Selenium	BDL	BDL	BDL	BDL	BDL	BDL
Nickel	BDL	BDL	BDL	BDL	BDL	BDL
Ammoniacal Nitrogen	34	42	59	52	61	38
Nitrate	0.11	0.14	0.14	0.16	0.14	0.12
Total Coliform (MPN/100 ml)	4200	4200	9600	12000	9284	4700
Faecal Coliform (MPN/100 ml)	2700	2600	7100	8100	5000	4000

❑ Water Quality Parameters

pH and alkalinity

Water quality of river Damodar generally exhibited neutral to an alkaline range between 7.0-8.0 and remained within the acceptable limit. This indicates that decomposition of organic matter contributed by industrial and human wastes has not suppressed the pH all along the study area (Cummins 1966). The alkalinity values are also found to be higher in the range of 78 to 143 mg/1 at all the sampling stations in comparison to the prescribed limit of 50 mg/1 for river water (IS : 2296, 1974).

Turbidity and Suspended Solids

Turbidity expresses the clarity of water and other insoluble inorganic matter present in the water system. Damodar river water all along the study area has been observed very turbid with a lot of suspended matter and appeared blackish to brownish blackish in colour with hardly any transparency. This is mainly due to discharge of coal washery and thermal power plant effluents containing a considerable load of suspended matter which comprises mainly coal fines and fly ash.

Three stations viz. III, IV and V which are sited on the downstream of the river showed comparatively low values of turbidity during mansoon season due to dilution of river water. On the other hand stations I and II reflected highest values of turbidity during the same period indicating that dilution by rain water did not suppress the turbidity levels due to discharge of more fly ash by Chandrapura Thermal Power Plant supplemented by uncontrolled discharge of effluents from Dugda Coal washery aggravating pollution on downstream of the river. The turbidity and suspended solids profile along various stations is given in Fig. 2 Highest values of turbidity observed during summer period at stations III, IV and V which is mainly due to the decrease in volume of river water coupled with the load of suspended solids.

One of the serious problems of Damodar river pollution is of high concentration of suspended solids. Transport of solids by flowing water in suspended form is attributed to the effluents coming from coal washeries containing high amount of suspended solids as well as erosion in the catchment area, and other industrial effluents including domestic sewage. Suspended solids at stations I and II were observed at maximum levels of 1652 mg/1 and 4382 mg/1, respectively during August 1992. During this period Dugda Coal washery effluents were found to contain suspended solids of 4672 mg/1 and in conjunction with the suspended solids (fly ash) discharged from Chandrapura Thermal Power Plant on U/S, Station II reflected highest suspended solids of 4382 mg/1. All along the stretch of the river under the study area suspended solids were considerably higher against the permissible limit of 100 mg/1.

Dissolved Solids and Hardness

Dissolved solids and hardness content in general, are found increasing with decreasing values of suspended solids. This is mainly due to the strong adsorption capacity of fly ash and coal mines. Dissolved solids content at all the five sampling stations during the entire study period ranged from 100 mg/1 to 284 mg/1 which were below the permissible limit of 500 mg/1 for drinking purposes. Dissolved solids and hardness profile are depicted in Fig. 3

Hardness varies from 92 to 190 mg/1 all along the stretch of river of the study area indicates that the river water is moderately hard. Maximum hardness was observed near Dugda Coal washery which were ranging from 96.8 mg/1 to 170 mg/1. The value of alkalinity at all the five sampling stations is less than the total hardness values which showed the hardness mainly of carbonate type.

Chloride and Sulphate

Chloride and sulphate contents were found quite lower and are in the desirbale range. These lower contents do not pose any implications for Damodar river water utilisation for various purposes.

Value of chloride varies from 15.8 mg/1 to 49.6 mg/1 along the study area. The concentration of sulphate all along the river starting from station I to II are in the range of 21 to 86 mg/1. Except at a few places, concentrations of both chloride and sulphate are not much higher showing that river water is very much suitable for agricultural purposes (Gurdeep Singh and Ram 1989).

DO, BOD and COD

Dissolved Oxygen (DO), Biochemical Oxygen Demand (BOD) and Chemical Oxygen Demand (COD) have been considered very important parameters to be assessed to know the extent of organic pollution in the river. It has been found that in summer season DO level of the river goes down. It is due to the fact that the waste volume entering into the river was more or less same but volume decreases and also increasing temperatures of effluents as well as river water result in reduction of DO level during summer season. DO profile of the river along the various stretches is given in Fig. 4. Along the entire study area DO level ranges from 5.2 mg/1 to 7.2 mg/1. BOD values range from 1.0 mg/1 to 4.1 mg/1 at all five stations. With these results of DO and BOD oxygen sag curve is viewed together for an appropriate interpretation of the Pollution profile of the river because even for any two equal DO values in the streams the DO sag would correspond to different BOD values. The BOD and COD profile of river along the various stretches as give in Fig. 5 shows that at sampling stations III (Telmucho Bridge) and station V (Damgarh Water Works) BOD values were comparatively higher which indicate that human and industrial wastes are affecting river water. In these zones of the river, available oxygen got utilised. Starting from sampling station I to sampling station V and increasing trend of BOD and decreasing trend of DO have been observed which clearly indicates increasing pollution load towards downstream. Organic pollution load is not significantly imparted by effluents from coal washeries and thermal power plants. Domestic effluents in general cover a long distance along the various nallah, streams and are biodegraded along the natural pathway by mixing with water, mine water discharge etc. and also land treatment of these occurs naturally and their organic load is not significantly exerted on Damodar river.

The portion of organic matter which is oxidisable by the activity of the micro-organisms is measured by the BOD test and the organic content inclusive of that measured by the COD test. The overall COD load was found to be in the range of 46-1760 mg/*l*, maximum COD was observed at sampling station II. These observed COD values were comparatively higher due to discharge of suspended solids from coal washeries and power plants effluents which contain considerable reducible matter. However, at all the sampling stations of Damodar river, it exceeded the threshold limit value of 25.0 mg/*l* (IS: 2296, 1974).

Total Coliforms and Faceal Coliforms

The fairly high values of BOD coupled with high values of Total coliforms and Faecal coliforms are indicative of incrasing pollution of the river by organic means particularly through the discharge of sewage and domestic effluents into the river. Total coliforms show gradually increasing trend from station I to station V.Faecal coliforms are also increasing in the same fashion as more and more domestic effluents find their way along downstream of Damodar river.

Total coliforms range from minimum of 1200 org/100 ml to a maximum of 12000 org/100 ml whereas faecal coliforms vary from 430 MPN/100 ml to 8100 MPN/100 ml. The higher values of these at stations III, IV and V indicate that river water is grossly polluted by sewage.

At Domgarh Water Works, ammoniacal nitrogen has been found ranging from 34 mg/*l* in May to 61 mg/*l* in September. Ammonical nitrogen at all other sites has not been found. These are found at Domgarh Water Works because of discharge of Sindri fertiliser complex effluents containing more ammonia compounds.

Trace Elements

Trace elements as analysed by GBC-902 AAS were observed either at below detection limits (BDL) or at insignificant concentration levels and as such do not pose any major water pollution problem.

Damodar River Water Quality Profile

Damodar river water quality profile with respect to specific criterion (CBPCWP 1987) as evolved on the basis of this study is presented below.

Sanitation and Quality

The higher coliforms at stations III, IV and V show that river is organically polluted. River water has been all along unsatisfactory for drinking purposes. Health hazards cropping out of chemically and bacteriologically polluted water can not be ruled out. There have been occasional outbreaks of acute gastroenteritis cholera, dysentry, diarrhea and vomitting at several pockets of this region of coalfield time to time.

Aesthetics

The criteria can be best described qualitatively. The surface water should be free of floating debris, scum and oil. Objectionable organoleptic features of colour, odour and taste should not be present (Berkson 1974). From general observation it can be seen that the water at the location I, II, III, IV and V is not suitable from aesthetic point of view.

Criteria of Raw Water Sources for Public Water Supply

Site IV and V are used for public water supply. On an average at these two points river water conforms to both the Indian and WHO standards for pH, Chlorides and DO. But all stations exceed the tolerance limits for BOD and COD.

River water here falls under the category 'C' and is only suitable for drinking with conventional treatment followed by disinfection. Further after treatment of Damodar river at these Water Works, back wash (Waste sludge) generaged during the treatment is again pushed back to the river with no safe disposal and subsequent utilisation taken care of.

Criteria for Bathing

International Joint Commission standards (1973) indicate that wastes with coliforms counts of 1000 MPN/100 ml or less are acceptable for bathing. The Indian standards for bathing ghats are the same as the standards for the raw surface water (IS: 2296, 1974), i.e. 5000 MPN/100 ml. The existing coliforms counts in all the stations do not meet standards for body contact attributed to the surface water bathing ghats and give rise to associated skin diseases.

Criteria for Oxygen Demanding Wastes

All forms of organic matters are basically undesirable in the water though to a certain level is tolerable. This may lead to serious trouble like eutrophication and depletion of the oxygen content of the system and thus impairing its health.

Indian standards set 3 mg/l of DO as the limit. The DO level at all the sampling stations was more than 3 mg/ *l*. This perhaps indicates that the self-purification capacity of this river, can assimilate the existing load. High bacterial contamination and oxygen demanding waste characterise river Damodar and figure shows the overall quality status for the utlity aspects of the river Damodar in the stretch from station I to station V. The present population size and distribution, the different water utility patterns and industrial wastes disposal all have resulted in considerably degraded river water in Damodar.

❑ REFERENCES

APHA. 1980. *Standard Methods for the Examination of Water and Wastewater*. 15th edition, Washington, D.C.

Bandopadhyay P. 1994. Processing and utilisation of minerals and pollution of water resources. Proceedings second National Seminar on Minerals and Ecology, ISM, Dhanbad, Jan. 1994.

Central Board for the Prevention and Control of Water Pollution 1985. Operation Control of Damodar River. Programme objective Series : PROBES / 23/1984-85.

Cumins K.W. 1966. A review of stream ecology with special emphasis on organisms substrate relationships Spl. Publ. Pymatuning Lab. Field Biol. 4 : 2-51.

Gupta Ravi K. and Gurdeep Singh. 1995. Water pollution profile of coal washeries. *Poll. Res.* 14(2) : 203-213.

Gurdeep Singh. 1990. Status of water quality in a coal mining environment- a case study in the Jharia coalfield. *Jr. Ind. Poll. Cont.* 6(2) : 61-67.

Gurdeep Singh and Bigan Ram. 1989. Pollution of Damodar river in coalbelt of Dhanbad. *Minetech* 10(1) : 73-78.

National Environmental Engineering Research Institute, 1988. *Manual on Water and Wastewater Analysis*, NEERI, Nagpur.

Rawat N.C., Saxena A.K., Gurdeep Singh and Sundriyal A.K. 1981. Physico-chemical characteristics of underground mine waters and x-ray analysis of corrosion products. *J. Mines, Metals & Fuels*, 108 : 114.

Authors :

1. **Binay K. Gupta :** Bihar State Pollution Control Board, Dhanbad Regional office, Dhanbad, Bihar, India.

2. **Gurdeep Singh :** Centre for mining Environment Indian School of - Mines, Dhanbad, Bihar, India.

■ ■

5

River Pollution in Central Gujarat

❑ INTRODUCTION

The natural water bodies, both lotic and lentic, are most important source of water to sustain life. These resources need special attention for conservation, development and management for optimal and sustainable utilization. The study of river ecology has gained immense importance because of multiple use of river water. Gross water pollution is clearly an immediate hazard to aquatic environment. The macro and micro communities in a natural water body play an important role in keeping the water clean and acceptable for various purposes. The life of an aquatic organism depends directly on physical and chemical characters of aquatic environment. The changes in physico-chemical composition of water may lead to drastic change in the community of aquatic biota, some of which may exploit the increased nutrients, while others may deminish. Such changes in structure and function of biota form the basis of water quality assessment. Various mode of pollution monitoring programme using biota as indicators of degree of pollution have been compiled recently (Rana, 1995). Assessment of pollution using both physico-chemical and biological methods are more reliable than chemical or biological methods alone. A good amount of work has been carried out by many Indian scientists on the ecology of various rivers including polluted rivers (Venkateswarlu, 1969 a, b, Prasad and Saxena, 1980, Gunale and Balkrishna, 1981, Venkateswarlu, 1981, Venkateswarlu and Sampathkumar, 1982, Nandan and Patel, 1985, Palria and Rana, 1985, Sengar and Sharma, 1987, Rana and Palria, 1988, Trivedy, 1988, Shaji and Patel, 1991, Rana and Jameson, 1995, Kargupta and Siddiqui, 1995).

In the present article ecology of certain polluted rivers of Central Gujarat region has been reviewed with reference to their physico-chemical, phycology, pollution indices and bioassay parameters.

On industrial map of India, Gujarat is number one among industrially advanced States. With liberalization of economic policy, Gujarat has received highest amount of financial investment for industrial development. About 90 thousand crore rupees investment is likely to come to Gujarat, out of which one thousand crore is to be invested in chemical industries alone.

Traditionally Gujarat is known for textile industry, however, large number of diverse industries such as soda ash, polyester fibers, dyes and dye intermediates, pharmaceuticals, oil and petrochemicals, refineries, fertilizers, power generation, telecommunication and other core sector industries are now well established. Gujarat Industrial Development Corporation (GIDC) has developed more than two hundred industrial estate throughout the State, out of which twelve are large water intensive estates. These water intensive industrial estates have more than 8000 small or medium size water polluting industrial units. The individual units have almost no waste water treatment plants and in many of the above estates combined waste water treatment facilities either do not exist or are partly available. The facilities of the treatment of industrial and urban community wastes are meager and therefore large quantities of untreated waste find their way into natural water bodies. Virtually, no water body lotic or lentic are free from pollution. Almost all the rivers of the State like Mahi, Sabarmati, Narmada, Tapi, Khari, Vishwamitri, Shedhy, Vatrak, Ambica etc. have become polluted either by sewage or industrial waste or both.

Study Area

The Gujarat State is situated on the west coast of India between 20^0 1' and 24^0 1' north latitudes and between 38^0 4' and 74^0 4' east longitude. Several rivers flow across the state and meet Arbian sea, Gulf of Khambhat and Kutch. The rivers like Sabarmati, Mahi, Narmada and Tapi are the longest among other rivers of Gujarat. The natural drainage of the State is provided by some major rivers with their tributaries. In the present review pollution status of following rivers which mainly receives domestic or industrial or both types of wastes have been discussed.

1. The Sabarmati - This river rises in the South-Western spurs of Arvali hills. It traverse through Sabarkantha, Ahmedabad, and Kheda districts and finally discharges into the Gulf of Khambhat (Cambay). The river has been studied at various places from Ahmedabad to the Gulf of Khambhat. It is mainly a dumping ground of domestic, textile, chemical, dye and industrial wastes of units located on or near the river bank.

2. The Khari - The river Khari rising in a cluster of hills about 16 km to the north-east of Himmatnagar in Sabarkantha districts passes through the extreme west of

Kheda district and finally merges with the Sabarmati near village Radhu. The river receives large quantities of industrial effluents from Vatava and other industrial estates of Ahmedabad.

3. The Vatrak - The river Vatrak with which the water of river Shedhy unites at Kheda rises from the hills of Dungarpur in Rajasthan. It runs Parallel to the Mahi and passes through Sabarkantha and Kheda districts and merge with the Shedhy river near Kheda town.

4. The Shedhy - This river originates from hills of Panchmahal district and traverses through Kheda district to merge with river Vatrak near Kheda town.

5. The Mini - A small tributary flows between industrial estate of Nandesari and petrochemical industry of Indian Petrochemical Corporation (IPCL), Ranoli near Baroda. The river receives effluents from large number of industrial units of Nandesari industrial area.

6. The Vishwamitri - The river flows into north-south direction through Baroda city and mainly receives domestic wastes of the city.

Physico-chemical parameters

The physico-chemical parameter of samples collected from the study sites and analysed on monthly/seasonal basis as per standard methods (APHA 1989, Trivedy and Goel, 1984) have been reported here.

The Sabarmati

River Sabarmati has been studied to its entire length by various workers. Shaji and Patel (1991) studied the river around Ahmedabad city. The city has many urban and industrial establishement which discharge their wastes through about 32 drains. Earlier, this river was studied for its hydrobiological qualities by Venkateswarlu and Jayanthi (1968). Shaji and Patel (1991) divided eight km. stretch of river into three sampling sites. The results have been summarized in Table 1. The result indicated that pH was always in alkaline range. The alkalinity was remarkably high and D.O. content was very low at all the three sites.The water was rich in chloride, phosphate, nitrate and COD indicating its polluted nature.

The downstream of river Sabarmati in Kheda region was studied by Jameson (1993). The river froms a complex of many rivers at village Vautha, where more than seven rivers confluences to discharge the water ultimately to the gulf of Khambhat. About 10 km. stretch of the river divided into three sampling site was studied in Kheda region. (Table 2) Among the three sites, second and third sites revealed high values of many parameters. The first site showed alkaline pH, whereas site two and three were acidic. The DO content of second and third sites was negligible. Other parameters like total solids, chloride, phosphate, COD and.

BOD were very high. However, even the contents of first site were very high compared to clean water river (Jameson and Rana, 1996).

Table 1
Physico-chemical parameters of River Sabarmati at Ahmedabad
(After Shaji & Patel, 1991)

S. No.	Parameters	Site I		Site II		Site III	
		Range	Average	Range	Average	Range	Average
1.	Temperature (^{0} C)	13-29	24	13-29	24.2	13-29	24.1
2.	pH	7.6-9.0	8.1	7.6-9.1	8.3	7.6-8.9	8.2
3.	Total Alkalinity	200-780	485	160-760	560.5	200-800	564,5
4.	DO	0-6	1.7	0-6	1.1	0-6	1.0
5.	Total Hardness	116-360	258.2	120-380	286.4	116-380	289.8
6.	Calcium	28.8-104.2	68.6	28.8-88	65.1	28.8-80	63.0
7.	Chloride	42.6-738.4	446.3	39.7-721.3	427.9	42.6-583.2	397.4
8.	Silica	1.7-34.4	18	1.7-34.4	16.6	1.7-34.4	15.9
9.	Nitrate	0.07-1.53	0.4	0.01-1.1	0.33	0.09-1.08	0.5
10.	Phosphate	0.133-6.25	2.7	0.16-5.24	1.9	0.26-5.24	2.4
11.	Sulphate	14.9-519	160.4	14.9-386.2	149.4	14.9-414.0	142.4
12.	COD	80-650	372.5	100-600	330.6	80-650	373.1

All parameters except temperature and pH are in mg/*l*

Table 2
Physico-chemical parameters of river at Kheda region

S. No.	Parameters	Range	Average
1.	Temperature (^{0} C)	17-34	27.2
2.	pH	7.1-8.8	7.9
3.	Total Solids	1800-6800	3758
4.	DO	3.1-5.1	4.1
5.	Total alkalinity	440.8-850	603.4
6.	Hardness	130-410	277.8
7.	Chloride	286-514	409.4
8.	Phosphate	3.1-9.6	5.5
9.	Sulphate	0.9-16.9	11.4

10.	Nitrate	0.1-1.1	0.5
11.	Silica	17.3-46.8	36.2
12.	Calcium	31.6-86.4	48.8
13.	BOD	140-212.8	180.2
14.	COD	240-450	343.8

All parameters except temperature and pH are in mg/*l*.

Further, downstream of river Sabarmati near Khambhat region was studied by Harish Kumar (1994). He divided the river into three sections. The results tabulated in Table 3 indicate a decreasing trend of many parameters as river proceed towards station two and three. Water becomes near alkaline from acidic range at stations two and three, however, DO content remains negligible throughout the river. Other parameters like total solids, nitrate, phosphate, sulphate, calcium and COD also get reduced to significant level. Chloride content, however, increased two-folds from station one to station three.

Table 3

Mean seasonal values of physico-chemical parameters of river Sabarmati at Kheda region. (After Harishkumar, 1994)

S. No.	Parameters	Site I			Site II			Site III		
		A	B	C	A	B	C	A	B	C
1.	Temperature (^{0} C)	24	21	28	23.5	19	32.1	28.7	23.4	26.3
2.	pH	7.3	5.1	3.8	7.2	7.1	6.3	7.9	7.5	8.1
3.	DO	-	-	-	1.9	0.8	-	1.4	1.3	0.7
4.	Total Alkalinity	240.6	315.7	418.7	214.7	262.8	283.4	315.7	318.8	435.3
5.	Nitrate	1.31	1.57	1.63	0.97	0.78	0.65	0.37	0.51	0.65
6.	Phosphate	11.4	11.8	11.7	11.9	11.7	12.0	8.1	4.3	7.7
7.	Sulphate	38.7	35.7	43.8	25.3	22.1	20.1	18.7	19.3	15.3
8.	Calcium	50.4	58.1	63.7	38.1	35.1	34.3	25.9	27.6	28.3
9.	Chloride	1117	1317	987	1813	1737	1631	2538	2438	2438
10.	Hardness	268	313	438	181	183	231	313	438	493
11.	Total Solids	2201	2309	2514	2077	2017	1981	931	637	531
12.	COD	1948	2314	2555	988	975	1371	335	531	631

All parameters except temperature and pH are in mg/*l*.

A = Monsoon, B = Winter, C = Summer

The Khari

The Khari river water is highly acidic and DO content was almost absent. (Table 4) Other parameters like total solids, chloride, phosphate, sulphate, calcium, BOD and COD were very high. The effluent received from number of industrial units of Vatava industrial estate makes this river highly polluted (Rana and Jameson, 1995).

Table 4

Physico-chemical parameters of river at Khari

S. No.	Parameters	Range	Average
1.	Temperature (^{0} C)	16.0-32	26.9
2.	pH	1.8-5.3	2.4
3.	Total Solids	4500-10,000	8370
4.	DO	Nil	Nil
5.	Total alkalinity	-	-
6.	Hardness	590-1910	1172.3
7.	Chloride	702-2634	2060.3
8.	Phosphate	10.4-36.8	23.5
9.	Sulphate	11.3-120	55.3
10.	Nitrate	0.25-2.5	0.95
11.	Silica	35.6-158.6	91.2
12.	Calcium	65.1-392.5	190
13.	BOD	235-625	418.9
14.	COD	970-2880	2275

All parameters except temperature and pH are in mg/*l*.

The Vatrak

The physico-chemical parameters of river water collected from nearby kheda town before its merger with Shedhy river has been shown in Table 5. The river water was alkaline in nature with moderate DO content. Comparing to Sabarmati and Khari rivers, many of the physico-chemical parameters were moderately low. All the parameters like dissolved solids, phosphate, nitrate, chloride, BOD and COD were in low quantity.

The Shedhy

The sample water of river Shedhy was collected from Kheda town before its merger with river Vatrak. The river water was in alkaline range with high DO content. Other parameters were also in moderate range indicating low pollution load in the river (Table 6).

Table 5

Physico-chemical parameters of river at Vatrak

S. No.	Parameters	Range	Average
1.	Temperature (^{0} C)	20-31.5	26.2
2.	pH	7.5-8.5	7.9
3.	Total Solids	600-1300	875
4.	DO	7.6-9.2	8.5
5.	Total alkalinity	140-350	272.5
6.	Hardness	172-388	249
7.	Chloride	69.5-88	77.2
8.	Phosphate	0.1-0.8	0.3
9.	Sulphate	1.8-2.1	1.9
10.	Nitrate	0.21-0.31	0.2
11.	Silica	11.5-18.6	15.3
12.	Calcium	15-28.4	19.8
13.	BOD	29-35	31.2
14.	COD	40-48	432

All parameters except temperature and pH are in mg/*l*.

Table 6

Physico-chemical parameters of river at Shedhy

S. No.	Parameters	Range	Average
1.	Temperature (^{0} C)	19-32	25.2
2.	pH	7.1-8.6	7.7
3.	Total Solids	500-2000	1190
4.	DO	3.4-12	9.3
5.	Total alkalinity	108-380	303
6.	Hardness	72-320	186
7.	Chloride	65.3-364	208.9
8.	Phosphate	0.2-0.9	0.5
9.	Sulphate	0.6-6.4	3.3
10.	Nitrate	0.1-0.3	0.19
11.	Silica	10.0-16.8	13.2
12.	Calcium	8.6-47.2	26.5
13.	BOD	31.5-65	48.8
14.	COD	30-80	60.2

All parameters except temperature and pH are in mg/*l*.

The Mini

The physico-chemical parameters were studied by Shaji (1990). The river water was alkaline in nature and contains negligible quantity of DO. It showed high amount of chloride, sulphate, calcium and COD indicating high degree of pollution (Table 7).

Table 7

Physico-chemical parameters of river Mini (After Shaji, 1990)

S. No.	Parameters	Range	Average
1.	Temperature (^{0} C)	21-29	24.7
2.	pH	6.8-8.5	7.4
3.	Total alkalinity	110-610	149.5
4.	DO	0-1.6	0.58
5.	Hardness	230-1400	731.3
6.	Calcium	40-328	117.3
7.	Chloride	198-1391	777
8.	Nitrate	0.12-1.15	0.46
9.	Phosphate	0-0.38	0.14
10.	Sulphate	39.7-1529	559.7
11.	Silica	9.7-37.9	20.6
12.	COD	1025-2130	1584

All parameters except temperature and pH are in mg/*l*.

The Vishwamitri

The river was studied around Baroda city by dividing it into three sampling sites by Nandan (1983). The physico-chemical parameters revealed that there was progressive pollution from station one onwards in downstream (Table 8). The river water was alkaline and had moderate DO. The other parameters like chloride, phosphate, hardness, sulphate, COD etc. were in high range compared to clean water rivers (Nandan and Patel, 1985).

Phycological Characteristion

Algal samples collected from the study sites of various rivers on monthly or seasonal basis were analysed. The algal samples preserved in 4% formalin were identified by drawing camera lucida diagrams. The algal taxa were identified using standard texts, monographs and published literature. Further, to use algae as pollution indicators, Palmer (1969) index was scored for many river water algae. A brief summary of the work has been discussed here.

Table 8
Mean values of physico-chemical parameters at three station of Vishwamitri river (After Nandan and Patel, 1985)

S.No.	Parameters	Station		
		I	II	III
1.	Temperature (^{0}C)	28.5	28.7	28.9
2.	pH	7.7	7.5	7.6
3.	DO	8.4	6.8	5.9
4.	Total Alkalinity	240.6	281.2	278.7
5.	Nitrate	0.17	0.17	0.18
6.	Phosphate	1.49	1.8	1.95
7.	Sulphate	50.4	58.7	67.0
8.	Calcium	127.2	139.9	159.5
9.	Chloride	38.5	50.0	55.5
10.	Hardness	260.8	2817	283.9
11.	Total Solids	920	988	1067
12.	COD	154	172	208

All parameters except temperature and pH are in mg/*l*.

The Sabarmati

Three parts of river Sabarmati studied by three groups of workers (Shaji, 1990, Jameson, 1993 and Harish Kumar, 1994) recorded number of algal forms from various sampling sites. A selective list of three parts has been given in Table 9. Shaji and Patel (1991) recorded 187 taxa belonging to 62 genera of four major groups viz., Cyanophyceae, Chlorophyceae, Euglenophyceae and Bacillariophyceae from the river around Ahmedabad. Phytoplankton assembalage were mainly made up of Cyanophyceae and Euglenophyceae. Cyanophycean members were most dominant and all the twenty most pollution tolerent genera were recorded (Palmer, 1969) from these sites. The more common forms were *Oscillatoria Microcystis, Chroococcus, Merismopedia, Ankistrodesmus, Scenedesmus, Pediastrum, Cyclotella, Navicula, Nitzschia, Synedra, Euglena, Phacus etc.*

The second site studied by Jameson and Rana (1996) recorded 83 taxa belonging to 35 genera. The Cyanophyceae was represented by 28 taxa belonging to 10 genera, Chloprophyceae by 13 taxa having 6 genera, Bacillariophyceae by 30 taxa of 16 genera and Euglenoids were represented by 8 taxa of 3 genera. The river was dominated by Cyanophyceae. The third phase of the river also recorded all the major forms of algae. The early part of the river was dominated by Cyanophyceae followed by mixed population of Cyanophyceae, Chlorophyceae and Bacillariophyceae.

Table 9

Name of the species	S1	S2	S3	S4	S5	Kh	Vi	Mi
Cyanophyceae								
Microcystis aeruginosa	+	-	-	-	-	-	+	-
M. flos-aquae	+	+	+	+	+	-	-	-
M. elongata	-	-	-	-	-	-	+	-
Synechocystis aqutilis	-	-	-	-	-	-	-	+
S. pevalekii	+	-	-	-	-	+	-	+
Chroococcus minor	-	+	-	-	-	-	-	-
C. minutus	-	+	+	+	+	+	+	-
C. montanus	-	-	-	-	-	+	-	-
Synechococcus aeruginosa	-	+	-	+	+	-	-	-
S. elongatus	-	-	-	-	-	-	-	+
Merismopedia minima	+	+	+	+	+	+	-	-
M. punctata	+	-	-	-	-	-	-	-
M. tenussima	-	+	+	+	-	+	-	-
Arthrospira jenneri	-	+	+	+	+	-	-	-
A. platensis	+	+	-	-	-	+	+	-
Spirulina major	-	+	-	+	+	-	-	-
S. subsalsa	+	+	+	+	-	-	-	-
S. gigantea	-	-	-	-	-	-	+	-
S. laxissima	+	-	-	-	-	-	-	-
Oscillatoria accuminata	-	+	+	+	+	-	-	-
O. boryana	-	+	-	-	-	-	-	-
O. chlorina	-	+	-	+	-	+	+	+
O. formosa	-	+	+	-	-	-	+	-
O. lemmermanni	-	+	-	+	-	-	-	-
O. limosa	+	+	+	+	+	-	-	-
O. proteus	-	+	-	-	-	+	+	-

Name of the species	S1	S2	S3	S4	S5	Kh	Vi	Mi
O. schultzii	-	+	-	+	+	-	-	-
O. subtilissima	-	+	-	+	-	-	-	-
O. curviceps	-	-	+	-	-	-	+	-
O. princeps	+	-	+	-	-	-	+	-
O. rubescens	+	-	-	-	-	-	+	-
O. brevis	-	-	-	-	-	+	+	-
O. tenuis	+	-	-	-	-	-	-	-
O. salina	+	-	+	-	-	-	-	-
O. chalybea	+	-	-	-	-	+	-	-
O. perornata	+	-	-	-	-	-	-	+
O. subbrevis	-	+	+	-	-	-	+	-
O. pseudogeminata	+	-	-	-	-	-	-	+
O. amphibia	+	+	-	-	-	-	-	+
O. foreauii	-	-	-	-	-	-	-	+
O. trichlorides	-	-	-	-	-	-	-	+
Lyngbya sp.	-	-	-	-	-	+	-	-
Phormedium sp.	-	+	+	+	+	+	+	+
Anabaena aphanizomenoides	+	-	-	-	-	-	-	-
A. constricta	+	-	-	-	-	-	-	-
A. spiroides	+	-	-	-	-	-	-	-
A. arnoldii	+	-	-	-	-	-	-	-
Anabaena sp.	-	+	+	+	+	-	-	-
Anabaenopsis sp.	-	+	-	+	-	-	+	-
Raphidiopsis meditaronea	+	-	-	-	-	-	-	-
CHLOROPHYCEAE								
Pediastrum duplex	-	-	+	+	+	-	-	-
P. simplex	+	-	+	+	+	-	+	-
P. tetras	+	-	-	+	-	-	-	-
Pandorina morum	+	-	+	+	+	-	+	-
Eudorina elegans	+	-	-	-	-	-	+	-
Chlorella vulgaris	+	+	-	+	+	-	-	-
Ankistrodesmus convolutus	-	+	-	-	-	+	+	-
A. falcatus	-	+	+	-	-	-	+	-

Name of the species	S1	S2	S3	S4	S5	Kh	Vi	Mi
Schroederia planctonica	-	-	-	-	-	-	+	-
S. setigera	+	-	-	-	-	-	-	-
Actinstrum hontzschii	+	-	-	-	-	-	-	-
Dictyosphaerium pulchellum	+	-	-	-	-	-	-	-
Golenkinia radiata	+	-	-	-	-	-	-	-
Micractinium pusillum	+	-	-	-	-	-	-	-
Oocystis crasa	+	-	-	-	-	-	-	-
Tetradron minimum	+	-	-	-	-	-	-	-
T. muticum	+	-	-	-	-	-	-	-
Coelastrum microporum	+	+	-	-	-	-	+	-
Staurodesmus cuspidatus	-	-	-	-	-	-	+	-
Chlamydomonas conferta	+	-	-	-	-	-	-	-
C. reinhardii	+	-	-	-	-	+	-	-
C. mutifilis	+	-	-	-	-	-	-	-
C. umbonata	-	-	-	-	-	-	-	+
Chlamydomonas sp.	-	-	-	-	-	+	-	-
Scenedesmus armatus	-	+	+	-	-	+	+	-
S. benardii	-	+	-	+	-	-	-	-
S. bidijitatus	-	-	-	+	+	-	-	-
S. bijugatus	-	-	-	+	-	-	+	+
S. dimorphus	+	+	+	+	+	-	+	-
S. obliquus	+	+	+	-	-	-	-	-
S. protuberans	-	+	-	+	-	-	-	-
S. quadricauda	+	+	+	+	+	+	+	-
S. acuminata	-	-	-	-	-	-	+	-
Scenedesmus opoliensis	-	-	-	-	-	-	+	-
S. ascuatus	-	-	-	-	-	-	-	+
S. abundans	+	-	-	-	-	-	-	-
Scenedesmus sp.	+	+	+	-	-	-	-	-
Chlorogonium euchlorum	+	-	-	-	-	-	-	-
Pyrobotrys elongata	+	-	+	-	-	-	-	-
P. gracilis	+	-	-	-	-	-	-	-
Closterium dianae	-	-	-	-	-	-	+	-

Name of the species	S1	S2	S3	S4	S5	Kh	Vi	Mi
C. gracile	+	-	+	-	-	-	-	-
C. trilobetum	-	-	-	-	-	-	+	-
Closterium sp.	+	+	+	+	-	-	+	-
Closteridium lunuta	-	-	-	+	-	-	-	-
Cosmarium bicardia	+	-	+	+	+	-	-	-
C. laeve	-	-	-	+	+	-	-	-
C. lundellii	+	-	-	+	+	-	-	-
C. subcostetum	+	-	-	+	+	-	-	-
Coelastrum microporum	-	+	-	-	-	-	-	-
Staurastrum tetracerum	+	-	-	-	-	-	-	-
Ulothrix zonata	-	-	+	-	-	-	+	+
U.subtilissima	-	-	-	-	-	-	+	-
Oedogonium sp.	-	-	-	+	-	-	+	-
Spirogyra sp.	-	-	-	+	+	-	+	-
Hydrodictyon sp.	-	-	-	+	+	-	-	-
Schizomeris leibleinii	-	-	-	-	-	-	+	-
EUGLENOPHYCEAE								
Euglena gracilis	+	+	+	-	-	-	+	-
E. agilis	+	+	+	+	+	-	+	-
E. acus	+	+	+	-	-	-	+	-
E. caudata	+	+	+	-	-	-	+	-
E. pascheri	+	+	-	-	-	+	-	-
E. tripteris	+	-	-	-	-	-	-	-
E. viridis	+	+	+	-	-	-	+	-
E. spirogyra	+	-	-	-	-	-	-	-
E. granulata	+	-	+	-	-	-	+	-
E. minima	+	-	-	-	-	-	-	-
E. anabaena	-	-	-	-	-	-	+	-
E. proxima	+	-	-	-	-	-	-	-
E. variabilis	+	-	-	-	-	-	+	-
E. ratornata	-	-	-	-	-	-	-	+
Phacus orbicularis	-	+	-	-	-	-	-	-
P. acuminatus	+	-	-	-	-	-	-	-

Name of the species	S1	S2	S3	S4	S5	Kh	Vi	Mi
P. pyrum	+	-	-	-	-	-	+	-
P. glober	-	-	-	-	-	-	+	-
P. longicauda	+	-	+	-	-	-	+	-
P. agilis	+	-	-	-	-	-	-	+
Phacus brachykentron	+	-	-	-	-	-	-	-
P. circumflexum	+	-	-	-	-	-	-	-
P. indicus	+	-	-	-	-	-	-	-
P. pleuronectes	+	-	-	-	-	-	-	-
Trachelomonas hispida	-	+	-	+	+	-	+	-
T. intermedia	-	-	-	-	-	-	+	-
T. volvocina	+	-	-	-	-	-	+	-
Lepocinclis fusiformis	+	-	-	-	-	-	-	-
BACILLAROOPHYCEAE								
Cyclotella meneghiniana	+	+	+	+	+	-	+	+
C. striata	-	-	-	-	-	-	+	-
Melosira granulata	+	-	-	-	-	-	-	-
Melosira sp.	-	+	+	+	+	-	-	-
Fragillaria gracillima	-	-	-	+	-	-	-	-
F. brevistriata	-	-	-	-	-	-	+	-
Fragillaria sp.	-	-	+	+	+	-	-	-
Synedra labulata	-	+	-	-	-	-	+	-
S. ulna	+	-	-	-	+	-	-	-
Synedra sp.	-	+	+	-	-	-	+	-
Achanthus karyophila	-	-	-	+	-	-	-	-
A. exigua	-	+	-	-	-	-	-	+
Cocconeis Placentula	-	+	-	+	+	-	+	-
Cocconeis sp.	-	-	+	-	-	-	+	-
Diploneis sp.	-	+	-	-	-	-	-	-
Anomoeoneis sphaerophora	-	+	-	+	+	-	-	+
Navicula anglica	-	+	-	+	-	-	-	-
N. cryptocephela	+	-	+	-	-	-	-	-
N. cuspidata	-	+	+	-	+	-	+	+
N. pupula	+	+	-	-	-	-	+	-

Name of the species	S1	S2	S3	S4	S5	Kh	Vi	Mi
N. salinarum	-	+	+	-	-	-	+	-
Navicula sp.	-	-	+	-	-	-	-	+
Pleurosigma itium	-	-	-	+	-	-	-	-
Gyrosigma sp.	+	-	+	+	+	-	-	-
Gomphonema sp.	-	+	+	+	-	-	+	+
Amphora acutiuscula	-	+	-	-	-	-	-	-
A. angusta	-	+	-	+	-	-	-	-
A. granulata	-	+	-	-	-	-	-	-
Amphora sp.	-	+	+	+	-	-	-	-
Nitzschia amphibia	-	+	-	+	-	-	+	-
N. filiformis	-	+	+	-	-	-	-	-
N. intermediata	-	-	-	-	+	-	-	+
Nitzschia macilanta	-	-	-	+	-	-	-	-
N. amphioxiodes	+	-	+	-	-	+	-	+
N. microcephela	-	+	-	-	-	-	-	+
N. palea	+	+	-	-	-	-	+	+
N. pygma	-	-	-	-	-	-	-	+
N. reversa	-	-	-	-	+	-	-	-
N. subtilis	-	+	-	-	-	-	-	-
N. thermalis	-	+	-	-	-	-	-	-
N. Fasciculata	-	-	-	-	-	-	+	-

S1 = Sabarmati, S2 = Sabarmati 2, S3 = Sabarmati 3, Sh = Shedhy, Va = Vatrak, Kh = Khari, Vi = Vishwamitri, Mi = Mini.

+ = Present, - = Absent.

The Khari

River Khari being highly polluted no indigenous algae were recorded (Rana and Jameson, 1995). However, Shaji and Patel (1990) rocorded 17 taxa of pollution tolerant forms from this river near Laly village. These forms were recorded from the soil enriched with effuent during rainy season when effluent was diluted. The important forms recorded were species of *Chroococcus, Synechocystis, Merismopedia, Arthrospira, Oscillatoria, Ankistrodesmus, Scenedesums, Nitzschia and Euglena.*

The Vatrak

The river was studied near Kheda town before it mixes with Sabarmati. Algal samples were collected from various spots of the river. In all 53 taxa belonging to

31 genera were recorded from this river. Cyanophyceae was represented by 15 taxa, Chlorophyceae by 20 taxa of 8 genera, Bacillariophyceae by 17 taxa of 12 genera and Euglenophyceae were represented by two species. The more common forms recorded were *Merismopedia, Arthrospira, Oscillatoria, Anabaena, Pediastrum, Scenedesmus, Clostredium, Cosmarium, Spirogyra, Fragillaria, Nitzschia and Euglena.*

The Shedhy

The algal samples collected from the river before it merges with river Vatrak near Kheda town. In all 68 taxa belonging to 37 genera were recorded. River Shedhy was represented by 19 taxa of Cyanophyceae, 32 taxa of Chlorophyceae, 15 taxa of Bacillariophyceae and two species of euglenoids.

The Mini

The river water being highly polluted it could sustain only a few algae taxa. In all 33 taxa of 18 genera were recorded. The Cyanophycean members were represented by 10 taxa of 4 genera mainly of Oscillatoria sp. Chlorophyceae members were poorly represented by 4 taxa of 3 genera containing *Scenedesmus* and *Ulothrix*. The diatoms were represented by 17 taxa of 9 genera. The more common diatoms were *Cyclotella, Achenthis, Navicula, Gomphonema, Nitzschia and Cymbella.*

The Vishwamitri

The river Vishwamitri was studied by dividing into three parts by Nandan (1983). In all 148 taxa belonging to 49 genera of four classes viz., Cyanophyceae, Chlorophyceae, Bacillariophyceae and Euglenophyceae were recorded. The Cyanophycean members were represented by 34 taxa of 8 genera. The more common members were *Chroococcus, Microcystis, Phormidium, Arthorospira, Spirulina, Oscillatoria and Raphidiopsis.* The Chlorophyceae was represented by maximum number of 57 taxa of 26 genera. The more common were *Scenedesmus, Micractinium, Ankistrodesmus, Pediastrum, Schroederia, Closterium, Cosmarium and Spirogyra.* The diatoms were represented by 23 taxa of 11 genera. The more common forms were *Navicula, Nitzschia, Cyclotella, Synedra and Gomphonema.* The euglenoids were represented by 34 taxa of four genera and mainly belonging to species of *Euglena and Phacus.*

Pollution Index

Large number of biological pollution indices are in use for detection and assessment of pollution in water bodies (Shannon and Weaver, 1949, Nygaard, 1949, Palmer, 1969, Cairns and Dickson, 1971, Rana and Nirmal Kumar, 1993). In the present study Palmer's pollution index (Palmer, 1969) was scored for the evaluation of pollution status of various rivers. Palmer (1969) listed algal genera and species and rated them on the basis of their pollution tolerant genera and rating was given to them. A score of 20 or more for a water body indicates gross organic pollution, while a score of 15-19 indicates probable evidence of organic pollution and lower figure indicate low organic pollution. Similar rating was given to algal species tolerating organic pollution. On the basis of occurrence of algal genera and species in various rivers a table indicating the score of Palmer's index was preparted (Table 10). The score of genera and species indicated that all the rivers were highly polluted scoring more than 20 points to show confirmed high organic pollution.

Table 10

Palmer's pollution index of Various Rivers

S.No.	Rivers	Score	
		Genera	Species
1.	Sabarmati (Ahmedabad)	38	37
2.	Sabarmati (Kheda)	30	29
3.	Sabarmati (Khambhat)	31	-
4.	Khari	23	-
5.	Vatrak	30	16
6.	Shedhy	30	17
7.	Mini	28	-
8.	Vishwamitri	30	-

Pollution Monitoring

Bioassays or toxicity tests are commonly used methods for monitoring wastes or toxic chemicals using standard test organisms (APHA, 1989). Toxicity test using different test organisms have been conducted by many workers and reviewed by Shubert (1984) and Rana (1995). River Khari receives enormous amount of untreated industrial discharges, particularly from Vatava industrial estates of Ahmedabad. The toxicity of this river water was assessed by preparing a dilution series and using different test organisms. Toxicity test was carried out under laboratory conditions using *Anacystis nidulans* (Cyanobacteria), *Daphnia* sp. (Cladoceron), *Puntius* sp. (Fish) and seeds of Bajra, Radish and Mustard. Maximum 96 hrs. LC 50 value was calculated for the first three test organisms. The LC 50 value found to be 4, 11, and 12 percent for *A. nidulans, Daphnia and puntius* respectively. The seed germination test of three test crops were conducted in a dilution seried of Khari river. It was found that seeds germinated upto in 75 percent sample water. However, the other parameters like fresh weight, dry weight, root-shoot length and root-shoot ratio of seedlings were severely affected above

25 percent sample water in all crops (Jameson and Rana, 1995).

Discussion

Many Indian rivers polluted due to industrial effluents have variety of pH ranging from acidic to alkaline (Mitra, 1982, Rana and Palria, 1982, Palria and Rana, 1985, Trivedy, 1988, Shaji and Patel, 1991). The rivers receiving organic wastes have low D.O. content and affects the quality and quantity of biota living in such habitats (Verma et al., 1978, Goel et al., 1980, Rana and Palria, 1988, Venkateswarlu, 1986). Chloride content is an important pollution parameter and known to increase the degree of eutrophication (Goel et al., 1980 Palria and Rana, 1985, Sinha, 1986). Phosphate and nitrate, two important nutrients for the proliferation of algae are contributed by sewage and many industrial wastes (Rana and Palria, 1985, Sinha, 1988, Shyamsundar, 1988, Bharti and Murthy, 1990). B.O.D. and C.O.D. commonly used parameters to indicate pollution load of industrial wastes and receiving water bodies have been investigated by many workers (Kumar et al., 1974, Verma et al., 1978, Rana and Palria, 1988, Venkateswarlu, 1986, Sudhakar et al., 1992). Phycology of many polluted waters including sewage, industrial wastes, lakes, ponds and rivers have been investigated by many workers (Rana, 1977, Prasad and Saxena, 1980, Rai et al., 1981, Nandan and Patel, 1983, Rana and Palria, 1988, Gunale, 1991, Sudhakar et al., 1991, Rana et al., 1995). Cyanophyceae is known to dominate in such polluted water bodies. Palmer (1969) compliled a list of pollution tolerant genera and species and rated them on the basis of their occurrence. Many such genera and species have been found to occur in polluted water bodies of Indian habitats (Rai, 1978, Govindan and Sundarsan, 1979, Rai and Kumar, 1980, Hosmani and Bharti, 1980, Nandan and Patel, 1983, Rana and Palria, 988, Sengar and Sharma, 1987, Sudhakar et al., 1992, Shaji and Patel, 1991 and Rana et al., 1995). Many species of chlorocoocales, diatoms and euglenoids are known to occur and occupy selective polluted habitats and can serve to indicate the ecological conditions (Rai and Kumar, 1983, Shubert, 1984, Round 1991). Number of polluted rivers of Central Gujarat investigated in the present review indicated different degree of pollution as revealed by its physico-chemical parameters. The occurrence of wide range of algal species in these habitats indicate their wide tolerance to different types of pollutants. Pollution indices such as Palmers index provides useful information about the pollution load in the water bodies.

Toxicity tests of various types of wastes using different test organisms are very useful for monitoring the pollution status of such habitats. A multi-species toxicity test covering various components of the ecosystem is very important, since the stability of ecosystem rests upon the healthy existence of its various components (Caims, 1983). The present data of toxicity test indicated that Khari river water is highly toxic to all the test organisms and the discharge of such toxic industrial effluents into the rivers should be stopped.

Systematic studies of various pollutants and their co-relation with biota may prove useful for the pollution biology and can be exploited usefully for managing and restoring many damaged ecosystems.

Acknowledgement
The present work was carried out under the financial help provided by the UGC, New Delhi.

❑ REFERENCES

APHA, 1989. *Standard Methods for the Examination of Water and Waste water* (17th edn). American Public Health Association, New York.

Bharti S.G. and Murthy S.R.K., 1990. Effect of industrial effluent on river Kali around Dandeli, Karnataka. Part I. physico-chemical complexes. *Indian J. environ. Hlth. 33* : 167-71.

Cairns J. Jr., 1983. Are single species test alone adequate for estimating hazards ? *Hydrobiologia 137* : 271-278.

Cairns J. Jr. and Dickson K.L., 1971. A simple methods for the biological assessment of the effects of waste discharges on aqatic bottom dwelling organisms. *J. Water Poll. Conrol Fed. 43* : 755-772.

Goel P.K., Gopal B. and Trivedy R.K., 1980. Impact of sewage on freshwater ecosystem-II physico-chemical characteristics of water and their seasonal changes. *Int. J. Ecol. Environ. Sci., 6* : 97-116.

Govindan V.S. and Sundaresan B.B., 1978. Seasonal succession of algal flora in polluted region of Adyar river. *Indian J. Environ. Hlth. 21* : 131-142.

Gunale V.R., 1991. Algal communities as indicators of pollution. *J. Environ. Biol. 12* : 223-232.

Gunale V.R. and Balkrishnan M.S., 1981. Biomonitoring of eutrophication in the Pavana, Mula and Mutha rivers flowing through Poona. *Indian J. Environ. Hlth., 23* : 316-322.

Harish Kumar, 1994. Screening and characterzation of pollution strategies of river Sabarmati, Gujarat, India. *Geobios 21* :283-289.

Hosmani S.P. and Bharti S.G. 1980. Algae as indicators of organic pollution. *phykos. 19* : 23-26.

Jameson J. 1993. *Ecological studies of certain polluted Rivers of Gujarat*. Ph.D. Thesis, S.P. University, Vallabh Vidyanagar, Gujarat.

Jameson J. and Rana B.C. 1995. Seed germination and seedling growth of certain plants in industrially polluted Khari river water. *Geobios*. 22 : 39-41.

Jameson J. and Rana B.C. 1996. Pollution status of the river complex of Sabarmati at Kheda region of Gujarat. I - physico-chemical characters. *Pollution Research, 15* (1).

Jameson J. and Rana B.C., 1996. Pollution status of the river complex of Sabarmati at Kheda region of Gujarat. II- phycological characters. *Poll. Res.15* (2) :

Kargupta A.N. and Siddiqui E.N., 1996 (eds) *Algal Ecology* : An overview. International Book Distributors, Dehradun, India.

Kumar H.D., Bisaria G.P., Bhandari L.M., Rana B.C. and Sharma V. 1974. Ecological studies on algae isolated from effluents of an oil refinery, fertilizer factory and brewery. *Indian J. Environ. Hlth. 16* : 247-255.

Mitra A.K. 1982. Chemical characteristics of surface water at selected guaging stations in the river Godavari, Krishna and Tungabhadra. *Indian J. Environ. Hlth., 24* : 165-179.

Nandan S.N. 1983. Algal Flora of polluted Waters of Gujarat. Ph.D. Thesis, S.P. University, vallabh vidyanagar, Gujarat.

Nandan S.N. and Patel R.J. 1983. Study of algal communities in river Vishwamitri, Baroda, as indicators of organic pollution. *Indian J. Ecol., 10* : 11-15.

Nandan S.N. and Patel R.J. 1985. Study of phytoplankton and physico-chemical parameters in Vishwamitri river, as indicator of eutrophication. *Res. J. Pl. Environ., 2* : 23-27.

Nygaard G. 1949. Hydrobiological studies in some ponds and lakes II - the quotient hypothesis and some new or little known plankton organisms. *Kgl. Danske. Vendenskselsk Biol. Skrifler 7* : 1-293.

Palria S. and Rana B.C. 1985. Ecological studies of river polluted by dyeing and textile factories waste. *Comp. physiol. Ecol., 10* : 235-237.

Palmer C.M. 1969. A composite rating of algae tolerating organic pollution. *J. phycol., 5* : 78-82.

Prasad B.N. and Saxena M. 1980. Ecologycal study of Blue-green algae in river Gomati. *Indian J. Environ. Hlth., 22* : 151-168.

Rana B.C. 1977. Algae of sugar factory waste. *Geobios*. 4 : 121.

Rana B.C. (ed) 1995. *Pollution and Biomonitoring*. Tata Mc Graw Hill Publishing Co., New Delhi, pp. 425.

Rana B.C. and Nirmal Kumar J.I. 1993. A composite rating of trophic status of certain ponds of Gujarat, India. *J. Environ. Biol., 14* : 111-118.

Rana B.C. and Jameson J. 1995. Evaluation of pollution status of river Khari near Kheda region of Gujarat. *Geobios new Reports, 14* : 146-148.

Rana B.C. and Palria S. 1982. Ecological studies of certain heavy metal pollution in Rajastan. In : S.P. Banerjee (ed) *Proc. Nat. Sem. Minerals and Ecology* VI/5 : 1-5. Indian School of Mines, Dhanbad, India.

Rana B.C. and Palria S. 1988. Phycological and physico-chemical evaluation of the river Ayad, Udaipur. *Phykos 27* : 211-217.

Rana B.C., Nirmal Kumar J.I. and Sreenivas S.S. 1995. phytoplankton ecology of certain inland water bodies of Central Gujarat, India. pp. 101 - 129. In : A.N. Kargupta and E.N. Siddiqui (eds) *Algal Ecology : An Overview*. International Book Distributors, Dehradun, India.

Rai L.C. 1978. Ecological studies of algal communities of the Ganges river at Varanasi. *Indian J. Ecol., 5* : 1-6.

Rai L.C., Gaur J.P. and Kumar H.D. 1981. Phycology of heavy metal pollution. *Biological Review, 56* : 99-151.

Rai L.C. and Kumar H.D., 1980. Algae and the problem of water pollution. In : K.S. Bilgrami and K.M. Vyas (eds). *Biology of Microorganisms,* pp. 127-140.

Round F.E. 1991. Diatoms in river water monitoring. *J. Apl. phycol., 3* : 129-145.

Sengar R.M.S. and Sharma K.D., 1987. Role of algae in the assessment of pollution in river Yamuna. *J. Indian Bot. Soc., 66* : 325-334.

Shaji C. 1990. *Studies on Algal Flora and Ecology of some Polluted Waters in Gujarat* Ph.D. Thesis, S.P. University, Vallabh Vidyanagar, Gujarat.

Shaji C. and Patel R.J. 1990. Chemical and biological assessment of pollution in the river Khari, Gujarat. *Indian J. Ecol., 17* : 6-8.

Shaji C. and Patel R.J. 1991. Chemical and biological evaluation of pollution in the river Sabarmati at Ahmedabad, Gujarat, *phykos, 30* : 91-100.

Shannon C.E. and Weaver W. 1949. *the Mathematical Theory of Communication*. University of Illinois Press, Urbana, pp. 117.

Shyam Sunder, 1988. Monitoring the water qualiy in a stretch of river Jhelum, Kashmir. In : R.K. Trivedy (ed). *Ecology and pollution of Indian Rivers.* Ashish Publishing House, New Delhi.

Sinha M.P. 1986. Limnobiotic study of trophic status of a polluted freshwater reservior of coalfield area. *Poll. Res., 5* : 13-17.

Sinha M.P. 1988. Effect of waste disposal on water quality of river Damodar in Bihar. Physico-chemical characteristics. In R.K. Trivedy (ed.) *Ecology and pollution of Indian Revers.* Ashish Publishing House, New Delhi.

Shubert L.E. 1984 (ed.) *Algae as Ecological Indicators.* pp. 434. Academic Press, London.

Sudhakar G., Jyothi B. and Venkateswarlu V. 1991. Metal pollution and its impacts on algae in flowing waters in India. *Arch. Environ. Cont. Toxicol., 21* : 556-566.

Sudhakar G., Jyothi B. and Venkateswarlu V. 1992. Impact of paper mill effluents on the distribution of cyanobacteria in the river Godavari, India. *Water Management, 11* : 263-270.

Trivedy R.K. (ed.) 1988. *Ecology and Pollution of Indian Rivers.* Ashish Publishing House, New Delhi.

Trivedy R.K. and Goel P.K. 1984. *Chemical and Biological Methods for Water Pollution Studies.* Environmental Publications, Karad, India.

Venkateswarlu V. 1969. An ecological study of the algae of the river Moosi, Hyderabad (India) with special reference to water pollution - I. physico-chemical complexes. *Hydrobiologia, 33* : 117-143.

Venkateswarlu V. 1969. An ecological study of the algae of the river Moosi, Hyderabad (India) with special reference to water pollution - II. Factors influencing the distribution of algae. *Hydrobiologia, 33* : 352-363.

Venkateswarlu V. 1981. Algae as indicators of river water quality and pollution. *Cent. Bd. Prev. Cont. Water Poll.* Osmania University, 93-100.

Venkateswarlu V. 1986. Ecological studies on the rivers of Andhra Pradesh with special reference to water quality and pollution. *Proc. Indian Sci. Acad. 96* : 495-508.

Venkateswarlu V. and Sampath Kumar P.T. 1982. Chemical and biological assessment of pollution in the river Moosi, Hyderabad (A.P.), *India, Biol. Bull. India, 4* : 23-30.

Venkateswarlu T. and Jayanthi T.V. 1968. Hydrobiological studies of the river Sabarmati to evaluate water quality. *Hydrobiologia 31* : 442-448.

Verma S.R., Tiagi A.K. and Dalela R.C. 1978. physico-chemical and biological characteristics of Kadrabad in U.P. *Indian J. Environ. Hlth., 20* : 1-13.

Authors :

1. * **B.C. Rana and J. Jameson :** Department of Biosciences, Sardar Patel University, Vallabh Vidyanagar - 388120. Gujarat (India.)

■ ■

6

A Pollution Profile of Damodar River

❑ INTRODUCTION

Damodar river is the lifeline of the thickly and densely populated and highly industrialised Jharia coalfield. This river is the chief source of water for industrial use, power generation, irrigation and human consumption in the region and simultaneously acts as the main sink of the wastes effluents released from the different industries. As no systematic study on water quality monitoring has been carried out in Damodar river of Dugda-Sindri industrial stretch of Jharia coalfield the present study has been undertaken with the following objectives

To Characterise the effluents from various industries including domestic effluents being discharged into the Damodar river and to study the pollution profile of Damodar river.

Methodology

As the study is restricted to certain selected stretches of the river only eight sampling stations were selected from Dugda to Sindri. Industrial effluents of various industries situated in between these stations were also monitored. Monitoring was done monthwise. The sampling points are U/S Damodar river before Dugda coal washery D/S Damodar river after Dugda coal washery, Telmochu bridge before Garga river and after Garga river, Jamadoba water works, Premnager basti Domgarh water works & Nimatand basti. Effluents of Dugda coal washery Moonidih coal washery, Coke oven plants (Loyabad, Bhowra, Bararree, and Lodna), U/S Dumrijore nalla near Jamadoba coal washery, Jamadoba coal washery, D/S Dumri Jore nalla, Sudamdih coal washery, Chasnalla coal washery, F.C.I. Sindri were collected from their final discharge.

Dugda coal washery
This coal washery is situated at 2 km. away from Damodar river. Effluents of this coal washery is allowed to settle in a series of ponds & overflows collects in a pond where a pump is installed through treated effluent is recycled to the plant but when pump is not operating or when pond is filled up by slurry effluent goes in Damodar river. The sample has been collected and flow was measured in M^3/ Sec.

Moonidih coal washery
The unit has settling tanks for recovery of coal fines from the slurry but inadequate supernatent effluent is discharged to river Damodar. Final effluent leading to Damodar river near CISF training centre has been collected.

Coke oven plants
There are four coke oven plants viz. Lodna, Bhawra, Bararree & Loyabad. Their effluents are discharged in a nalla which meet river Damador.

U/S Dumri jore nalla
This nalla contains effluents of Jamadoba power house No-3 and domestic efluents of the area.

Jamadoba coal washery
This washery has a few kuccha and pucca settling tanks. There is a pump installed here which recycles the treated effluent. The overflow from the tank goes to Dumri jore nalla.

D/S Dumri jore nalla
This sampling point gives quality of nalla water after mixing & dilution of washery overflow, power house & domestic effluents.

Sudamdih coal washery
Effluents from this washery after passing through a few settling tanks overflow to Damodar riser.

Chasnalla coal washery
Final effluent is discharged to a nalla which finally meets river Damodar.

F.C.I. Sindri
Effluents of FCI Sindri collected from Bara Domgarh contains powerhouse effluents which meets Damodar after traversing about 6 km. Main F.C.I. effluent were collected from Sirsakuri bridge and flow rate was measured.

Sampling Procedure
River water and effluents were collected as per IS: 2488, 1966, part 1 NEERI manual & APHA. Samples were collected in cleaned polyethylene bottles which were rinsed

initially with a portion of sample. For oil & grease sampling a pre-cleaned small wide mouth glass container was used because this particular parameter requires whole sample for the analysis. For bacteriological sampling pre-sterilised (at 121 ^{0}C for 15 minutes) paper wrapped BOD bottles were used.

Frequancy of Sampling

Monitoring was done on monthly basis starting from January 1995 to December 1995. composite samples were collected at one hour interval from 7 AM to 5 PM representing 10 Hr. period. Grab samples thus collected at various intervals were mixed to make one final composite samples (APHA).

Flow Measurement of Waste Water from Industries

Where industrial effluent is flowing through a nalla flow measured by measuring velocity of flow, width of nalla and depth of the nalla.

Flow in M^3/sec = v(m/s) * depth (m).

Preservation of Samples

Samples collected were analysed immediately in the lab., however for some of the parameters like ammoniacal nitrogen & trace metals sample was preserved. Samples for ammoniacal nitrogen were preserved by adding 0.8 ml conc. sulfuric acid and samples was refrigerated. For trace metals 100 ml filtered samples were taken in a 100 ml. small plastic bottles and to it was added 5 ml. 6N nitric acid. These bottles were refrigerated till analysis was done.

Analysis

Analysis of temp., pH, total dissolved solids were done in the field by using centuary ck. 710 water Analyser kit. Dissolved oxygen was also determined at the spot by iodometric method by using alkali azide modification method. Other parameters like T.S., TSS, Turbidity, BOD, COD, Ammoniacal nitrogen, trace metals. Total and faecal coliforms were done as per APHA methods.

Flow rate of the River Damodar at the sampling points in (M^3/Sec.)

Sampling Point	Jan.	Feb.	Mar.	Apr.	May	June	July	Aug.	Sept.	Oct.	Nov.	Dec.
U/S Damodar before Dugda coal washery.	28.2	26.5	14.4	15.5	20.2	30	31.2	30.4	30	28	25	25.2
D/S Damodar after Dugda coal washery.	30.2	28.4	22.3	24.2	22.1	25	29.2	30.0	31	32	30	25
Telmuchu bridge	27.8	35.2	36.5	37	36.2	36.4	39	40.9	39	35	32	33

Jamadoba water works.	30.2	31.4	26.8	25.2	26.4	25.2	30.4	31.2	37	32	31	30
Premnagar Basti (u/s FCI)	31.1	30.8	28.2	26.0	25.2	26.4	31.8	32.0	38	32	31.9	32
Domgarh water works.	26.4	26.2	25.4	26.2	24.1	22.3	29.4	30.0	32	32.4	30.	31
Nimatand Basti (d/s FCI)	32.0	31.7	30.8	31.4	28.2	27.6	28.6	31.0	32.7	32.5	30	31.7

Average flow rate in MLD in year 1995

U/S Damodar before Dugda coal washery.	2193.091
D/S Damodar after Dugda coal washery.	2371.68
Telmuchu bridge	3153.6
Jamadoba water works.	2568.672
Premnagar Basti (u/s FCI)	2630.88
Domgarh water works.	2404.0831
Nimatand Basti (d/s FCI)	2435.04

Average flow rate of industrial effluent in MLD in 1995

Dugda coal washery	466.56
Moonidih coal washery	368.064
Loyabad coke oven	6588.00
Lodna coke oven	557.28
Bararree coke oven	504.576
Bhowra coke oven	298.08
U/S Dumrijore nalla	912.384
Jamadoba coal washery	177.12
D/S Dumri jore nalla	989.28
Sudamdih coal washery	582.336
Chasnalla coal washery	318.816
FCI, Sindri (power house)	1412.64
FCI, Sindri (Sirsakuri Bridge)	1234.68

Analysis resuits of the effluents of Industries & nalla.

Parameters (mg/1)	Dugda Coal washery	Moonidih Coal washery	U/S Dumri Jore Nalla	Jamadoba Coal washery	D/S Dumri Jore Nalla
pH	6.8-7.8	6.8-7.8	7.2-8.4	6.8-7.8	7.2-8.5
Temp. (A/W)	30/29-38/31	30/29-38/31	30/29-36/31	30/29-36/31	30/29-36/31
Turbidity (NTU)	132-1700	104-168	115-228	858-1068	826-1194
TSS (mg/1)	200-4672	126-298	250-605	1268-1626	1268-1958
TDS (mg/1)	240-600	584-716	198-282	225-308	200-262
Alk (mg/1)	190-472	100-145	128-148	68.6-102	148-182
Hard. (mg/1)	112-290	98-168	128-180	68.6-102	148-182
Ca- (mg/1)	27.5-17.8	36-58	22.0-38.0	18.0-28.0	48-58
Mg- (mg/1)	12.6-136	22-28	20.4-30.4	6.8-18.2	20-32
Cl - (mg/1)	17-68	36-40	62.8-70.8	20.6-36.2	68-90
SO4-2 (mg/1)	12-377	27.46	28.0-64.0	25-48	34-64
DO (mg/1)	3.0-6.5	2.5-3.2	5.8-6.8	3.8-4.6	4.2-5.8
BOD (mg/1)	1.0-3.9	1.0-1.2	1.0-1.8	1.0-1.8	1.2-1.8
COD (mg/1)	168-2100	198-280	308-432	248-548	628-900
Oil & grease (mg/1)	9.5-65	7.8-16.8	3.6-9.8	5.2-9.8	3.8-9.8
Pb (mg/1)	BDL	BDL	BDL	BDL	BDL
Total Cr	BDL-0.002	BDL	BDL	BDL	BDL
Cu (mg/1)	BDL-0.004	BDL	BDL	BDL	BDL
Zn (mg/1)	0.004-0.5	0.28-0.5	BDL	0.4-0.5	Nil-0.2
Fe (mg/1)	BDL-0.04	BDL-0.3	BDL-0.03	BDL-0.03	BDL-0.03
Mn (mg/1)	BDL	BDL	BDL	BDL	BDL
Se (mg/1)	BDL	BDL	BDL	BDL	BDL
Ni (mg/1)	BDL	BDL	BDL	BDL	BDL
Ammoniacal Nitrogen	Nil	Nil	Nil	Nil	Nil
Nitrate	Nil	Nil-0.12	Nil-0.21	Nil	Nil

Parameters (mg/1)	Sudamdih Coal washery	Chasnalla Coal washery	F.C.I. Bara Domgarh, Sindri	F.C.I. Srsakuri Bridge	F.C.I. Sindri Pond No.-9
pH	7.2-8.4	7.2-8.0	7.1-8.3	6.5-7.8	9.2-12.5
Temp. (A/W)	26/23-38/35	25/22-37/31	26/23-35/32	25/23-35/32	26/24-36/32
Turbidity (NTU)	60-164	112-214	105-208	358-568	89-124
TSS (mg/1)	300-379	726-1878	150-205	68-626	68-98
TDS (mg/1)	85-868	412-528	138-268	215-328	200-526
Alk (mg/1)	184-195	142-168	138-178	68.0-100	1202-2123
Hard. (mg/1)	142-190	128-172	138-170	108-120	68-120
Ca- (mg/1)	38.5-48	36-48	32.0-38.0	18.5-38.0	18-21.4
Mg- (mg/1)	22.8-34	23-38	17.4-27.4	3.8-8.2	2.3-5.87
Cl - (mg/1)	68.6-80.5	56-70	52.8-60.8	29.6-34.2	168-324
SO4-2 (mg/1)	25.8-57.8	27-66	28.0-54.0	22-44.5	14-24
DO (mg/1)	2.2-4.2	2.5-3.7	5.2-6.8	4.8-6.6	4.2-5.6
BOD (mg/1)	1.0-1.2	1.1-3.2	1.0-1.8	1.0-1.8	1.2-1.6
COD (mg/1)	408-572	2298-3489	108-170	148-348	228-320
Oil & grease (mg/1)	4.8-9.2	3.8-8.8	3.6-9.8	2.2-6.8	12.5-18.9
Pb (mg/1)	BDL	BDL	BDL	BDL	BDL
Total Cr	BDL	BDL	BDL	BDL	BDL
Cu (mg/1)	BDL	BDL	BDL	BDL	BDL
Zn (mg/1)	0.5	0.3-0.5	BDL	0.2-0.5	Nil-0.2
Fe (mg/1)	BDL-0.02	BDL-0.4	BDL-0.03	BDL-0.04	0.5-1.5
Mn (mg/1)	BDL	BDL	BDL	BDL	BDL
Se (mg/1)	BDL	BDL	BDL	BDL	BDL
Ni (mg/1)	BDL	BDL	BDL	BDL	BDL
Ammoniacal Nitrogen	Nil	Nil	Nil	Nil	112.7-214.9
Nitrate	Nil	Nil	Nil	7.84-14.8	Nil

Analysis results of coke oven plants for the year 1995.

Parameters (mg/1)	Lodna Coke oven	Bhawra Coke oven	Bararee Coke oven	Loyabad Coke oven
pH	8.2-8.4	7.9-8.5	7.0-7.8	6.9-7.7
Temp. (A/W)	24/23-36/33	25/22-36/31	26/23-35/32	25/24-36/32
Turbidity (NTU)	160-964	122-198	125-238	158-468
TSS (mg/1)	2300-3379	726-978	950-1245	968-1265
TDS (mg/1)	125-9868	312-428	538-1187	715-876
DO- (mg/1)	Nil	3.5-4.7	3.0-3.8	4.8-5.4
BOD (mg/1)	2454-4578	1.1-1.2	0.1-1.8	0.9-1.2
COD (mg/1)	8345-10987	298-489	808-970	2342-6543
Ammonia	1465-2312	45-54.5	34.8-49.2	8.4-10.2
Phenol	1286-2109	0.9-0.98	1.1-1.3	1.1-1.7
Cyanide	12.4-23.8	0.02-0.03	0.2-0.5	0.76-0.8

Results and Discussion

There is a wide variation in the water quality parameters as analysed during Jan. to Dec. 1995 of Damodar river water of all seven monitoring stations along Dugdha-Sindri industrial belt in Jharia coalfield. The water quality of Damodar river ranges between 6.6 to 8.8, i.e. neutral to alkaline. This indicates that decomposition of organic matter contributed by industial & human wastes has not suppressed the pH all along the study area (Cummins, 1966). The alkalinity values are also found higher in the range of 64.0 to 142 mg/1 at all points which is higher than the limit of 5 mg/1 (IS : 2296, 1974). Turbidity ranges 112 to 1986 NTU. Biochemical oxygen demand and chemical oxygen demand have been considered very important parameters to assess the extent of organic pollution in the river. It is due to the fact that waste volume entering into the river was same but river water volume decreases and thus increasing effluent and river water temperature . As a result DO level during summer season. Along the entire study area DO level ranges from 4.2 to 7.8 mg/1 BOD value ranges from 0.5 to 5.8 mg/1. With these results of DO and BOD oxygen sag curve is viewed together for an appropriate interpretation of any two equal DO valus in the stream the DO sag would correspond to different BOD values. The BOD and COD profile of river along the various stretches shows that at sampling stations Telmucho Bridge BOD is higher & COD near Dugdha coal washery were higher which indicates that human & industrial wastes are affecting river water. In these zones of the river, available oxygen got utilised. The chlo-

Analysis results of coke oven plants for the year 1995.

Parameters (mg/1)	Dugdha (U/S)	Dugdha (D/S)	Telmucho bridge (before garga)	Telmucho bridge (After garge)	Jamadoba Water Works	Premnagar Basti	Domgarh Water Works	Nimatand Basti
pH	6.8-7.8	7.0-7.9	7.1-7.6	7.3-7.8	7.4-8.8	6.6-7.8	7.2-8.8	7.6-8.5
Temp. (^C)	28-35	28-35	29-35	29-35	28-34	23-33	23-34	22-35
Turbidity (NTU)	543-986	745-1986	112-450	825-1945	312-550	225-684	625-884	250-640
TSS (mg/1)	868-1678	1125-2309	221-542	544-1289	245-567	344-647	688-1158	96-244
TDS (mg/1)	145-204	112-254	102-198	102-298	68-122	97-128	114-165	75-99
Alk (mg/1)	92-142	88-136	82-98.2	76-92	66.2-98	76.2-97	64-89.9	67.9-98
Hard. (mg/1)	105-142	105-134	108-168	102-88	102-126	98-121	45.8-98	67-87
Ca- (mg/1)	10.5-28.6	12.5-38.6	10.4-28.6	22-39.5	12-21	12-28.5	19.2-27.3	24-77.3
Mg- (mg/1)	29.5-50.8	18.5-23.8	14.1-22.8	14.8-19.9	5.6-8.8	6.0-7.6	4.6-9.9	11.8-19.9
Cl - (mg/1)	18.2-22.6	21.4-41.5	14.4-31.6	31.8-78.9	21.8-48.9	21.6-48.9	24.8-32.4	24.8-32.4
SO-4 (mg/1)	22.8-66.4	34.8-68.7	34-43.7	42.2-69.2	24.8-54.2	42.2-54.2	25.8-65.7	22.2-35.8
DO (mg/1)	5.0-7.2	6.1-7.1	6.4-7.7	4.2-5.8	6.4-7.8	7.0-7.8	5.4-6.8	5.4-6.7
BOD (mg/1)	1.0-2.3	2.8-2.6	0.8-1.2	3.0-5.8	3.0-4.5	0.5-1.9	1.8-2.9	1.8-2.9
COD (mg/1)	354-1685	334-1785	42-48	66-28.8	96-188.8	44-56	148-345	231-354
Oil & grease (mg/1)	0.6-2.4	1.1-8.2	0.2-0.4	0.1-1.2	0.4-1.2	0.3-1.0	0.8-1.7	0.8-1.7
Pb (mg/1)	BDL	BDL	BDL	BDL	BDL	BDL	BDL	BDL
Total Cr	BDL	BDL	BDL	BDL	BDL	BDL	BDL	BDL
Cu (mg/1)	BDL	BDL	BDL	BDL	BDL	BDL	BDL	BDL
Zn (mg/1)	0.4-0.5	0.4-0.5	0.4-0.5	0.4-0.5	0.1-0.2	Nil -0.2	0.1-0.2	Nil - 0.2
Fe (mg/1)	BDL-0.001	BDL-0.003	BDL-0.001	BDL-0.001	BDL-0.001	BDL - 0.003	BDL -0.001	BDL -0.003
Mn (mg/1)	BDL	BDL	BDL	BDL	BDL	BDL	BDL	BDL
Se (mg/1)	BDL	BDL	BDL	BDL	BDL	BDL	BDL	BDL
Ni (mg/1)	BDL	BDL	BDL	BDL	BDL	BDL	BDL	BDL
Ammoniacal Nitrogen	Nil	Nil	Nil	Nil	Nil	Nil	11-43	11-43
Nitrate	Nil	Nil - 0.3	Nil - 0.11	0.16-0.9	Nil	Nil	Nil	Nil
Total Coliform (MPN/100ml)	1200-1875	1430-2400	1432-2100	6600-12100	2854-8900	2400-8100	2100-4200	2100-4200
Faecal coliform (MPN/100ml)	400-850	450-760	850-1660	2456-8654	2192-6960	560-1250	830-1750	830-7950

ride and sulphate contents were found quite low and are in desirable range. These lower contents do not pose any implications for river Damodar utilisation for various purposes.

The fairly high values of BOD coupled with high values of Total coliforms and Faecal coliforms are indicative of increasing pollution of river by organic substances particularly through the discharges of sewage and domestic effluents into river. At Telmucho bridge coliforms were higher which gradually decreases towards Nimatand. However, at all the sampling points of river Damodar Coliforms were high showing that river is organically polluted. River water has been all along unstatisfactory for drinking purposes. Health hazards cropping out of chemically and bacteriologically polluted water can not be ruled out. There has been occasional outbreaks of acute gastroenterites, cholera, dysentry, diaarrhea and vomitting at several pockets of this region of coalfield time to time. Asthetically river water is also not suitable. Site V & VII are used for public water supply. On an average these two points conform to both the Indian and WHO standards for pH, DO and BOD.

River water here falls under the categories of 'C' as and is only suitable for drinking with conventional reatment followed by disinfection.

International Joint commission Standards (1973) indicate that waters with coliforms counts of 1000 mpn/100 ml or less are acceptable for bathing. The Indian standards for bathing ghats are the same as the standards for the raw surface water (IS:1974) i.e. 5000 MPN/100 ml. The existing coliforms counts in all the stations do not meet standards for body contact attributed to the surface water bathing ghats.

All forms of organic matters are basically undesirable in the water though to a certain level is tolerable. This may lead to serious troubles like eutrophication and depletion of the oxygen content of the system and thus impairing the health.

Indian standards (IS:1974) set 3 mg/1 DO as the limit. The DO level at all the sampling stations was more than 3 mg/1. This indicates that the self-purification capacity of this river can assimilate the existing load.

Summary and conclusion

From general observation it can be seen that the river water starting from Dugdha to Sindri is not suitable from aesthetic point of view. River water all along the study area appears brownish-black colour and quite turbid with hardly any transparency.

The most polluting among the various sources of pollution are coal washeries. Analysis of effluents from washeries revealed that they contain a lot of suspended solids, significant levels of COD and oil and grease and impart colour on the surface of the river Damodar.

Waste effluents of coke oven plants are highly toxic in nature though their presence has not been confirmed in the river. This is because sources of discharge are situated away from Damodar river and also effluents have found some degradation pathways while getting mixed with water discharge and domestic effluents. Thermal power plants are adding much of the suspended solids near Dugdha at station 1 thus increasing TSS and Turbidity here. Further domestic effluents such as sewage and garbage contains lots of organic matter and other undersirable matters without any treatments. Near Telmucho bridge Garga river meets which contains lots of domestic effluents of Bokaro steel city thus aggravating bacteriological contamination.

Analysis shows River water all along is unsatisfactory for drinking purposes. Surface water can be seen full of debris, scum and oil. Thus aesthetically river water is not suitable. River water all along the study area falls under the category 'C' as per classification made by Central Pollution Control Board except at points Telmucho bridge (after meeting Garga river) Jamadoba water works, Domgrah water works where coliforms are very high River water is only suitable for drinking with conventional treatment followed by disinfection. But as far as bathing is concerned, the existing coliforms counts in all the stations do not meet standards for contact attributed to the surface water bathing ghats.The DO level at all the points are more than 3 mg/1. This perhaps indicates that the self-purification capacity of this river, can assimilate the existing load.

Acknowledgements
Aurthors are grateful to Prof. H. Roy, Head of Botany Department, Binoba Bhawe University for his valuable suggestions and to Sri. M.K. Prasad, R.O. B.S.P.C.B.Dhn.

❑ REFERENCES

Anil Kumar De, Asit Kumar Sen, Deb. P. Modak, Tapash K. Mishra and Nihar R. Sengupta 1991 Toxic metals in effluents and sediments in Durgapur region, *J. of PHU, India,*3.

Anil K. De, Asit K. Sen and Deb P. Modak 1980 *Environmental International, Vol. 4. pp. 101-105.*

Anil K. De, Asit K. Sen and Deb. P. Modak 1981 The Damodar River Water Quality- Upstream and Downstream of Durgapur Barrage, *J. of the IPIIE India,* 3.

Anil K. De, Asit K. Sen and Taoesh K. Mishra 1981 Pollution Profile of Effluent and Damodar River bed Sediments in Durgapur region, *J. of the IPHE India. Vol. 1981.* No. 3.

Anil K. De, Asit K. Sen, Deb P. Modak, Tapash K. Mishra and Nihar R. Sengupta, *J. of the IPHE India. Vol. 1982.* No.3.

A Report on the Status of pollution in Durgapur - Asansol Region, 1983.

Anil K. De, Asit K. Sen, Deb. P. Modak and Tapash K. Mishra. The Impact of fertiliser Effluents and their Sediment on Damodar River, *J. Inst. chemistry (India), Vol. 55, January 1983.*

Anil K. De, Asit K. Sen, D.P. Modak and Md. R. Karim, Some industrial waste effluents in Asansol - Raniganj region and their impact on the Damodar river water quality, *J. of the Geological Sociaty of India. Vol. 30,* No. 5, 1987.

APHA, Standard Methods for examination of wastes and wastewater, 15th edition, washington, D.C., U.S.A. (1980)

Berkson H., Water quality criteria In : Env. Engg. *hand book vol. I - Water pollution* (Ed. Liptack, B.G.) Chilton Book co., Radnor, Pennsilvania (1974).

Bandopadhyay P. and Rao T.C., Dewatering characteristics of Indian Coal Fines; Int. Symp. on Fine Particle Sociaty, Honululu, Aug., 1983.

Cairney T. and Frost R.C. (1975) - A case study of water quality deterioration mainsforth colliery, country Durham, *J. of Hydrology,* 25, 275-293, 1975.

Choudhury S.K., Sodimonts characterietics of the river Ganga in Sultanganj - Bhagalpur Region, Bihar, Env. & Ecology 9(1), 249-251, 1991.

Central Board For The Prevention and Control of Water Pollution Operation Control In Damodar River. Programme Objective Series : PROBES /23/1984-85.

Clear E.J., Persepiective on river quality diagnosis J. Awowa, 69, 522, 1977.

Cumins K.W., A review of stream ecology with special emphasis on organism substrate relationships, spec, Publi. Pymatuning Lab. *Field Biol. 4,* 2-51, 1966.

Davis D.G. A practical approach to selting effluent standards, water pollution control, 14, 902, (1975.)

D.K. Saikia, R.P. Mathur, S.K.Srivastava, Heavy Metals in Water and Sediments of Upper Ganga, *Indian J. Env. Hlth., Vol. 31, No. 1,* 11-17, (1988).

Dora M.M. and Roy N.N, Investigation of water quality of Subarnarekha river for irrigation., *Indian J. Environ. Hlth, vol. 29, No. 4,* 292-298. (1987).

Fish H., "Water conservative by Effluent reuse." year book of association of river authorities of river authorities, P. 91 (1965).

Fuller W.S. In : Sewage disposal, Mc Graw Hill Book Co., New Your, 1912.

Goel P.K., Gopal B. and Trivedy R.K., Impact of Sewage on freshwater ecosystems II. Physico-chemical characteritic of water and their seasonal changes. *Int. J. Ecol. Environ. Sci.,* 116, 1980.

Gurdeep Singh, A survey of corrosivity of underground mine waters from Indian Coal mines Metals and Fuels, 169-172, July, 1980.

Gurdpeep Singh, Status of Jharia coalfield, International Symposium on Environmental Mangement, Istanbul Turkey, June 5-9, 1986.

Authors :

1. **Dr. Gurudeep Singh** : CME, ISM, Dhanbad, Bihar, India.
2. **Binay Kumar Gupta** : Bihar State Pollution Control Board, Regional Officer, Dhanbad, Bihar, India.

■ ■

7

Damodar River Pollution Caused By Coke Plant Effluent and The Low Cost Technology for its Safe Disposal

❑ INTRODUCTION

Coke plants are producing coke for metallurgical purposes on one hand and also producing gas on the other hand. These are responsible for entire supplies of benzene, toluene, anthracine, napthane and coal tar products which constitutes the raw materials for the manufacture of synthetic dyes, drugs and high explosives. Large quantities of water are used for quenching of hot coke and also for washing gas produced from ovens. Effluents generated from quenching of coke mainly contain suspended matter of coke breeze. Water used for washing gas comes out as strong ammoniacal liquor, which contains high concentration of ammonia, phenol, cyanide and other toxic surfaces. These effluents are causing serious surface water pollution in the area.

The agricultural potential of coke oven effluents is not being utilized in our country. It is, however, a pressing need and we must not over look the reuse of this nutrient rich valuable water source in irrigated agriculture. For the abatement of surface water pollution due to coke oven effluents land application system may be considered as a low cost technology for the treatment and disposal of coke oven offluent. It will give a good nutrient removal from the coke effluents for raising useful crops, provide primary, secondary and tertiary treatment of the effluents. It will not only control the surface water pollution but also maintain ecological balance of the ground water. The challange here is to utilize the chemical physical and biological properties of soil as an acceptor of the water with minimum unwanted effects to the crops that are to be grown, to the characteristics of the soil and to the quality of ground water and surface water run off. Under controlled application rates of waste water, it is also possible to maint within the soil.

There are four coke oven plants running under BCCL. They are producing a huge quantity of effluents and discharge into the river Damodar without any proper treatment. One large coke oven has been selected for detailed study. A suitable land near to the plant under study has been indentified for the collection of soil samples for pilot study of land application system of coke oven effluents.

Effluent produced from coke plant & its impact on surface water quality
Coal cannot be used as such in blast furnace. It is transformed by carbonisation into a hard, porous mass, derived of volatile matter called coke. Carbonisation of coal consists of heating bituminous coal in the absence of air to final temperatures of 900^0 C to 1100^0 C in an oven or a retort, driving of volatile products that are recovered as tar, light oil, and gas. The coking operation is carried out in a battery of 10 to 100 individual ovens designed to provide relatively uniform production of finished products and to recover heat in a way that minimizes fuel requirement of the 315 to 630m^2 of gas produced per ton of coal charged. About 40 to 50% is returned as fuel to the coke oven gas.

The waste water produced during the carbonization and gasification of fuel may be classified in three basic groups.

a. Water used for quenching the coke discharged from the retorts or ovens.

b. Wastes formed during cooling and washing of the gas.

c. Waste formed furing the purification of by-products.

Relatively large amounts of water are used when producer gas is washed, however, as it is only slightly centaminated, it may be recycled. The most heavily polluted liquors from coking plants are the wastes from the ammonia stills, where the condensate from the gas coolere is distilled, Ammonia, phenol, cyanides and sulphides are toxic to aquatic life. Chipman (1934) has shown that ammonia is toxic to fishes, phenolic compounds can affect fresh water fishes adversely by lowering the available oxygen. The insolubel pollutants (EIFAC 1973), especially tar give rise to the formation of a surface layer which hinders the access of oxygen from the air, where as larger particles of suspended matter settle on the bed of the stream thus causing further pollution. Fine suspended particles clog the respiratory organs of fish, making it impossible for them to live in such a stream. If the degree of dilution is insufficient, not only the phenols but also thiocyanides, free lime and pyridine basis may cause serious difficulties in self purification process. A considerable consumption of dissolved oxygen is caused by sulphur compounds, mainly sulphides, in addition to their harmful effects, on fish life in the stream. Cyanide is actually toxic to most fishes even at very low concentration ranging from 50 to 200 g/1. Cyanide toxicity is essentially an inhibition of oxygen metabolism, that renders the tissue incapable of exchanging oxygen. The degree of hazard exhibited by sulphide to aquatic and animal life. Dependent on the temperature. pH and dissolve oxygen. Oil and grease unless trapped from the effluent will lead to the formation of ugly oil slicks and indescent colours, all leading the diffusion of oxygen from air to water and effects the self purification of water.

Land treatement of waste water

Chemical treatment methods are costlier. In some old coke plants sufficient space may not be available for biological treatment. Land disposal system may be chosen as it will be cheap and does not require waste water treatment plant, requiring high initial and maintenance cost. It is a combination of primary, secondary and tertiary treatment of waste water and it will also control surface water pollution. It will give a very good nutrient removal from the coke oven effluents for raising useful crops in the region. The cost of fertilizer will be minimized. It will be economical and safe as available irrigation water is small in quantity in the area. Water of irrigation canal can be saved when this system will be practised. The ground water table will be recharged by recycling the waste water by irrigation. In old coke plants where sufficient area is not available for making treatment plants, land disposal system may solve the problem.

Details of the Coke Plant Under Study

The coke plant under study is situated at 10 km from Dhanbad, 1155 km from Delhi and 269 km from Calcutta within latitude 23^0 38'-13^0 40^0 N and longitude 86^0 22' -86^0 30' E. It was installed in 1915 and is well connected by metal roads. This coke plant takes water from Ekra Jore Nallah and discharges its waste water again to the Nallah which finally meets the River Damodar. Detailed information of the coke plant is given below :

1.	Design	:	Coopers underfiring, Regeneration Cross Section, vertical flumes		
2.	No. of batteries	:	2 nos. and 40 ovens per battery		
3.	Coking hours	:	26/28 hours.		
4.	Yield of coke	:	7 MT per oven		
5.	Yield of gas	:	60,000-70,000 Nm^3/day		
6.	Products and its quantity per annum	:	Hard coke	=	75-8000 MT
			Coal tar	=	1250-1300 MT
			Coke	=	2000-2500 MT
7.	Water consumed for industrial and domestic purpose	:	700 KL each/day		

8. Waste water generation from industry: 175-200 KL/day
9. Area : 20.60 acre
10. No. of persons working : 600

METHODS

Sampling & Analysis of Coke oven Effluent

The effluent samples from the coke plant under study were coliected in each week on Monday and Friday. The effluents were poured into the colum lysimeter on the same day at 17.00 hours i.e. two irrigation in a week with 5 cm flooding was maintained. It was continued for 3 months. The raw effluent and leachate samples collected were analysed as per standard methods.

Collection of Soil Samples & analysis

A suitable land near the coke plant was selected on which coke plant effluent may be disposed for irrigation. The sampling site for digging the soil upto 2km depth was determined by means of suger boring. At site the land was marked for a dimension of 4 ft x 4 ft. The located plot was cleaned, i.e. grasses, stones and other unwanted materials were removed. The soils from various depths, i.e. 0-30 cm, 30-90 cm and 90-160 cm were collected in clean bags after proper mixing at each consecutive depth and brought to laboratory for the preparation of column lysimeters. Disturbed and undisturbed soil samples were collected for physico-chemical analysis.

Preparation of Column Lysimeters

Three PVC pipes each of 200 cm length having 20 cm diameter were selected in the laboratory. In each pipe three ports at given heights were driven inside it. One end of the port was closed by plug and the other end connected by polyethylene pipe of 1.27 cm (1/2") dia used for collecting the leachate from specified depth. The top surface of the port was made perforated and just above it gravel and sand were placed so that leachate through the port may be collected easily, as shown in plate-1. The soil collected from the agricultural land near the coke plant was used in the pipes and was carefully placed and compacted to duplicate the original land in situ condition pertaining to startification and density.

Disposal of coke oven effluent & Analysis

In lysimeter No. 1 only raw effluents of the coke oven were applied.Two irrigation in each week with 5 cm flooding and the quantity of raw effluent added each time was noted.

In lysimeter No. 2 diluted effluents of coke oven were applied over the soil.In each case two irrigation in each week with 5 cm flooding was maintained and quantity of water added each time was noted.

The leachates from lysimeter No. 1 and 2 wee collected through polythyne pipes. The collected leachates were analysed. After analysis of the leachate collected from lysimeter No. 2, it was observed that if the raw effluent when diluted to 1:1, it would not contaminate the ground water quality. Two irrigations per week with 5 cm flooding was continued for three months. Then the soil of different layers of each lysimeter were analysed 10 to 11times to assess the impact on soil quality due to the disposal of coke plant effluent.

RESULTS AND DISCUSSION

The temperature of the outcoming effluent were found in the range of $28^{0}7$ C to 32°C with and average of 30.4°C. The pH value was found to be in the range of 7.1 to 8.5 with an average of 8.28. DO values were found to be within 1.52 to 3.17 mg/1 with an average of 2.52 mg/1. COD values were found to be high, ranging from 525. 39 mg/1 to 810.30 mg/1 with an average value of 692.11 mg/1. Average value of BOD, oil and grease, TDS, suspended solids were found to be 80.60 mg/1, 9.61 mg/1, 1122.65 mg/1 and 1804.17 mg/1 respectivaly.Ammoniacal nitrogen was found to vary from 16.93 mg/1 to 22.30 mg/1 with an average of 18.95 mg/1 and nitrate nitrogen from 43.29 mg/1 to 55.39 mg/1 with and average of 49.30 mg/1. Phenol, cyanide, total hardness were found tobe 0.39 mg/1., 0.03 mg/1 and 462.76 mg/1 respectively.

In the leachate of the bottom port of lysimeter No. 1, oil and grease were found to be from 0.2 mg/1 to 0.09 mg/1 with an average value of 0.06 mg/1. Whereas oil and grease as per IS 10,500 for drinking water should be nil. The dissolved solids with an average value of 541.27 mg/1 also found to exceed lightly the tolerance limit of 500 mg/1. In case of nitrate nitrogen the value was found to be increased even from the raw effluent value. It may be due to the conversion of ammoniacal nitrogen to nitrate nitrogen. After three month irrigation the texture of the soil in each lysimeter was sandy loam as was before irrigation. The percentage of organic matter in soil of lysimeter No.1 was found to be more as compared to lysimeter No. 3. Water holding capacity was also found to be increasing the leaching losses of available nitrogen in lysimeter No. 1 was found to be less as compared to lysimeter No. 3. Only a very small amount of losses of phosphorous were found in the soil. The leaching losses of potassium was also found in a considerable amount but in lysimeter No. 1 was found to be less. The pH was found to be increased as the increase in depth of the soil in each lysimeter. The conductivity of soil was found in decreasing order as increase in depth of the soil. The cation exchange capacity of soil at various depth were found to change more or less. The trace metals such as Zn, Pb, etc. were found to be changed to a lesser extent.

To overcome the contamination of ground Water due to oil and grease, high TDS and nitrate nitrogen, the experiment in lysimeter No.2 was performed with diluted effluents. When dilution ratio was increased to 1:1 (1 part water and 1 part raw effluent), leachate analysis of the bottom port were found to be within the tolerance limit. TDS were found in the range of 465.21 to 470.59 mg/1 with an average of 467.55 mg/1. Nitrate nitrogen were found in the range of 34.09 to 37.27 mg/1 with an average of 35.85 mg/1.

The percentage of organic matter in soil of lysimeter No.1 was found to be more as compared to lysimeter No.2 and 3. Hence the water holding capacity was also found to be increased. The leaching losses of available nitrogen was found to be less in lysimeter No. 1 as compared to other two. A very small amount of losses of phosphorous were found in the soil during 3 months flooding. The leaching losses of potassium was found to be in considerable amount in each lysimeter. No. 1. The cation exchange capacity of soil at various depth were found to be changed more or less as pH change. The conductivity of soil was found in decreasing order as increase in depth of soil in each lysimeter after and before irrigation of soil. The trace metals were also found to be changed after three months irrigation but to lesser extent.

Conclusion

Renovation efficiency of living filter (RELP) at bottom port of lysimeter No. 2 for COD was found to be 86.92%, for TDS - 53.22% and for oil and grease - 99.37%. It has also been calculated that 12,400 m^3/ha can be applied within a period of three months and about 2.73 hactares of land can be irrigated with the effluents produced from the coke plant under study.

Table 1

Effluent Characteristics of the Coke plant (Average results)

Parameters	Minimum	Maximum	Average	Maximum Tolerance limit (IS : 2490)
pH	7.1	8.5	8.28	5.5 - 9.0
Temperature (^{0} C)	28.7	32.0	30.40	40^0 C
Dissolved oxygen	1.42	3.17	2.39	-
C.O.D.	525.39	810.30	692.11	250
B.O.D.	63.9	92.30	80.60	30
Oil and grease	6.92	14.23	9.61	10
Dissolved solids	945.0	1287.82	1122.65	2100
Suspended solids	1521.0	2031.0	1804.17	100
Ammonical nitrogen	16.93	22.30	18.95	50
Nitrate nitrogen	43.21	55.39	49.30	–

Total hardness	440.21	495.39	462.76	–
Ca-hardness	72.0	96.0	88.35	–
Mg-hardness	352.00	406.07	371.91	–
Phenol	0.20	0.60	0.39	1
Cyanide	0.01	0.07	0.03	0.2

All parameters except pH are expressed in mg/1 unless otherwise mentioned.

❑ REFERENCES

Bone W.A.,1980 Coal and its Scientific uses, F.R.S. 1980.

Fisher C.W., 1965 Coke and Gas, Edited by Fredgurnham, C 1965, *Industrial Waste-water Control,* Academic press, New York, San francisco, London.

Koziorowska B. and Kucharshi J., 1972 *Industrial Waste Disposal,* Pergamon press, Oxford, New York, Toronto, Sydney.

Chipman (Jr.) W.A., 1934 *The Role of Ammonium Compounds,* Ph.D. Thesis, University Columbia.

EIFAC, 1973 Water Quality Criteria for European Fresh Water Fish Report on Monohydric phenols and Inland Fisheries *Water, Res., 2:249.*

Metcalf & Eddy, INC. 1990. *Wastewater Engineering, Treatment, Disposal and Reuse,* McGraw-Hill, INC, New York.

Ghose M.K. & Sinha D.K., 1990. Surface Water *Quality Monitoring Programme* and status of water quality in Coal Mining Area *Ind. Jr. of Env.* Health 110, 6, 459-461.

Standard Methods for the Examination of Water and Waste Water 1985 APHA, AWWA and WPCF, Washigton D.C., 16th Edition.

Black C.A. 1968 Soil Plant Relationship. John Wiley and Sons, Inc, (2nd Edition), New York (1968).

Ghose M.K. 1989 Ecologically Balanced Waste Renovation and Control of Water Pollution Using Local Soil-Napier Grass Combination as a Recycling Media, *Ind. Jr. of Env.* Hlth 9, (9).

Authors :

1. **M.K. Ghose :** Asst. Professor, CME, ISM, Dhanbad-826 004.

2. **S. Roy :** Bihar State Pollution Control Board, Dhanbad-826 004. Bihar (India)

■■

8

Assessment of Water Quality of Some Indian Rivers based on Factor Analysis

❑ INTRODUCTION

Developing countries are undergoing a transition period from a largely agrarian economy to intensive industrial activity. People and establishments congregate at certain areas and their activities produce external effects, whether beneficial or adverse on the environment that supports them. One of the most vital components of the physical environment is water. Because of a growing global awareness in the maintenance of a "Clean World", public and private agencies have come to realize the importance of surface water to a Nations economy. Knowledge of water quality thus plays a significant role in the development of water quality control and management strategies.

An index is a number, usually dimensionless, which expresses the relative magnitude of some complex phenomenon of condition. A Water Quality Index also is a "Communication tool for transfer of water quality". It provides an instrument of tool that appropriately consolidates and presents as a single number, the values of multiple parameters selected to enter into the index formulation. The composite influence of significant physical, chemical and biological parameters is reflected in the index.

Some possible uses of Water Quality Index are for resource allocation, location ranking, standards enforcement, trend analysis, public information and scientific research. Horton (1965) first proposed the concept of indices to represent gradation in water quality. The head for such a reading understood evaluation tool was ultimately realized, and several authors have developed their own rating schemes (Brown et al. 1970, Dinius 1972, Lohani and Mya 1982).

Methodology Used-Factor Analysis

The method proposed to solve the water quality index is factor analysis. Factor analysis was first orginated to aid in the explanation of psychological data but has been widely applied to other fields. Its best feature lies in the persimonius reduction of data (Prati el al. 1971.) Factor analysis seeks to express a large number of variables in terms of a smaller, more manageable number of factors based on linear relationship between the original variables. These derived components or factors may be regarded as underlying influences of dimensions that, on further measurement can be substituted for the more numerous variables (Horton 1965).

TABLE : 1

Physical, Chemical and Biological Parameters of Important Rivers in India

River, Location, State	Temperature (°C)	pH	Conductivity (mmhos/cm)	Turbidity (JTU)	Alkalinity (mg/lt)
	1.	2.	3.	4.	5.
1. Bhima, Takali, Maharashtra	27.9	7.8	696	54.7	142.2
2. Brahmaputra, Pandu, Assam	24.3	7.4	120	59	75
3. Cauvery, KRS Dam, Karnataka	26.2	7.9	188.5	17	78
4. Cauvery, Satyagalam, Bridge Karnataka	26.6	8.0	250.8	42.7	109.2
5. Godavari, Asthi, Maharashtra	24.6	7.7	293.7	35.7	120.8
6. Godavari, Dhalegaon, Maharashtra	24.9	7.7	293.7	35.7	120.8
7. Godavari, Mancherial, A.P.	28.3	8.0	538.9	27.2	187.3
8. Godavari, Polavaram, A.P.	29.0	7.9	218.6	105.9	100.1
9. Krishna, Vijaywada, A.P.	29.2	8.2	432.5	22.3	130.2
10. Pennar, Nellore, A.P.	28.8	8.1	862.1	42.7	189.4

11. Tungabhadra, Ullanuru, Karnataka	27.3	8.0	470.7	31.2	129.7
12. Periyar, Alwaye, Kerala	30.7	6.5	1220.4	30.7	72.8
13. Periyar, Kalady, Kerala	30.8	7.1	129.4	9.8	23.4
14. Chaliyar, Koolimadu, Kerala	30.0	7.1	135.5	14.0	38.8
15. Chaliyar Kollapalli, Kerala	30.4	7.1	244.8	24	68.3
16. Sabarmati, Dharojdam, Gujarat	26.4	7.9	313.4	17.2	142
17. Sabarmati, Ahmedabad, Gujarat	29.3	7.8	1677.8	53.9	460.2
18. Mahi, Sevalia, Gujarat	27.3	8.2	281.8	25.7	127.8
19. Mahi, Vlsad, Gujarat	27.5	8.2	347.1	22.9	193.9
20. Narmada, Gurudeshwar, Gujarat	26.9	8.1	271.8	146.7	166.1
21. Tapi, Nepanagar, M.P.	26.0	8.0	313.1	130.9	200.8
22. Tapi, Burhapnur, M.P.	26.7	8.0	499.4	275.2	240.1
23. Sutlej, Amritsar, Punjab	22.5	7.8	1980	56	75
24. Beas, Goindwal, Punjab	22.0	8.3	2190	57	86
25. Mahanadi, Sambalpur, Orissa	27.5	7.7	302.8	590.6	94
26. Subarnarekha, Ranchi, Bihar	25.3	7.2	296.2	239.3	72.2
27. Subarnarekha, Jamshedpur, Bihar	28.5	7.8	250	59	65.1

28. Musi, Hyderabad, A.P.	28.9	8.0	700	35	124
29. Ganges, Kannauj, U.P.	24.9	7.7	300	123	167
30. Ganges, Allahabad, U.P.	26	8.3	370	130	166
31. Yamuna, Allahabad, U.P.	25.9	8.3	480	144	192
32. Yamuna, Karnal, Haryana	19.8	7.6	920	156	71
1. Bhima, Takali, Maharashtra	2322	6.3	5.4	34.2	1.220
2. Brahmaputra, Pandu, Assam	9296	7.5	0.8	5.6	0.02
3. Cauvery, KRS Dam, Karnataka	765	7.1	1.3	13.9	0.001
4. Cauvery, Satyagalam, Bridge Karnataka	101.0	7.0	1.5	16.6	0.001
5. Godavari, Asthi, Maharashtra	3176	5.9	5.8	42.9	0.250
6. Godavari, Dhalegaon, Maharashtra	1425	6.5	4.7	32.4	0.240
7. Godavari, Mancherial, A.P.	34	8.1	2.4	21.9	0.520
8. Godavari, Polavaram, A.P.	31.0	7.3	1.6	23.6	0.580
9. Krishna, Vijaywada, A.P.	155	8.2	2.2	23.6	0.78
10. Pennar, Nellore, A.P.	106.0	8.7	3.2	30.2	0.6
11. Tungabhadra, Ullanuru, Karnataka	1006	7.1	1.7	27.8	0.01

12. Periyar, Alwaye, Kerala	1342	5.8	1.4	36.4	4.23
13. Periyar, Kalady, Kerala	4038	7.6	1.2	26.8	0.89
14. Chaliyar, Koolimadu, Kerala	2695	7.2	1.2	25.2	0.64
15. Chaliyar Kollapalli, Kerala	4147	6.1	2.4	26.2	2.5
16. Sabarmati, Dharojdam, Gujarat	6203	9.6	2.2	6.9	0.4
17. Sabarmati, Ahmedabad, Gujarat	17E6	1.0	73.3	193.6	1.18
18. Mahi, Sevalia, Gujarat	26E6	8.9	2.0	6.9	1.31
19. Mahi, Vlsad, Gujarat	15E5	8.4	1.6	7.6	1.06
20. Narmada, Gurudeshwar, Gujarat	31E5	8.0	2.4	16.5	0.54
21. Tapi, Nepanagar, M.P.	1156	7.3	2.4	16.8	0.47
22. Tapi, Burhapnur, M.P.	1578	7.2	4.3	32.6	0.49
23. Sutlej, Amritsar, Punjab	2944	7.8	2.5	3.6	2.4
24. Beas, Goindwal, Punjab	3497	7.7	3.1	3.9	2.733
25. Mahanadi, Sambalpur, Orissa	9659	8.2	2.3	26	0.28
26. Subarnarekha, Ranchi, Bihar	15E3	4.7	7.2	41.9	2.05
27. Subarnarekha, Jamshedpur, Bihar	6312	7.2	2.5	7.6	1.89
28. Musi, Hyderabad, A.P.	3894	7.1	1.6	124	217

29. Ganges, Kannauj, U.P.	12990	7.2	2.3	8.9	0.908
30. Ganges, Allahabad, U.P.	9719	8.3	2.1	19.6	1.749
31. Yamuna, Allahabad, U.P.	8492	8.2	1.3	1.656	11
32. Yamuna, Karnal, Haryana	1800	5.4	3.5	15.1	0.325

1. Bhima, Takali, Maharashtra	92.08	75.8	32.4	49.6	151.1	187.4
2. Brahmaputra, Pandu, Assam	6	120	2	45	20	66
3. Cauvery, KRS Dam, Karnataka	15.2	30.1	9.7	13.1	46.3	60.4
4. Cauvery, Satyagalam, Bridge Karnataka	19.2	29.5	13.6	14.8	62.9	84.2
5. Godavari, Asthi, Maharashtra	41.2	16.3	15.3	34.4	100.1	131.1
6. Godavari, Dhalegaon, Maharashtra	72.2	69.7	20.7	38.2	113.6	171.0
7. Godavari, Mancherial, A.P.	37	44.2	79	66.4	82.4	148.5
8. Godavari, Polavaram, A.P.	15.3	11.9	19.3	44.9	29.4	74.9
9. Krishna, Vijaywada, A.P.	45.3	34.7	47.6	49.6	44.1	101.4
10. Pennar, Nellore, A.P.	113.6	77.6	135.2	82.9	76.5	155.4

11. Tungabhadra, Ullanuru, Karnataka	49.6	32.1	40.7	40.1	71.4	104.4
12. Periyar, Alwaye, Kerala	866.3	71.8	367.9	108.1	102.3	409.9
13. Periyar, Kalady, Kerala	1147.9	9.2	68.6	21.9	42.3	64
14. Chaliyar, Koolimadu, Kerala	39.7	4.2	16.4	18.6	13.4	35.1
15. Chaliyar Kollapalli, Kerala	307	25.9	135.3	69.8	82.6	161.3
16. Sabarmati, Dharojdam, Gujarat	39.6	13.4	32.5	75.7	54.1	128.6
17. Sabarmati, Ahmedabad, Gujaat	300.9	113.3	354.9	121.6	114.4	238.1
18. Mahi, Sevalia, Gujarat	28.4	7.9	34.6	54.2	66.1	119.5
19. Mahi, Vasad, Gujarat	47.7	10.7	53.4	60.2	90.1	151.2
20. Narmada, Gurudeshwar, Gujarat	27.3	12.6	35.4	67.1	95.6	171.8
21. Tapi, Nepanagar, M.P.	21.8	17.6	30.7	74.6	68	146.6
22. Tapi, Burhapnur, M.P.	48.5	37.4	39.5	110	96.5	208.8
23. Sutlej, Amritsar, Punjab	32	26	3.0	30.0	23.0	92.0
24. Beas, Goindwal, Punjab	39	49	2.0	46	29	97

25. Mahanadi, Sambalpur, Orissa	21	4	29	50	16	64
26. Subarnarekha, Ranchi, Bihar	17.3	19.8	11.8	51.4	40.5	89.5
27. Subarnarekha, Jamshedpur, Bihar	38.9	12.5	16.4	47.8	30.9	80.1
28. Musi, Hyderabad, A.P.	32	12	30	62	53	126
29. Ganges, Kannauj, U.P.	8	16	2	88	61	151
30. Ganges, Allahabad, U.P.	60	13	6	92	70	153
31. Yamuna, Allahabad, U.P.	18	12	7	98	88	174
32. Yamuna, Karnal, Haryana	61	14	12	58	38	96

Data Analysis

Data was obtained from CPCB, New Delhi, Table (1) for 32 sampling stations located on twenty different rivers in India. The following distribution was based for selection of variables : (I) Parameters commonly included in water quality indices for general use of water; (II) availability of data for as wide a range of sampling period as possible, and (III) by allowing compression with water quality indicators.

By using the Diagonal method of the factor analysis, the correlation matrix from the data matrix has been obtained (Prati et al.). Using this correlation matrix, we find the factor loadings for the first factor. Similarly, other factor loadings are calculated. The factors along with their factor loadings, communality, eigen values and cumulative variance are listed in Table 2. The results of principal factor analysis with met the objective and were employed for index building Rotated factors in principal factor analysis always have 100% cumulative variance. The last column of Table (4) only depicts the cumulative variance explained by a set of factors if higher factors are ignored.

Factors with eigen values higher than 1.00 were considered significant. Eigen values represent the sum of the sequences of the factor loading across all variables on that factor. Column entries in the Table (2) denote factor loadings, which are measured of how much each variable contributes to the explanation of the extracted factors. Variables that load heavily on a factor are most representative of that factor. There is, however, no strict criteria on what consititutes a high loading. In order to interpret the factor loading matrix some degree of subjectivity must be used. The communality is a measure of the portion of a variable's total variance attributable to other variables in the set. The greater the communality, the greater the interrelationships within the set of variables. In this analysis variables having factor loadings higher than or equal to 0.500 are considered significant.

Table 2

Rotated Factor Loading Matrix from Factor Analysis

Parameters	Stations					commu-nality
	Factors					
	1	2	3	4	5	h^2
1. Temperature	0.392	-0.224	0.262	0.090	0.103	0.291
2. pH	-0.456	0.216	0.338	0.088	0.227	0.428
3. Conductivity	0.489	0.189	0.174	0.313	0.139	0.422
4. Turbidity	-0.199	0.189	0.130	0.252	0.297	0.244
5. Alkalinity	0.644	0.446	0.201	0.201	0.134	0.777
6. Coliforms	0.366	0.370	-0.198	0.223	0.115	0.373
7. DO	-0.627	0.317	0.186	0.142	0.177	0.586
8. B.O.D.	0.687	0.830	-0.146	0.352	0.111	1.322
9. C.O.D.	0.671	0.665	0.123	0.362	0.254	1.103
10. Nitrogen	0.474	-0.231	0.302	0.196	0.209	0.451
11. Chloride	0.785	-0.147	0.475	0.262	0.308	1.027

12. Sulphates	0.535	0.148	0.180	0.090	0.258	0.415
13. Sodium	0.928	-0.175	0.081	0.070	0.126	0.919
14. Calcium	0.656	0.076	0.294	-0.256	0.198	0.627
15. Magnesium	0.714	0.251	0.255	0.191	0.134	0.692
16. Hardness	0.792	-0.267	0.286	0.248	0.283	0.907
Eigenvalue	6.047	2.004	0.970	0.804	0.754	_
Percentage of Variance	51.02	27.4	11.1	6.4	4.1	_
Cumulative % of Variance	51.02	78.4	89.5	95.9	100	_

Table 3

Eigenvalues and cumulative percentage of variance of the five factors

Factors	Eigenvalues variance	Cumulative % of
1	6.047	51.02
2	2.004	78.4
3	0.970	89.5
4	0.804	95.9
5	0.754	100.0

Water quality Index Development

In factor analysis, once the loadings or weights for each variable are known, sampling index in which each variable is weighted proportionately in its involvement in the patterns may be constructed. The greater the involvement, the higher the weight. An index I(1), calculated from the first factor may be expressed as

$$I(1) = \sum_{j=1}^{n} \frac{a(1i)\ y(i)}{(\lambda 1)}$$

where,

$(\lambda 1)$ = eigenvalue for the first factor;
$y\ (i)$ = standardized form of known variable j and
$a\ (1i)$ = factor loading of variable j on first factor

An index calculated from the foregoing formulation is valid only if the first factor extracts a large proportion of total variance among variables. None of the first factors had a cumulative variance, exceeding 0.70 and this limits the construction of a single index. Thus separate indices were calculated for factors having eigen-values greater than 1.00. The general form is

$$I\ (i) = \sum_{j=1}^{n} \frac{a\ (1i)\ \ y\ (i)}{(\lambda 1)}$$

where,

$(\lambda 1)$ = eigenvalue of factor;
$y\ (i)$ = standardized value of the variable j and
$a\ (1i)$ = factor loading of variable j on factor i

Table 4

Calculation of Water Quality Index of 32 Stations

River, Location, State	$\Sigma aiiyi$	$(\Sigma aiiyi/\lambda 1)$	Area WQI. 1 as a (%)	Rating
1. Bhima, Takali, Maharashtra	4.3527	0.7198	71.98	9
2. Brahmaputra, Pandu, Assam	2.061	0.3408	34.08	26
3. Cauvery, KRS Dam, Karnataka	1.2995	0.2142	21.42	31
4. Cauvery, Satyagalam, Bridge Karnataka	1.2345	0.3199	31.99	27
5. Godavari, Asthi, Maharashtra	4.584	0.5781	75.81	8

6. Godavari, Dhalegaon, Maharashtra	3.9901	0.6598	65.98	12
7. Godavari, Mancherial, A.P.	3.9912	0.6601	66.01	11
8. Godavari, Polavaram, A.P.	1.412	0.2334	23.32	30
9. Krishna, Vijaywada, A.P.	2.693	0.4454	44.54	23
10. Pennar, Nellore, A.P.	5.241	0.8668	86.68	5
11. Tungabhadra, Ullanuru, Karnataka	2.846	0.4706	47.06	22
12. Periyar, Alwaye, Kerala	5.8624	0.9695	96.95	2
13. Periyar, Kalady, Kerala	1.8725	0.3097	30.97	28
14. Chaliyar, Koolimadu, Kerala	0.7314	0.1209	12.09	32**
15. Chaliyar Kollapalli, Kerala	4.8141	0.8086	80.86	7
16. Sabarmati, Dharojdam, Gujarat	3.0392	0.5026	50.26	21
17. Sabarmati, Ahmedabad, Gujaat	5.9638	0.9862	98.62	1*
18. Mahi, Sevalia, Gujarat	2.4194	0.4001	40.01	24
19. Mahi, Vasad, Gujarat	5.4257	0.8973	89.73	4
20. Narmada Gurudeshwar, Gujarat	3..4276	0.5668	56.68	18

21. Tapi, Nepanagar, M.P.	4.9395	0.5678	56.78	17
22. Tapi, Burhapnur, M.P.	5.4748	0.9053	90.53	3
23. Sutlej, Amritsar, Punjab	3.5359	0.5847	58.47	16
24. Beas, Goindwal, Punjab	4.2941	0.7101	71.01	10
25. Mahanadi, Sambalpur, Orissa	1.8288	0.3024	30.24	29
26. Subarnarekha, Ranchi, Bihar	4.9345	0.816	81.60	6
27. Subarnarekha, Jamshedpur, Bihar	2.7204	0.3757	37.57	25
28. Musi, Hyderabad, A.P.	3.3773	0.5585	55.85	19
29. Ganges, Kannauj, U.P.	3.9379	0.6512	65.12	13
30. Ganges, Allahabad, U.P.	3.8667	0.6394	63.94	15
31. Yamuna, Allahabad, U.P.	3.9196	0.6482	64.82	14
32. Yamuna, Karnal, Haryana	3.1945	0.5283	52.83	20

* Most Polluted ** Least Polluted

The calculated indices are then transformed to values in the "area under the normal curve (Table 4). The body of the Table 4, may be regarbed as degree of pollution values. The proportion of total area column is this value, normalized to a linear variable. Values thus range from 0-100, with zero representing good water quality and progressively becoming poorer as the value tends towards 100. With these values, the rating has been done for 20 rivers at the 32 values, the rating has been done for 20 rivers at the 32surface station (Table 4). The station which is ranked 1 is the most polluted, while the station ranked 32 is the least polluted among the 32 stations. In Table 4, only the WQI for the first factor has been calculated. Similarly, WQI can be calculated for other factors also. A bar graph, surface stations Vs. water Quality Index percentage, has been drawn (Fig. 2) which shows the comparision in the extent of pollution between the 32 surface station of the 20 rivers.

Conclusion

From the WQI-1 (Table 4), we can conclude that the river Sabarmati at Ahmedabad in Gujarat is the most polluted. It is followed by the river Periyar at Alwaye in Kerala and so on as the ratings given. The least polluted river is the Chaliyar at Koolimadu in Kerala, perceded by the Cauvery at KRS Dam in Karnataka. All the other rivers are moderately polluted.

With the availability of statistical packages, factor analysis provides a quick, simple and analytical method of quality evaluation. The factor scores may be further utilized for the develpment of composite index for water quality assessment. Further different processes may be calculated. The development of index may be summarzed as : data collection, test for normality, missing value estimation, cluster analysis, factor analysis, and transformation to normal curve index values.

❑ REFERENCES

Ball R.O. and Church R.L., "Water Quality Indexing and Scoring", ASCE Journal of the Environmental Engineering Division, Vol. 106, No. EE4, 1980, pp. 757-771.

Brown R. et al., "A Water Quality Index - Do We Dare?", Water and Sewage Works, Oct., 1970, pp. 339-343.

Dinius S.H., "Social Accounting System for Evaluationg Water Resources," Water Resources Research, Vol. 8, No. 5, 1972, pp. 1159-1177.

Horton R.K., "An Index Number System for Rating Water Quality", Journal Water Pollution Control Federation, Vol. 37, No.3, 1965, pp. 300-306.

Lohani B.N. and Mya M.S., "An Indexing Approach for Classification of Water

Uses in the Chao Phraya River", Proceedings of the Third Congress of the Asian and Pacific Regional Division of the International Association for Hydraulic Research, Aug. 24 - 26, Indonesia, 1982.

Prati L., Pavahello R., and Pesarin F., "Assessment of Surface Water Quality by a Single Index of Pollution", Water REsearch, Vol. 5, 1971, pp. 741-751.

Singh R.K. and Anand H., (1996), "Water Quality Index of Some Indian Rivers" Indian Journal of Environmental Health, Vol. 38, No. 1, pp. 21-34.

Authors :

1. **Ram Karan Singh :** Lecturer, Department of Civil Engg. BITS, PILANI - 333 031.
2. **Dr. Shiv Prasad :** Lecturer, Department of Mathematics, BITS, PILANI-333 031.
3. **N.S.J. Naidu :** Undergraduats, students, BITS, PILANI-333 031.

■■

9

Assessment of the Water Quality of Purna River, Valsad (Gujarat) with special Reference to the Heavy metal pollution

❑ INTRODUCTION

The indiscriminate urbanization, and industrial development can effect substantial changes in the water quality of rivers, which receive effluents, from various sources. Among the pollutants heavy metals play an important role as they are biologically non-degradable and tend to accumulate in the biosphere. Heavy metal pollution of aquatic environment has been studied in detail in many countries. However very few reports from India have contributed to this subject Mhatre *et al.*, (1980) Patil *et al.*, (1983) Zingde, *et al.*, (1979, 1980, 1981, 1987) Hasan *et al.*, (1986) Narvekar (1987).

In the present study four metals of ecotoxicological importance, copper, cadmium, lead and zinc were determined by atomic absorption spectro photometer, from Purna river along with other physico-chemical parameters. Purna river originate from sahiadrain water shed and drains entire length of Valsad district east to west. Enroute this, Purna collects domestic sewage and discards from industries in and around Navsari city.

METERIALS AND METHODS

Water samples were collected for four seasons in 1989-90. Samples were collected in 2 litre polythene bottles, previously soaked with 6N nitric acid and subsequently washed with distilled water. The samples for heavy metal analysis were collected separately and acidified immediately after collection with 6N HNO_3 (8ml/L). Physico-chemical analysis were carried out as specified in APHA (1975). Copper, Cadmium, Lead and Zinc in water were estimated by Atomic absorption spectrophotometer, after preconcentration by evaporation of unfiltered samples in the presence of HNO_3 and H_2SO_4 mixture (APHA, 1975) and expressed the results as total. Algal samples were collected from river by using plankton net and preserved immediately after bringing to the laboratory in 4% formalin and Lugol's iodine.

For identification standard monographs and journals were followed. (Desikachary, 1959 Husted, 1930 and Sarode and Kamat 1984).

RESULTS AND DISCUSSION

Physico-chemical characteristics of Purna river have been summarized in Table-1. The water, irrespective of season was always muddy due to the flushing character of the river. Temperature was varying seasonally. pH was alkaline throughout the collection. Dissolved Oxygen was very low except in Spring (5.270 mg/L). Salinity varied from 3.618% in Summer to 6.82% in spring. Nitrite, Nitrate, Ammonia contents were very low. The high value of phosphate and silicate revealed the amount of industrial effluents released into the rivers. Copper content varied from 0.046 to 0.094 ppm. Cadmium level was comparatively low (0.012 to 0.04 ppm). The alarmingly high level of lead (0.2 to 1.08 ppm) was due to the burning of the gasoline, or by the petrol operated vehicles or industries (Hasan *et al.* 1986). Zinc concentration was relatively low (0.15 - 0.25 ppm).

Table 1

Physico-chemical characteristics of Purna River
In Four Seasons in the Year 1989-90

Parameter		Winter	Spring	Summer	Autum
Colour		Muddy	Muddy	Muddy	Muddy
Temperature 0°C					
	Atmospheric	30°C	26°C	36°C	27°C
	Water	28°C	25°C	35°C	26°C
pH		7.88	8.0	8.16	8.32
Dissolved Oxygen mg/L		3.243	5.270	2.838	3.648
Nitrite	mg/L	0.085	--	0.039	0.025
Nitrate	mg/L	0.708	--	1.240	--
Ammonia	mg/L	--	--	0.039	--
Phosphate	mg/L	0.18	0.28	0.22	0.10
Silicate	mg/L	3.8	10.2	13.6	5.0
Copper	ppm	0.046	N.D.	0.046	0.094
Cadmium	ppm	0.012	0.04	0.012	0.026
Lead	ppm.	1.08	0.61	0.4	0.2
Zinc	ppm	0.21	0.25	0.18	0.15

The algal flora present in the river have been summarized in Table 2. Diatoms were dominated in phytoplankton community.

Table 2
Phytoplankton Population in Purna River

I.	*CYANOPHYCEAE*	
1.	*Chroococus Sp.*	Recorded in Winter
2.	*Microcystis Sp.*	Recorded in Winter and Autumn
3.	*Oscillatoria Sp.*	Recorded in all Seasons
4.	*Lyngbya Sp.*	Recorded in all seasons
II.	*CHLOLOPHYCEAE*	
1.	*Spirogyra Sp.*	Recorded in Winter
III.	*BACILLARIOPHYCEAE*	
1.	*Melosira*	Recorded only in Winter
2.	*Cyclotella meneghiniana*	Recorded only in Winter
3.	*Cosinodiscus excentricus*	Recorded in all seasons
4.	*Skeletonema Sp.*	Recorded in Winter
5.	*Fragilaria brevistriata*	Recorded in Winter
6.	*Synedra acus*	Recorded in all seasons
7.	*Synedra ulna*	Recorded in all seasons
8.	*Cocconeis maharashtrensis*	Recorded in Winter
9.	*Achnanthes affinis*	Recorded in Winter
10.	*Gyrosigma scalprodies*	Recorded in Winter and Summer
11.	*Navicula minuta*	Recorded in all seasons
12.	*Navicula confervacea*	Recorded in all seasons
13.	*Navicula papula*	Recorded in Winter
14.	*Navicula sabhamulata*	Recorded in Winter and Summer
15.	*Pinnularia acnosphaeria*	Recorded in all seasons
16.	*Pinnularia dolosa*	Recorded in all seasons
17.	*Cymbella Chandolensis*	Recorded in Winter
18.	*Cymbella gracillis*	Recorded in Winter and Spring
19.	*Cymbella gracile*	Recorded in Winter
20.	*Gomphonema constrictum*	Recorded in Winter
21.	*Gomphonema gracile*	Recorded in Winter and Spring
22.	*Nitzschia amphibia*	Recorded in Winter
23.	*Nitzschia closterium*	Recorded all seasons
24.	*Nitzschia intermedia*	Recorded in Winter
25.	*Nitzschia palea*	Recorded in Winter and Summer
26.	*Surirella biseriata*	Recorded in Winter
27.	*Suriella ovata*	Recorded in Winter

❑ REFERENCES

APHA 1985 *Standard Methods for Examination of Water and Waste Water,* Washington, DC.

Dean J.G., Bosqui F.L. and Lanouctts K.H., 1972. Removing heavy metal from Waste Water, *Envir. Sci. Technol.* G. 518-522.

Desikachary T.V., 1959. *Cyanophyta,* Indian Council of Agricultural Research, New Delhi.

Hasan M.Z., Pande S.P. and Saxent K.L., 1986. Heavy metals in Aquatic Environment, *Journal of the I.W.W.A., Vol.18* (2) : 155-159.

Husted F. 1930. Bacillariophyta, (Diatomaceae), In a pascher's Die Susswasser, *Flora Mittaleuropus 10.*

Mehtre G.N., Chaphekar S.B., Rao Ramani, I V, Patil M.R. and Haldar B.C., 1980 Effect of industrial pollution on the Kalu River Ecosystem. *Environ. Pollt.* Series A 23, 67-78.

Narvekar P.V. and Zingde M.D., (1987), Behaviour of Boron, Calcium and Magnesium in Purna and Auranga estuaries (Gujarat), West Coast of India, *Indian J. Mar. Sci. 16,* 46-50.

Patil M.R. and Patil A.R., (1983), Water Quality of Ulhas River with respect to Copper, Cadmium and Mercury, *Poll. Res.* 2(1) : 24-27.

Sarode P.T. and Kamat N.D., 1984. Freshwater Diatoms of Maharashtra, Saikripa Prakashan, Aurangabad.

Zingde M.D., Sharma R.V. and Desai B.N., 1979. Pollution in River Par and its abatement, *Indian J. Marine Sci.* 8 : 266-270.

Zingde M.D., Narvekar P.V., Sharma R.V. and Desai B.N., 1980. Water Quality of the River Damanganga (Gujarat), *Ibid.* 9, 94-99.

Zingde M.D., Chandar S., Rokade M.A. and Desai B.N. 1981. Base line Water Quality of the River Narmada (Gujarat) *Ibid. 10,* 161-164.

Zingde M.D., Mandalla A.V and Rokade M.A., 1987. Flashing Characteristics and Pollution assessment of Parna River Estuary, *Indian J. Marine Sci.,* 16 : 94-98.

Authors :

1. **Manoj E.** : Department of Zoology, M.S. University of Baroda, VADODARA Gujarat (India)

2. **G. Ragothaman** : Department of Aquatic Biology, South Gujarat University, SURAT - 395 007. Gujarat (India)

■ ■

10

Ecological Studies on River Panar of Araria (Bihar) with Emphasis on Its Biological Components

❑ INTRODUCTION

Water is a vital natural resource available on the earth, which occupies the key position in the life of organisms. It is the most essential component needed by men, plants and animals. Fresh water is essential for existence of life - human, animal and plant. River, ponds and lakes have attracted mankind since time immemorial as they are the chief sources of water needed for the development of an area. River water is used for the development of an area and all developmental activities like urbanization, industrialization, mining, construction and agriculture are dependent upon water. Rapid industrialization and urbanization and extensive use of fertilizers and pesticides have put severe strain on river ecosystems. Increasing pollution of rivers has become a matter of great concern over a period of last few decades.

The river ecosystem receives water primarily from their watersheds, marginal run off and domestic sources. These waters contain excessive organic matter, excess of nitrogen and phosphorus, suspended solids, chloride and pathogenic organisms. They get a lot of garbage, effluent and sewage. These control the nature of vegetation and fauna of the aquatic fauna (Hynes, 1975; Readdy and Gractz, 1981; Smart et al., 1985 and Warren, 1979). Pesticides and heavy metals from agricultural and industrial effluents are known to affect the living organisms. Pesticides are quite persistent and their extreme toxicity even to non-target organisms is due to fact that their metabolism and breakdown products especially those formed by oxidation are more stable and even the parent component (Holden, 1973, Heith, 1985, Rico et al., 1989, Gillom and Clifton, 1990). Heavy metals being conservative, can never be totally eliminated, once they contaminate water body and remain persistent in sediments from where they are slowly released causing serious hazards to aquatic life (Whitton, 1970, 1980, Rai et al., 1985, Stokes, 1983).

India has 14 major rivers along with their numerous tributaries. Most of the Indian rivers are polluted by domestic sewage and industrial wastes (Ajmal et al., 1984, Kar et al., 1987, Anwar and Siddiqui, 1968, Kulshreshtha, 1988 and Shukla et al., 1989).

The present study has been undertaken on the river Panar, which is considered as one of the most important tributaries of Koshi in Bihar. During its flow it receives diverse pollutants.

Materials and Methods

The present investigation was carried out in the river Panar at five sampling stations selected in the stretch of the 15 km. 1993 and Dec. 1993 from each sampling station in a polythene bottle of 3 ltrs capacity. Water temperature was recorded with an ordinary mercury thermometer graduated from 0^0 to 50^0 C. The analysis of pH, dissolved oxygen, free carbon dioxide, carbonate, bicarbonate, chloride, phosphate and sulphate was done according to the methods of APHA (1975) and Trivedy and Goel (1984).

For the plankton, 50 ltrs of river water was taken and filtered through the boilting silk plankton net (No.30) with 77 mesh/sq cm. The samples were taken from different areas of the stations regularly. The plankton concentrate on the plankton net was preserved with 5% formalin. Separation and counting the plankton was done by taking 1 ml of sub-samples into a Sedgwick Rafter plankton counting chamber of 1 ml capacity. All organisms were counted according to the procedure described by Welch (1952).

Result and Discussion

The physico-chemical analysis of the river water is shown in Table-1.

It is well known fact that temperature exerts direct as well as indirect influences on many abiotic and biotic components of the ecosystem. Many parameters such as density, viscosity, conductivity and specialization of nutrient salt etc. In the present investigation the water temperature varied with the current rise and fall of atmospheric temperature. In the present study the temperature varied between 20.1-39^0 C. The river was found alkaline throughout the year (pH 7.2-8.2). The highest value of pH during winter season appeared to be influenced by water level, large number of phytoplanktons and highest value of oxygen (Shyam Sundar, 1988, Pandey et al., 1993).

Table 1

Physico-chemical parameters of river Panar.

	Temp. W	Temp. A	pH	DO_2	Free CO_2	HCO_3	Cl	NO_3	PO_4	SO_4	Trans
Jan.	20.1	21.0	8.2	8.9	0.9	82.5	24.2	0.35	0.11	9.9	75.5
Feb.	20.5	24.8	8.1	8.6	0.8	84.6	22.2	0.36	0.90	9.5	65.1
March	21.6	29.9	7.6	8.1	0.1	86.3	23.1	0.39	0.12	9.2	54.2
April	23.9	30.4	7.7	7.9	1.2	95.6	26.7	0.45	0.15	12.2	49.4
May	26.5	33.5	7.5	7.7	1.3	100.5	29.5	0.51	0.17	14.1	46.2
June	30.5	34.5	7.4	7.6	1.5	68.5	30.9	0.54	0.18	16.8	39.1
July	30.9	34.9	7.2	7.4	1.9	58.5	31.9	0.65	0.80	13.5	32.6
August	28.9	33.2	7.3	7.1	1.8	54.6	18.4	0.75	0.70	18.2	30.1
Sept.	27.6	32.4	7.6	7.2	1.7	56.3	19.2	0.68	0.90	20.5	33.8
Oct.	25.4	30.9	7.7	7.0	1.2	58.9	20.2	0.61	0.12	21.9	53.1
Nov.	24.3	29.7	7.8	6.9	0.9	52.1	21.8	0.52	0.14	14.3	60.1
Dec.	22.1	24.6	7.9	7.6	0.7	72.2	22.1	0.50	0.12	14.2	64.3

The oxygen content was highest during winter season, which might be due to increased vegetation which took part in photosynthetic and consequence release of oxygen. The low value of oxygen during rainy season might be due to higher growth of bacteria, which utilised oxygen for their metabolic activities. The respiratory activity of aquatic organisms produces considerable amount of CO_2 in water, which dissolves partly in water as carbonic acid and a part of diffuses into atmosphere. CO_2 was present throughout the year. The high value of CO_2 was observed during rainy season which might be due to high rate of decomposition of organic matter by microbes resulting in rapid production of CO_2 (Sinha, 1988, Pandey et al., 1995). The lowest value of CO_2 during winter might be due to high growth rate. Phytoplankton and vegetation which utilize CO_2 in their photosynthetic activities. Chloride contents in the river under investigation was found highest during winter. The increased value of chloride during summer might be due to decreased amount of water and increased value of excreta laid by the various aquatic fauna (Mishra and Yadav, 1978; coley, 1979, Pandey et al., 1991). The level of chloride found in some of the polluted Indian river are 20-396 mg/L (Venkateshwarlu and Jayanthi, 1968) and 190-110 mg/L (Paramsivam and Sreenivasan, 1981) in river cauvery, 14-540 mg/L) Palharya and Malaviya, 1988) in Narmada river, 25-3324 mg/L (Shrotriya and Dubey, 1990) in river Kurel, 28.4-386.2 mg/L and 9.94-100 mh/L (Khatavkar, 1991) in river Krishna and Koyana and 17.5-279.0 mg/L.

Alkalinity in fresh water is due to the presence of three types of substances - bicarbonates, carbonates and hydroxides in the natural and polluted water. In the present investigation carbonate alkalinity was absent throughout the year. Bicarbonate alkalinity ranged from 52.1 to 100.5 mg/L. On the basis of average value of bicarbonate the river water can be classified under high category of nutrient type after the classification of Philipose (1959), as alkalinity values were>100 mg/L. On the basis of classification of Moyle's (1946) it can safely be termed as hard water type.

Nutrients play a major role in the metabolism of living beings and some of these have been reported to influence primary production (Goldman, 1964). Nitrogen is a very important element in water fertility. The nitrate content of the river ranged from 0.35 to 0.75 mg/L. The nitrate value was highest during monsoon (0.67) and lowest during winter (0.43). The high value of nitrate during monsoon might be attributed due to rain showers, decomposition of organic matters which get transferred by aerobic and anaerobic bacteria at high temperature into nitrogenous matter and influx of flood water (Rao and Govind, 1966). The low value of nitrate in winter might be due to high macrophytic and plankton production. Phosphorus in one of the most critical single factor in the maintenance of water fertility. The average phosphate value was 0.36 mg/L. In the present study the possibility of phosphorus leakage from rooted plants into water cannot be ruled out. On the basis of P and N the investigated water body may be categorised as of moderate trophic level. Sulphate is ecologically important for the growth of plants and its short supply may inhibit the development of plankton. Sulphr is also an important component in protein metabolism. The high value of sulphate during monsoon might be due to that surface run off brings more suspended solid along with organic and soluble salts (Sinha, 1986; Pandey et al.,1992). The low value of sulphate during winter might be due to higher phytoplankton population.

Phytoplankton

Phytoplankton play a key role in maintaining proper equilibrium between abiotic and biotic components of ecosystem. They have been regarded as the chief primary producers of natural ecosystem. Phytoplankton productivity is ofter the primary source of all the organic matter in an ecosystem. Being an index of tophic status phytoplankton reflects the over all environmental condition of the system and its potentiality (Agarwal et al., 1993). Their density have been reported to be effected by the quality of water (Bilgrami, 1979, 1985).

The phytoplankton communities of the river were presented mainly by four groups comprising 34 species of algae-chlorophyceas (37%), myxophyceae (28%), bacillariophyceae (25%) and euglenophyceae (10%), Table -2. Fig. 1.

Table 2

Monthly distribution of phytoplankton and zooplankton of the river Panar.

Months	Chloro phyceae	Myxo phyceae	Bacillario phyceae	Eugleno-phyceae	Rotifera	Cladocera	Copepoda
Jan	1055	695	610	250	245	120	110
February	1085	740	630	245	178	85	75
March	805	650	325	249	475	90	65
April	780	740	305	220	470	205	135
May	725	715	350	195	250	450	110
June	710	705	295	175	240	275	95
July	650	450	375	200	305	255	70
August	575	395	390	205	301	255	75
Sept.	560	340	350	190	270	175	68
Oct.	545	345	401	172	275	180	70
Nov.	705	495	445	154	305	195	95
Dec.	896	696	585	145	375	200	120

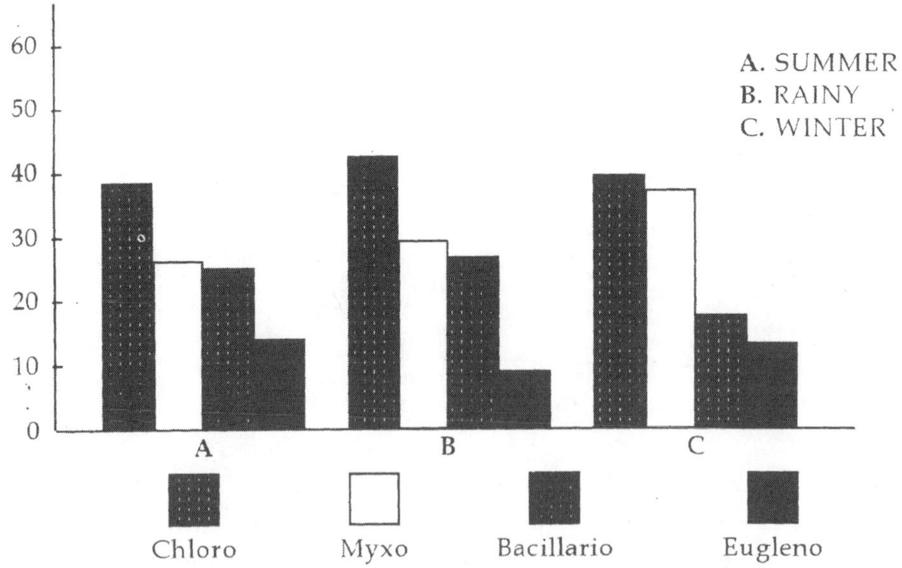

Fig. 1. Histogram Showing volume of phytoplanktonic Components in three different seasons.

A. **Chlorophycee** : *Chlamydomonas sp., Volvox, Clostridium s.p. Chara sp., Microspora sp., Chlorella sp. Oedogonium sp., Ulothrix s.p, Zygnema sp.,Spirogyra sp., ocystis sp. and Pediastrum sp.*

B. **Myxophyceae** : *Oscillataria limentica, Microcystis aeruginosa, M. robusta, Nostoc sp. Calothrix sp. Anbaena spherica, Rivlaria sp. and Spirulina sp.*

C. **Bacillariophyceae** : *Navicula phvgamea, Diatoma sp. Cymbella sp. Synedra acus, S. Splendenis, Fragilaria, Pinnularia sp. Cyclotella sp. Melosira granulata, Synedra ulna,Tabellaria sp. and Mastogolia sp.*

D. **Euglenophyceae** : *Euglena viridis, Phacus sp.*

Seasonal variations of phytoplankton alongwith temperature changes, may be due to oxygen and carbondioxide alongwith other chemical characteristics of water. Tripathy and pandey (1990) have reported that besides oxygen and carbondioxide variations in physico - chemical characteristic like temperature, pH. chloride, alkanity, calcium, magnesium, nitrate and sulphate in different seasons which affect the growth of diatom species. In the present investigation peak of the phytoplankton was observed during winter followed by summer and rainy. There was a decline in number with increase in temperature suggesting that phytoplankton preferred moderate temperature. Rainfall and high turbidity produced by high wind velocity during rainy season had a direct bearing on phytoplankton populatinon reducing them to minimum numbers. The phytoplankton population showed positive correlation with pH. DO, HCO_3, Chloride and transparencey. On the basis of palmer's algal index the river water is polluted as the algal species index is more than 20. Phytoplankton population showed positive correlation with pH. dissolved oxygen. bicarbonate, chloride (except Bacillariophyceae and transparency. Bacillariophyceae and Euglenophyceae are the only group showing positive correlation with phosphate (Table - 3). The river harbour many pollution indicators Diatoms like *Navicula, Cyclotella* and *Fragillaria* atc. Diatoms seem to represent populated condition. Rai and kumar (1979) and verma et al. (1978) have reported *Nitzschia* from highly populated waters in India. In the present investigation *Anabaena, Oscillateria, Microcystes, Chlamydomonas, Chlorella, Oocystis, Pediastrum* and *Euglena* were found in polluted condition.

Zooplankton

The zooplankton forms a link between phytoplankton and macro-invertebrates which in turn provides food to fishes and aquatic birds. They constitute an important link of secondary production in an aquatic ecosystem and play a vital role in the energy allocation in trophic levels. The knowledge of their seasonal qualitative and quantitative fluctuation has been considered always essential for a proper understanding of the factors influencing biological productivity and fisheries development.

Table 3

Correlation coefficient between plankton density and physico-chemical factors.

		Water Temp	pH	DO_2	Free CO_2	HCO_3	Cl	NO_3	PO_4	SO_4	Trans
Chloro	r	-0.776	0.780	0.892	-0.611	0.578	0.085	-0.875	-0.127	-0.825	0.801
Phyceae	r^2	0.6022	0.6081	0.7964	0.3731	0.3346	0.0072	0.7665	0.0161	0.6801	0.6419
Myxo-	r	-0.493	0.423	0.739	0.516	0.828	0.460	0.815	-0.400	-.754	0.527
Phyceae	r^2	0.2431	0.1793	0.5458	0.2667	0.6861	0.2114	0.6643	0.1598	0.5689	0.2776
Bacillario	r	-0.663	0.794	0.500	-0.370	0.045	0.326	-0.451	0.089	0.381	0.751
Phyceae	r^2	0.4398	0.6307	0.2497	0.1367	0.0020	0.1062	0.2036	0.0079	0.1452	0.5638
Eugleno-	r	-0.412	0.252	0.777	-0.272	0.534	0.040	-0.513	0.206	-0.636	0.205
Phyceae	r^2	0.1696	0.0637	0.6036	0.0738	0.2853	0.0016	0.2630	0.0423	0.4040	0.0419
Rotifera	r	-0.177	-0.157	-0.071	-0.380	-0.223	-0.017	-0.115	-0.402	-0.233	-0.026
	r^2	0.0313	0.0.248	0.0050	0.1445	0.0499	0.0003	0.0133	0.1617	0.0543	0.007
Cladocera	r	0.599	-0.559	-0.373	0.507	0.147	0.550	0.377	-0.145	0.297	-0.476
	r^2	0.3586	0.3125	0.1388	0.2567	0.0216	0.3029	0.1423	0.0210	0.0885	0.2147
Cope-	r	0.278	0.331	0.261	-0.142	0.528	0.302	-0.392	-0.569	-0.287	0.383
poda	r^2	0.0775	0.1093	0.0680	0.0202	0.2784	0.0913	0.1533	0.3242	0.0824	0.1465

The zooplankton community mainly comprises of rotifers (22) Cladocera (4) and Copepods (4), Comprising 27 species (Rotifera 51.09%, Cladocera 34% and Copepods 15%), Table - 2 and fig 2 .

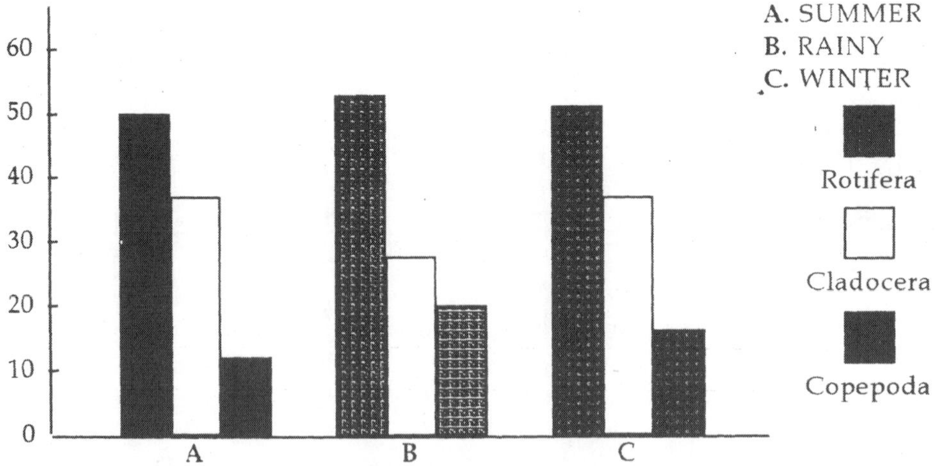

Fig. 2. Histogram Showing volume of zooplanktons Components in three different seasons.

The three groups of Zooplanktons were

A. **Rotifera :** *Brachionus calyciflorus, B. rubens, B. guadridentata, B. Plicatilis, B. falcatus, B. caudatum, B. angularis, keratella tropica, K. vulga, K. eanadensei, Horaella sp. Lapadella sp. Polyarthra sp. Filinia terminalis, Testudinella sp. Rotatoria sp. Monostyla lunaris, M. bulla, Cephalodella ferticula, Ophryoxus sp. and chydorus gibbus.*

B. **Cladocera :** *Moina sp. Macrothrix sp. Daphnia sp. and Ceriodaphnia sp.*

C. **Copepoda :** *Heliodiaptomus vidunus, Mesocyclops sp. Nauplius sp. and Diaptomus.*

Planktons depend on water quality, remains of dead and decaying vegetation as well as burnt and half burnt dead bodies. These results in the increase of organic matter and growth of bacterial population which increase the zooplankton population. The structure of zooplankton community is quite characteristic and is the product of the environmental conditions prevailing there Several zooplanktons species have been classified as indictive of polluted conditions like *Brachionus, Keratella, Adona, Cyclops, Lecone, Cephalodella, Bosminia, Orptolabris, Chydorus* etc. (Rawsen, 1956, Davis, 1961, Patalas, 1972 Tundisi and Tundisi, 1976, Widuto, 1977, Pennak, 1978 and Pandit, 1980)

In the present investigation out of 27 species, several are pollution to lerant forms. *Ceriodaphnia and chydorus* grow abundantly in the polluted condition. The genus *mesocyclops* can tolerate substantially low level of oxygen indicating the organic pollution (Aycock, 1942) *Ophryoxus* can occur in unpolluted as well as highly polluted conditions. *Keratella* sp. is as an indicator of eutrophication. as recorded by various workers (Sampath et al. 1978).

The presence of some tolerant indicator zooplankton species has also been reported in polluted condition of several other rivers in India. Bhattacharya and saha (1988) have identified *Keratella, Ceriodaphnia and Mesocyclops* from river Gomati in Uttar Pradesh and *Keratella* at the polluted stations (Raina et al. 1982).

Quantitative analysis of zooplankton during rainy season shows lower density which might be due to flood and fast current of river water. The species composition of zooplankton indicates that rotifers occur more pre-dominantly than cladocerans and copepods. It is presumed that rotifers utilize nutrients and phytoplanktons more rapidly to build up their population. This may be reason for the world wide distribution of the rotifers (Pennak, 1978) and their main characteristics inhabitation in the inland waters (Dumont and Green, 1980). pH and temperature are also the main factors in the appearance and abundace of different rotifers (Yousuf and Gadri, 1981). The next dominant group is cladocera. Cladoceran populations can be correlated with many physical parameters such as temperature, pH (Gadri and Yousuf, 1980) as well as magnitude of diatoms bloom and rotifer population (Harper and Ferguson, 1982, Wetzel, 1983).

In order to assess the importance of abiotic interaction an attempt was made to analyse the data statistically as shown in the Table 3. Rotifers abundance showed negative correlation with all the parameters. Cladocera exhibit positive correlation only with water temperature, free CO_2, bicarbonate, chloride, nitrate and sulphate. Copepods show positive correlation with pH dissolved oxygen. bicarbonate, chloride and transparency.

Macroinvertebrates

The macro-invertebrates are one of the major biological components of the aquatic ecosystems. They usually occupy a higher trophic level in a food chain. The role of macro-invertebrates in assessing river pollution has been emphasized by several workers (Kolwitz and marsson, 1908, 1909, Carpenter, 1926; Butcher, 1946, Pentelow et al., 1938, Back, 1955, Hawkes, 1979, Hynes, 1960 and chandler, 1970). The macro invertebrates by virtue of being relatively stationary are constantly exposed to changes undergoing in overlying water and hence respond very well to pollution. As macro invertebrates live in river bed benthos attached to some objects, they are considred best indicator for water quality.

Presence of bioindicators, *Chironomous* larvae and *Tubifex species* indicate the effect of pollution, Hawkes (1979) has reported that the members of oligochacta of the group Annelida are usually favoured by organic environment and remain dominant in severally polluted conditions. *Fristalis* sp. and *Tubifex* sp. occured in strong sewage pollution and anoxic waters. They are abundant at cremation site, as they feed on dead organic matter.

The molluscs usually do not tolerate pollution as high as do *Tubifex and Chironomids,* but a few like *lymnea* can tolerate very low oxygen conditions (Harman, 1974).

Macrozoobenthos

1. **Annelida**
   a) *Hirudinaria granulosa.*
   b) *Tubifex tubifex.*
   c) *Stylaria* sp.
   d) *Pheritma* sp.
   e) *Nais* sp.
2. **Coleoptera**
   a) *Hydrophilus* sp.
   b) *Laccophilus* sp.

c) *Aqabinus* sp.

d) *Haliplus* sp.

e) *Guignotus pradhani.*

f) *Berosus pluchellus.*

g) *B. indicus.*

h) *Enochrus* sp.

3. **Hemiptera**

a) *Micronecta proba.*

b) *Micronecta marpoe.*

c) *Corixa promontoria.*

d) *Lithocerus* sp.

e) *Notonecta* sp.

4. **Diptera**

a) *Chironomus* sp.

b) *Antocha* sp.

c) *Monopelopia* sp.

d) *Chaoborus* sp.

e) *Clinotanypus* sp.

f) *Eristalis* sp.

5. **Trichoptera**

a) *Hydropsyche* sp.

6. **Ephemeroptera**

a) *Ephemerella* sp.

b) *Caenis* sp.

c) *Heptagenia* sp.

7. **Odonata**

a) *Hagenius* sp.

b) *Enallagma* sp.

9. **Gastropoda**

a) *Lymnaea acuminata.*

b) *L. Klutsola.*

c) *Viviparus bengiensis.*

d) *Gyraulus convexiusculus.*

e) *Allocinma orcula.*

f) *Planispira proxima.*

10. **Bivalvia**

a) *Corhicula stiatella.*

b) *Lamellidens marginalis.*

Macro Invertebrates	Species	Percentage
Annelida	5	13.15
Coleoptera	8	21.05
Hemiptera	5	13.15
Diptera	6	15.78
Trichoptera	1	2.63
Ephemeroptera	3	7.89
Odonata	2	5.26
Gastropoda	6	15.78
Bivalvia	2	5.26
	38	

Macrophytes

Following macrophytes were listed in the present

Investigation :

1. *Hydrilla verticellate.*
2. *Ceratophyllum demersum.*
3. *Utricularia flexuosa.*
4. *Utricularia stellaris.*
5. *Pistia.*
6. *Azolla.*
7. *Limnophila* sp.
8. *Cyperus platystylis.*
9. *Jussiea repens.*
10. *Caesulia axillaris.*

11. *Eichhornia crassipes.*
12. *Neptunia oleracea.*
13. *Polygonum* sp.
14. *Lemna.*

Aquatic macrophytes are used for large scale nutrient removal and water purification. Among aquatic macrophytes, free floating plants create large number of problems particularly related to the use and management of water resources, they severely affect the hydrological regime and also create pollution problems after their death and decay. The floating mats of water hyacinth cuts the light reaching the water and severely retard photosynthesis. Most of the plants of the above mentioned serve as food for pigs and buffaloes. They choke up the surface of the water and serve as breeding ground for mosquitoes. During the tenure of the present investigation it was realized that Malaria, Dysentery. Typhoid, Filaria, Cholera and kala azar are most prevalent among the people of villages located near the river. Water hyacinth provide ideal site for the growth of molluscs which in turn choke water supply and also impart undesirable taste and odour to water. These molluscs are intermediate host of the Liver fluke which causes Liver rot in cattle leading to less production of milk. Liver rot is a very common disease in this region and algal growth and thereby disbalance the ecosystem.

Nitrogen fertilizers alter the pH of the soil, upsetting the eco-equilibrium devasting many natural microbes. In this contest prime importance is given to bio-fertilizers like *Azolla* which can substitute the chemical nitrogen fertilizers to a great extent, abat pollution and augment microbiol soil fertility (Singh et al, 1988). In addition, *Azolla* is found to improve the fertility of soil by providing gradually decaying biomass promoting the soil microbiol growth. Sevich and Madhusoodanan (1996) have found Azolla bio fertilizers to be a convenient and economic mode of sustainable in the paddy fields of kuttanad, Kerala.

Fish fauna

The Indian subcontinent has a large number of rivers. The total length of rivers in India is estimated at about 27360 Km. while all these rivers with their tributaries, canals and irrigation channels occupy roughly an area of 1,12,600 sp. km. (Jain, 1986, Samulel and kurup 1983) This area offered a wide variety of suitable habitats of fishes. There are several hundreds of species in the above mentioned area.

North Bihar is well known for its fish production (Particularly koshi zone). Koshi zone is well known for fish production which contributes 40% of the total fish output of Bihar During present investigation following fishes were found in the river Panar.

Summary

The paper deals with an ecological study with special reference to biological components on river Panar at Araria, Purnia (Bihar). A large proportion of the catchment is comprised of urbanized and human activity influencing the river.

The results revealed well defined seasonal variation. On the basis of physico chemical parameters the river appears to be moderate trophic level and hard type. The bulk of phytoplankton was shared by chlorophyceae (12). Myxophyceae (B), Bacillariophyceae (12) and Euglenophyceae (2). Many of the algal species out of the total 34 reported from river like *Euglena, Microcystis, Chlamydomonas, Anabaena and Spirulina* are recognised as pollution indicator. The bulk of zooplankton was shared by rotifera (19), cladocera (4) and copepods (4). In all 27 species of zooplankton were identified. Zooplankton also comprises of some pollution tolerant species like *Brachionus, Keratella and Cyclops* etc. The macro invertebrates fauna belonged to nine major groups with a total of 38 species. Presence of *Chironomids* and *Tubifex* indicated the build up of substantial quantity of organic matter in the sediments of certain places. The species diversity was found to be 2.34 indicating moderate quality of water. 14 species of macrophytes and 44 species of fishes were identified. Thus the study indicates that the river is polluted both chemically and biologically.

On the whole the river is getting polluted due to human activities, discharge of drainage water, agricultural run off as well as dumping of human excreta and burning of corpse. Incidence of diseases like asthma, respiratory tuberculosis, dysentry, kala azar, malaria and skin diseases were noted common in the population residing near the river. Due to high pollution fish mortality and crop damaged were also noticed which affect the economy of the local population.

❑ REFERENCES

Agarwal R., Bahura C.K. and Sexena M.M. (1993). Planktonic productivity of a sewage fish pond and adjoining oxidation pond at Vallabh Garden, Bikaner. Acta. Ecol. 15(1) : 58-61.

Ajmal M., Raji Uddin and Ashan Ullah Khan (1985). Physico-chemical aspects of pollution in Kali Nadi. IAWPC. TECH. ANNUAL. 12 : 106-114.

Anwar S., and Siddiqui M.S. (1988). On the distribution and dynamics of micro-invertebrate fauna of the river Kali in northern India. J. Environ. Bio. 9 (3 Suppl) : 333-341.

APHA (American Public Health Association) American Water Works Association and Water Pollution Control Federation (1975). *Standard methods for examination of water and waste water.* 14th edition. APHA, Washington, U.S.A.

Aycock D. (1942). Influence of temperature on size and form of Cyclops Vernalis Fisher. Y. Elisha Mitchell S.C. Soc. 58 : 84-93.

Back W.M. (1955). Suggested method for reporting biotic data *Sewage and Indust. Wastes,* 27 : 1193-1197.

Bhattacharya T. and Sqaha Ratan K. (1988). A comparative study of the physico-chemical properties and plankton density of two streams and tributaries of Gomti river in Tripura. Geobios, 15(50 : 201-206.

Bilgrami K.S. and Dutta Munsi J.S. (1979). Limnological survey and impact of human activities on the river Ganga (Barauni to Farakka). Allie Press, Bhagalpur, India.

Butcher R.W. (1946). the biological detection of pollution. J. Inst. Sew. Purif. 2 : 92-102.

Carpenter K.E. (1926). The lead mine as an active agent in river pollution ann. Appl. Biol. 13 : 395-401.

Chandler J.R. (1970). A biological approach to water quality management, Wat Pollut. Control. 4 : 415 -422.

Cole G.A. (1979). A text Book of Limnology 2nd Ed. The E.N. Mosley Co. London.

Davis C.C. (1961). The biotic community in the great lakes with respect to pollution. Proc. Conf. on water pollution and the great lakes, Depaul Univ. Chicago, 80-87.

Dumont H.J. and Green J. (1980). Rotatoria., *Hydrobiol.* 73 : 1-262.

Gillom R.J. and Clifton D.G. (1990). Organochlorine pesticides residues in bed sediments of the San Joaquin river, California. Water Resour. Bull. 26 : 11-24.

Goldman M.G. (1964). Primary productivity and micronutrients limiting factors in some North American and New Zealand lakes. Vern. Internat. Varn. Limnol. 15 : 365-374.

Haith D.A. (1985). Variability of pesticides loads to surface water. J. Water pollut. Cont. Fed. 57 : 1062-1067.

Harman W.N. (1974). Snails (Mollusca : Gastropoda) Pollution Ecology of Fresh-Water Invertebrates, Eds. C.W. Hart and S.L.H. Fuller, 274-312. Academic Press, New York and London.

Harper D.M. and Fergusson A.J. (1982). Zooplankton and their relationships with water quality and fisheries. Hydrobiol. 88 : 135-145.

Hawkes H.A. (1979). Eutrophication of lakes by domestic drainage, Ecology, 28 : 288-295.

Hawkes H.A. (1979). Invertebrates as indicator of River Water quality In (Biological Indicators of Water Quality) Ed. by A. James and Lillian Evison, John Wiley and Sons, New York.

Holden A.V. (1973). Effect of pesticides on fish. In : Environmental pollution by pesticides, Edwards C.A. (ed.) Plenum Press, London. pp. 213-252.

Hynes H.B.N. (1960). The Biology of polluted waters. Liv. uni Press, 202 pp.

Jain K.K. (1986). Inland Fishery Division, Central Institute of Fishery Education, Varsova, Bombay.

Kar G.K., Mishra P.C. and Das R.C. (1987). Pollution studies in river Ib III : Plankton population and primary productivity. Ind. J. Env. Hlth. 29(4) : 322-329.

Khatavkar S.D. (1991). Studies on the pollution of some fresh water reservoirs and rivers with special reference to bioindicators of organic pollution. Ph.D. Thesis, Shivaji University, Kolhapur.

Kolkwitz R. and Marsson M. (1908). Okologie der pflanzlichen saprobein. Ber. Dtsch. Bot. Ges. 26 : 505-519.

Kolkwitz R. and Marsson M. (1909). Okologie der pflanzlichen Saprobein. Int. Rev. Ges. Hydrobiol. 2 : 126-152.

Kulshreshta S.K., Adholia V.N. Khan Altaf A., Bhatnagar Anita, Baghail Manju (1988). Macrozoobenthose in river Chambal and its tributaries. Ind. Assoc. Water Poll. Tech. Ann. 15 : 39-41.

Mishra G.P. and Yadav A.K. (1978). A comparative study of physico-chemical characteristics of river and lake water in Central India. Hydrobiol.. 59(3) : 275-278)

Moyle J.B. (1946). Some indices of lake productivity. Trans. Am. Fish. Soc. 76 : 322-334.

Palharya J.P. and Malviya S. (1988). Pollution of the Narmada river at Hosahgabad in Madhya Pradesh and suggested measures for control. 55-56. In : Ecology and Pollution of Indian Rivers. Eds. R.K. Trivedy, Ashish Publ. House. New Delhi.

Palmer C.M. (1969). Composite rating of algae tolerating organic pollution. J. Physiol. 5 : 78-82.

Pandey B.N. and Mishra R.D. (1991). Hydrobiological features of River Saura. Bio. J. 2(2) : 337-342.

Pandey B.N. Lal R.N., Mishra P.K. and Jha A.K. (1992). Seasonal, rhythms in the physico-chemical properties of river Mahananda (Katihar), Bihar. Env. & Ecol. 10(2) : 354-357.

Pandey B.N., Mishra A.K., Das P.K.L. and Jha A.K. (1995). Studies on hydrological condition of River Saura in relation to its impact on biological health. Recent Researches in Aquatic Environment, (Eds. Ashutosh Gautam and N.K. Agarwal) Daya Publishing House, New Delhi.

Pandit A.K. (1980). Biotic factors and food chain structure in some typical wetlands of Kashmir, Ph.D. Thesis, University of Kashmir, Srinagar.

Paramsivam M. and Sreenivasan A. (1987). Changes in algal flora due to pollution in Cauvery River. Indian J. Environ. Hlth. 23 No. 3P. 222-238.

Patalas L. (1972). Crustacean plankton and eutrophication of St. Lawarence Great Lakes. J. Fish Res. Bd. Can. 20: 1451-1462.

Pennak R.W. (1978). Fresh water invertebrates of United States. John Wiley and Sons, New York.

Pentelow F.T.K., Butcher R.W. and Grindley U. (1938). An investigation on the effects of milk wastes on the Bristol Avon Fish Invest Lond. Series 1, Vol. 4 No.1.

Philipose M.T. (1967). Chloroccales, ICAR, New Delhi, pp. 365.

Gadri M.Y. and Yousuf A.R. (1980). Influence of the physico-chemical factors on the seasonality of cladocerans in lake Manasbal. Geobios., 7: 273-276.

Rai L.C., Gaur S.P. and Kumar H.D. (1981). Phycology and heavy metal pollution. Biol. Rev. 56: 99-151.

Raina, Rajni, Subla B.A. and Zutchi D.P. (1982). Water quality and plankton of Jhelum river Int. J. Ecol. Environ. Sci. 8 : 11-17.

Rao D. Subba and Govind B.V. (1966). Hydrobiology of Tungbhadra reservoir, Indian J. Fish 11 : 321-344.

Rawson O.S. (1956). Algal indicators of trophic lake types. Limnol. Oceanoger. 1: 18-25.

Reddy K.K. and Graetz D.A. (1981). Use of shallow reservoirs and flooded soil system for waste water tretement Nitrogen and Phosphorous transformation. J. Environ. Qual. 10: 113-119.

Rico M.C,, Hernandez L., Fernandez M., Gonzalez M.J. and Montero M. (1989). Organochlorine comtamination in water of Donane National Park, Wat. Res. 23: 57-60.

Sampath V. Sreenivasan A. and Ananathanaryaynam R. (1978). Rotifera as biological indicators of water quality in Cauvery river. Abst. Symp. Environ. Biol. p. 5.

Sevich P.J. and Madhusoodunan P.V. (1996). Ecological and Economical impacts of *Azolla rubra* as bio-fertilizer in the acidic paddy fields of Kerala. J. Ecobiol. 8 : 63-65.

Shrotriya L. and Dubey P.S. (1990). Ecological assessment of river Kurel, (Ratlam) In River pollution in India : Ed. : R.K. Trivedy, Ashish Publ. House, New Delhi.

Shukla S.C., Kant Rajani and Tripathi B.D. (1989). Ecological investigation on physico-chemical characteristics and phytoplankton productivity of river Ganga at Varanasi. Geobios. 16(1) : 20-27.

Shyam Sunder (1988). Monitoring the water quality in a stretch of river Jhelum.

Kashmir. In : R. K. Trivedy ed. Ecology and pollution of Indian Rivers. U.P. 131-162. Ashish Publishing House, New Delhi.

Singh A.L., Singh P.K. and Singh P.L. (1988). Comparative studies on the use of green manuring, organic manuring, use of Azolla and BGA bio-fertilizers to rice. J.: Agri. Sci. Camb. 110 : 337-344.

Sinha A.K., Srivastava S. and Srivastava K.N. (1989). Physico-chemical studies of river Ganga water at Kalakonklar (Pratapgarh). Indian J. Env. Prot. 9(3) : 194-197.

Sinha M.P. (1986). Limnobiotic study on trophic status of a polluted fresh water reservoir of coal field area. Poll. Res. 5(1) : 13-17.

Sinha M.P. (1988). Effect of waste distribution on water quality of river Damodar in Bihar. Physico-chemical characteristics In : R. K. Trivedy (ed.) Ecology and Pollution of Indian Rivers. Ed. pp. 219-249. Ashish Publishing House, New Delhi.

Smart M.M., Jones J. R. and Sebangh (1985). Stream watershed relations in the Missouri Ozark plateau province. J. Environ Qual. 14: 77-82.

Stoke P.M. (1983). Response of fresh water algae to metals. Progress in Phycological Research 2 (Round/Chapman eds.) Elsevier Sciene Publishers B.V. pp. 87-112.

Trivedy R.K. and Goel P.K. (1984). chemical and biological methods for water pollution studies. Environmental Publications, Karad, India.

Tripathy A.K. and Pandey S.N. (1990). Water Pollution, Ashish Publ. House, New Delhi.

Tundisi T.M. and Tundusi J.C. (1976). Plankton studies in lacustrine environment. Oecologia. (Berl.) 25 : 265-270.

Venkateshwarlu T. and Jayanthi T.V. (1968). Hydrobiological studies on river Sabarmati to evaluate water quality. Hydrobiologia. 443-448.

Warren C.E. (1979). Toward classification and rationale for watershed management and stream protection. EPA-600, 30-79-59 ((USEPA, Cornvalis, Oregan) 143.

Watanabe J., Roger P.A., Ladha J.K. and Van Hovec. (1992). Bio-fertilizer germplasm collections at IRRI Int. Rice Res. Inst. Philippines.

Welch P.S. (1952). *Limnology*. MC. Graw Hill Book Co. 2nd Edition 538.

Wetzel R.G. (1983). Limnology, IIed. Saunders College Publ.; New York. p. 153. Bristol Avon. Fish Invest. Lond. Series 1 Vol. 4 No. 1.

Whitton B.A. (1970). Toxicity of zinc, copper and lead to chlorophyta growing in flowing waters. Arch. Microbiol. 72 : 356-360.

Whitton B.A. (1980). Zn and plants in rivers and streams. In : Zinc in the environment part II. Health effects (Ed. I.D. Naraign) pp. 363-400. John Wiley and Sons Pvt. Ltd. New York.

Widuto J. (1977). Zooplankton in the Kortow Skil Lake after 18 years of the experimental restoration of the lake. Pol. Arch. Hydrobiol. 24(1) : 91-106.

Yousuf A.R. and Gadri M.Y. (1981). Seasonal abundance of rotifers in a warm monomictic lake. J. Ind Inst. Sci., 63:23-34.

Authors :

1. **B.N.Pandey, A.K.Gupta, A.K.Mishra, P.K.L.Das, A.K.Jha,** P.G. Deptt. of Zoology, Purnea College, Purnea - 054301. (Bihar) India.

11

Field study on the status of Pollution Caused By Industrial Effluents of the River Saraswati, West Bengal, India.

❑ INTRODUCTION

Aquatic ecosystem are facing constantly serious pollutional threat from various types of industrial effluents, agriculture and urban runoff containing various xenobiotic substances. Most of the industries of our country are situated on the side of major rivers and they discharge their effluents mostly in untreated condition into these rivers (Jhingran, 1979). These untreated effluents contain various types of toxicants which in mixture may produce various physiological abnormalities on the fish and fish food organisms (Panigrahi and Konar, 1989; Ghatak and Konar, 1992).

Field studies were conducted on Saraswati River, which is flowing through the populated area near Bansberia and ultimately mixed into the river Hooghly near Tribeni. Saraswati is one of the tributeries of river Ganga which receives huge amount of effluents from paper pulp industries regularly. The River also receives domestic sewage of the nearest town.

The objective of this investigation is to determine the toxic effects of paper mill effluents and sewage on aquatic ecosystem of Saraswati River, West Bengal.

Methods

The survey was made starting from 3000 meter upstream of Saraswati river to 2000 meter away from the meeting point of river Saraswati to Hooghly. Four spots were surveyed at surrounding area where the canal carrying effluents from paper mill meets the river Saraswati (Fig. 1). The fifth spot was on the river Hooghly (Fig.1). The details of the spots are given in Table -1 and Figure 1.

Table 1
Information about the sampling stations on the rivers Saraswati and Hooghly

Sampling Stations	Details	Distanc (km)
S1	Control zone at river Saraswati	0
S2	Saraswati river	3.10
S3	Meeting point of effluent carrying canal (from paper mill) near Bansberia, Hooghly.	5.24
S4	Saraswati river	6.84
S5	Hooghly river	7.00

The samples were collected during low tide at regular interval (30 days) throughout the period of investigation particularly to avoid contamination of several toxicants due to tidal influx. All water, soil and plankton samples were collected at random in five replicates at every station. For water quality criteria such as dissolved oxygen and free carbondioxide water was collected 3 meter below the depth of the surface water.

For alkalinity, hardness, phosphate and pH, water samples were collected from about 1-2 cm below the surface level of river water.

For plankton collection, water was collected from 45-60 cm below the water surface of the river with the help of one liter plastic mug which was concentrated to 200 ml for plankton analysis and then preserved in formalin. Bottom soil samples were collected from river bed where water remains throughout the year. Parameters of the samples were analysed as per APHA et al. (1980). Different schedules followed during investigations are give in Table 2. The data were subjected to statistical analysis using *t*, *d* and *F* tests and all variations were considered significant at 5% levels of probability (Campbell, 1967).

Results

The results obtained from the survey on the Saraswati river, West Bengal show that paper mill effluent as well as sewage has significant impact on the different physico-chemical and biological parameters of the receiving ecosystem.

Table 2

Various schedules followed during field studies on the river Saraswati

Schedule	Observation made on Days
Plankton analysis	0, 30, 60, 90, 120, 150, 180
Bottom biota recording	0, 30, 60, 90, 120, 150, 180
Water analysis	
Dissolved Oxygen	
Free carbondioxide	
Alkalinity	0, 30, 60, 90
Hardness	120, 150, 180
pH and phosphate	

Biotic Parameters

The planktonic populations (both zoo and phytoplanktons) and chironomid populations decreased significantly ($p < 0.05$) at all stations from S2-S5 (Table 3). Maximum reduction of chironomid population was observed at S3. Similarly there was significant ($p < 0.05$) reduction of molluscs at S3-S5. The degree of reduction of molluscan population at different stations was varied in the order of S3, S5, S4 (Table 3).

Table 3

Mean values of biological parameters (planktons and bottom organisms) at different sampling stations on river Saraswati. Statistical significance is shown at $p < 0.05$ *

Sampling	Planktons (No./l)		Bottom organisms (No./m^2)	
Stations	Zoo	Phyto	Chironomid larvae	Molluscs
S1	363.65	640.97	203.20	46.30
S2	132.55	343.35*	141.46 *	32.79
S3	26.87	115.79 *	45.51*	13.08*
S4	111.00*	279.64*	140.53*	20.36*
S5	119.48*	294.74*	76.37*	17.42*

Physico-chemical Parameters

Among the different physico-chemical parameters, the dissolved oxygen decreased significantly ($p < 0.05$) at all stations (S2-S5) but it was greatly decreased at the discharge point (S3) of raw effluent into the river. Free carbondioxide increased significantly at S2-S5. The alkalinity decreased significantly at S2 and S4. The hardness and phosphate concentration of water decreased significantly ($p < 0.05$) (Table 4). No marked change in pH of water was observed during the survey period.

Table 4

Mean values of some physico-chemical parameters of different sampling station on river Saraswati. Statistical significance is shown at $p < 0.05$ (*)

Sampling Stations	Free carbon dioxide	Dissolved oxygen (mg/1)	Alkalinity (mg/1)	Hardness (mg/1)	Phosphate (mg/1)	pH
S1	1.70	9.75	207.73	184.18	0.092	7.37
S2	2.90*	4.44*	126.44*	115.78*	0.061*	7.07
S3	5.98*	1.47*	175.60	128.59*	0.040*	7.69
S4	4.85*	2.45*	145.72*	112.86*	0.058*	7.18
S5	6.71*	6.20*	195.65	173.71	0.051*	7.76

DISCUSSION

The reduction of planktonic populations and bottom organisms might be due to the presence of brown color effluents in river water which decreases the penetration of light which ultimately reduced the growth of phytoplankton and various aquatic plants (Varma et al., 1980). No zooplankton was observed within 500 meters of effluent discharge point in the Hooghly estuary at Haldia during low tide (CIFRI, 1980). It was observed that waste waters of paper and pulp mill were toxic to planktons and benthic organisms (Ghosh and Konar, 1980). The significant reduction of molluscs at various sampling stations might be due to toxic effect of paper pulp and various industry effluents (Verma et al., 1980). So maximum reduction of molluscs indicates that the area is loaded by pollutants. The study of bottom fauna is very important for the determination of the status of water (Wiebe, 1928).

The physico-chemical parameters of the river Saraswati water was greatly altered due to the factory discharges. It was also observed reduced dissolved oxygen at various sampling stations of Hindan river due to the influence of paper pulp effluents (Varma et al., 1980). The domestic wastes contain a large amount of putrified organic matter which demand large quantities of oxygen for its degradation thereby leading to depletion of oxygen in the affected area (Ghosh and Konar, 1980).

The reduction of alkalinity and hardness at various stations depends upon the interaction of various factors. Taabta (1966) studied the toxic components in the effluents from ammonium base semi chemical pulping operations and the waste was found more toxic to Daphnia to alkaline water (pH 8.0) because of increase in toxicity of ammonia. It was investigated that municipal waste, paper pulp and paint industries contain phenol and phenolic compounds which contaminate the aquatic ecosystem (Mukherjee and Bhattacharya, 1977; Dauble and Barraclongh, 1983).

In the river Saraswati fish catches were nil in this region reported by the local fisherman. Maximum mortality of fishes, molluscs and crabs occurred during low tide as concentration of raw effluents increased at that time. Jhingran (1988) noted 100% mortality of *Rita rita* caged for 24 hours near the paper pulp and paper (soda process) mill outfall during summer and winter, decline in plankton population and primary production in the marginal stretches of the industrial zone.

However, the toxic action might be due to the interaction of various pollutants which get mixed together into the Saraswati river water and ultimately produce significant effect on riverine ecosystem (Ghatak and Konar, 1992).

So cautious steps should be taken so that Saraswati river water should not receive these industrial effluents, otherwise the river shall get polluted beyond repair.

Acknowledgement
We thank Head of the Department of Zoology, Kalyani University for providing necessary research facilities and ICAR, New Delhi for financial assistance.

❑ REFERENCES

APHA (American Public Health Association). 1980. Standard methods for the examination of water and waste water. 14th edition.

Campbell R.C. 1967. Statistics for biologists. Cambridge University Press, London, England.

CIFRI (Central Inland Fisheries Research Institute). 1980. Aquatic pollution in the Hooghly estuary with reference to Halidia oil refinery complex. Annual Report, project No. 20.14 CIFRI, Barrackpore, India.

Dauble D.D., S.A. Barraclongh, R.M. Bean and W.E. Fallen 1983. Chronic effects of coal liquid dispersion on fat head minnows and rainbow trout. *Trans. Am. Fish. Soc.* 112 :712-719.

Ghatak D.B. and S.K. Konar. 1992. Impact of various industrial effluents on Hooghly river ecosystem at Naihati, West Bengal, *Environ. Ecol.*

Ghatak D.B and S.K. Konar, 1992. Impact of various industrial effluents on the Hooghly river ecosystem at Naihati, West Bengal. *Environ. Ecol.* 1 : 144-150.

Ghosh T.K. and S.K. Konar, 1980. Toxicity of chemicals and waste waters of paper and pulp mills to worms, plankton and molluscs. *Indian J. Environ. Hlth.* 22 : 278-285.

Jhingran A.G. 1988. Aquatic pollution with special reference to the Ganga river system. *Nat. Symp. Environ. Imp. Anim. and Aqua.* May 14-16, 1988, p.6-19, Dept. of Zool., K.U., Nadia.

Jhingran V.G. 1979. Some aspects of capture and culture fisheries of Inland waters of India in relation to environmental pollution. Pages 183-190. in S.P. Roy Choudhuri and D.S. Gupta, editors. *Proc. Int. Symp. Env. Pollut. Toxicol.* Today and Tomorrow's printers and publsihers, New Delhi, India.

Mukherjee S. and S. Bhattacharya, 1977. Variations in the hepatopanereatic -amylase activity in fishes exposed to some industrial pollutants. *Water Res.* 11:71.

Panigrahi A.K. and S.K. Konar, 1987. Impact of industrial effluents on Hooghly estuary ecosystem with reference to oil refinery at Haldia, West Bengal. *Environ. Ecol.* 7 :57-61.

Tabata K., 1966. Synthetic studies on toxic components in industrial wastes with reference to the tolerance of aquatic lives. II. On acute toxic components in digested waste from ammonia based Sunichal Pulp mills. *Biol. Abs.* 47 :90682.

Verma S.R., I.P. Tonk and R.C. Dalela, 1980. Effect of thiox and malathion and their combination on the body composition of *Saccobranchus fossilis and Mystus vitatus. Int. J. Acad. Ichthyol.* 1 : 35-44.

Wiebe A.H., 1928. Biological survey of the upper Mississipi river with special reference to pollution *Bull. U.S. Bur. Fish.* 43 : 137-167.

Authors :

1. **D.B. Ghatak and S.K. Konar** Fisheries Laboratory, Department of Zoology, Kalyani University, kalyani - 741 235. (W.B.) India.

■ ■

12

Pollution studies on the Khan River At Indore

❑ INTRODUCTION

Years before, Khan was a flowing river. Due to urbanization of the city the river has become a wastewater disposal site. The growth of the city was fairly rapid during the past few years, as a consequence of which the waste entering into the river has increased at an enormous rate. At present it has been fully converted into a *'Nalah'*.

Practically no information is available on pollution of the Khan river. Hence a comprehensive study of its water quality has been undertaken. In the present investigation, certain physico chemical characteristics have been chosen to assess the pollutional levels of Khan river in a stretch of about 10 km.

STUDY AREA

Water samples were collected at 5 sampling locations, in a stretch of about 10 km of the Khan from Karbala, Harshiddhi, Krishnapura, Rambagh, Bhandari Mill Pool. The water samples were collected in Rainy Season for three months (July, Aug. & Sept. 1996).

Site A

The Station is situated at Karbala. Here washing of clothes is done by washermen.

Site B

This Station is near Harshiddhi. It is a slum area. Domestic waste is mixed in it. It is about two km from Karbala.

Site C

This site is located at Rajbada. It is also 2 km from Harshiddhi. Here also domestic waste is mixed.

Site D

This site is located at Rambagh. It is 1 km from Harshiddhi. A crematorium is also present here which also pollutes the river.

Site E

This site is located near Bhandari, Textiles Mill. It is about two km from Rambagh. Here waste from different factories gets mixed up in that water.

Table 1

Physico-chemical parameter of Khan (July 1996)
at Different sampling Location in mg/lit.

Sr. No.	Parameter	Karbala	Harshiddi	Krishnapura	Rambagh	Bhandari Mill
1.	Colour	Blackish	Blackish	Muddy	Blackish	Blackish
2.	Odour	Specific	Specific	Specific	Specific	Specific
3.	Temp (^{O}C)	31°	31°	31°	31°	31°
4.	pH	8.9	9.1	9.5	9.2	9.5
5.	Alkalinity as ($CaCO_3$)	170	180	230	180	240
6.	Chloride	180	160	200	170	230
7.	Total Hardness	300	420	400	450	470
8.	Ca Hardness	200	320	250	300	330
9.	Mg Hardness	100	100	150	150	140
10.	DO	2	2	2	2	1.6
11.	COD	256	206	225	218	356
12.	BOD	454	510	452	522	500
13.	Total Solids	4850	2520	1100	1660	2380
14.	Total Suspended Solids	750	470	140	280	610
15.	Total Dissolved Silids	4100	2050	960	1380	1870

Material and Methods

Water samples were collected form 5 sampling locations three times in a rainy season for a period of 3 months. (1 July 1996 to Sept. 1996).

Water samples were analysed for different physico-chemical parameters such as water temp., pH value, alkalinity, Chlorides, total suspended solids, dissolved solids, dissolved oxygen, biochemical oxygen demand (BOD), Total hardness, calcium hardness, magnesium hardness and COD. The temp. pH and the DO were, estimated in situ immediately on collection of the sample. The pH was determined electrometrically by using digital pH meter. Total dissolved solids and suspended solids were estimated by residue on evaporation of known volume of filtered water. Other tests were carried out in the lab by titration methods.

Table 2
Physico-chemical parameter of Khan (August 1996) at Different sampling locations in mg/lit.

Sr. No.	Parameter	Karbala	Harshiddi	Krishnapura	Rambagh	Bhandari Mill
1.	Colour	Blackish	Pale Black	Pale Yellow	Pale Yellow	Muddy
2.	Odour	Specific	Specific	Specific	Specific	Specific
3.	Temp (^oC)	26	26	26	26	26
4.	pH	8.1	8.2	8.5	8.2	9
5.	Alkalinity as ($CaCO_3$)	110	130	180	130	200
6.	Chloride	75	110	150	120	190
7.	Total Hardness	350	370	350	400	420
8.	Ca Hardness	210	220	230	240	250
9.	Mg Hardness	140	150	120	160	170
10.	DO	2.5	2.8	3	3	2.3
11.	BOD	358	385	326	400	450
12.	COD	200	180	190	200	268
13.	Total Solids	3960	1760	960	1200	1570
14.	Total Suspended Solid	640	200	240	80	170
15.	Total Dissolved Solid	3320	1560	720	1120	1400

Table 3
Physico-chemical parameter of Khan at
Different sampling Locations in mg/lit. (Sept. 1996)

Sr. No.	Parameter	Karbala	Harshiddi	Krishnapura	Rambagh	Bhandari Mill
1.	Colour	Pale Black	Muddy	Creamish	Creamish	Pale Yellow
2.	Odour	Specific	Specific	Specific	Specific	Specific
3.	Temp (^oC)	28°	28°	28°	28°	28°
4.	pH	7.6	7.7	7.9	7.6	8.2
5.	Alkalinity as ($CaCO_3$)	100	110	150	120	160
6.	Chloride	70	100	130	110	150
7.	Total Hardness	300	300	290	350	380
8.	Ca Hardness	170	160	170	200	220
9.	Mg Hardness	130	140	120	150	160
10.	DO	3.1	3.3	3.5	4.0	3.0
11.	BOD	302	333	298	356	410
12.	COD	178	154	182	188	230
13.	Total Solids	3750	1480	800	1100	1360
14.	Total Suspended Solids	550	160	190	120	180
15.	Total Dissolved Soilids	3100	1320	610	980	1180

RESULTS AND DISCUSSION

1. The maximum value of dissolved solids was recorded in july due to less flow of water and high impurities because rains did not pour much at that time. High value of dissolved solids were noticed (4100 mg/1) at sampling station Karbala and Bhandari mill. This may be due to stagnation.
2. The pH value of the water is above 8 at all the sampling stations. Higher pH value is in July (9.5). pH gets lowerer in August & September due to rainy water. pH at Krishnapura and Bhandari Mill is higher due to high concentration of sewage and industrial wastes.
3. The higher concentration of alkalinity was noticed at Krishnapura and Bhandari Mill due to sewage disposal (240 mg/1 as $CaCO_3$)
4. High value of chlorides was found because of the contamination of water due to sewage of the city. It is higher at Krishnapura, Harshiddhi & Bhandari Mill. It seems that it is heavily polluted water probably because of organic

matter. (230 mg/1).

5. The relative higher hardness was recorded at all the points (420 mg/1). It is due to sewage effluent and industrial waste. The hardness is higher in July & lower in August and September (250 mg/1).
6. The DO and BOD are the important parameters indicating the pollution level of Khan river. The higher BOD value was recorded in July. Higher BOD values at Harshiddhi and Rambagh were recorded (551 mg/1) due to less dilution and disposal of city waste water. In the August and September the BOD were less and DO is more (4 mg/1) due to dilution by rain water.
7. The COD found at Bhandari Mills is higher in comparision to other locations (356 mg/1). It is because of the mixing of different industirial waste.

SUMMARY AND CONCLUSION

The values of Physico-chemical Characteristics reveal that the water of the Khan river is highly polluted. The variations in various parameters in seaonal.

❑ REFERENCES

A manual of water analysis. NEERI, Nagpur.

Imhoff, K. 1956. Disposal of sewage and other water born waste. Butterworths Scientific Publications, London.

Shell G.L. and Burns, D.E. 1973. Advanced wastewater treatment methods. *Effluent and water treatment journal* 13 : 203.

Zayic, J.E. 1971. *Water Pollution.* Marcel Dekker Inc., New York

Authors :

1. **S.L.Garg, Miss Bum, Preeti, Miss Bharadia, Rama, Miss Rashmi, Sharma, Miss Minakshi, Jindal, Miss Kumudini, Joshi, Miss Savita Vyas :** Department of Chemistry, Govt. Girls Post Graduate College, Moti Tabela, Indore (M.P.) India.

■ ■

13

Comparative Evaluation of Water Quality of two rivers of Narkatiaganj, Bihar (India) with reference to Pollution, Fisheries and Diseases

❑ INTRODUCTION

Narkatiaganj a subdivision ot *West Champaran* district of *Bihar*, lies between latitue 27° 4.6'N and longitude 84° 28.8'E. The aquatic resources of this region consist of two river systems namely *Harbora* and *Pandai*, irrigation canals and their distributaries, paddy fields, ponds and water logged areas. The untreated effluents from the sugar factory and distillery of *Narkatiaganj* are discharged directly into the downstream of the river *Harbora* creating a serious problem of water pollution. These effluents caused severe damage to the flora and fauna of the water bodies. These units have an effluent treatment plant but not in working order as has been observed. Besides the effluents, man's influence on these river systems caused by rapid cutting of surrounding vegetation thus increasing silt and nutrient load, disposal of sewage, use for open defecation, cultural activity, agriculture around the water bodies using agrochemicals greatly increases the quality of nutrients and organic matter input deteriorates the quality of the river water leading to the death of fishes and various health hazards such as *Amoebic* and *Blood Dysentry, Malaria, Filaria, Worms diseases, Constipation, Skin diseases* etc.

The river *Harbora* and *Pandai* (Fig. 1) are hilly torrents, originating from Nepal, are famous for their swift current and meet the river *Sikrahana (Burhi Gandak)* and ultimately drain into the Ganges.

The water quality of freshwater bodies with reference to pollution, fisheries and diseases have been investigated by a number of workers in India (Agrawal et. al., 1976; Arora et al., 1973; Bilgrami and Datta Munshi, 1966, 1979; David A. and Ray P., 1966; Datta et al., 1988; David 1956; Hakim, 1981; Kudesia and Verma, 1985; Shanker et al., 1986; Sreenivasan and Sounder, 1967; Trivedi 1979; Trivedy,

1990 etc.) but there has not been a single comparative evaluation of water quality of two river systems of *Narkatiaganj (Bihar)* with reference to pollution, fisheries and diseases though the water of these rivers are used by inhabitants of the area under investigation in a variety of ways, hence the present investigation was undertaken.

MATERIALS AND METHODS

In the present investigation four sampling stations were taken for observation in the river *Harbora* and one in the river *Pandai*. In the river *Harbora* station I was upstrem of the river, II - the sugar factory effluent discharge zone, III - distillery effluent discharge zone, IV - downstream of ther river *Harbora*. The water and effluent samples were collected once in a month from all sampling sites. Samples were analysed for their physico-chemical characteristics according to APHA-AWWA-WPCF (1976), *Trivedy* and *Goel* (1986) and *Welch* (1948) respectively for a period of one year (September 1994 - August 1995). The fishes from the river *Pandai* and *Harbora* were collected with the help of different types of castnets. The specimens were fixed in 3% and later in 8% formalin solution. The identification of fishes are based on standard methods (*Day*, 1878; *Hamilton*, 1822; *Jayaram*, 1981 and *Talwar* and *Jhingran*, 1991). Classification of fishes followed are that of *Greenwood et al* (1966). The nomenclature of various species of fishes are based on recent revisions. The diseases prevailing amongst inhabitants of the area under investigation was surveyed by moving door to door and obtaining data from the pathological clinics.

OBSERVATIONS

Different atmospheric records of *Narkatiaganj* have been presented in Table-1. The physico-chemical characteristics of upstream, downstream, sugar factory effluent, distillery effluent samples and *Pandai* river water samples have been presented in Table-2, 3 & 4 respectively. List of tolerent species of fishes collected from the polluted zone of the river *Harbora* has been presented in Table-5. The atmospheric temperature ranged between 39.3 to 12.5^0 C and water temperature of the river *Harbora* and *Pandai* was found ranging between 20-35^0 C and 18-29.4^0 C respectively. Humidity and rainfall was recorded between 75-88.7% and 0-19.6" respectively during the period of investigation. *Harbora* and *Pandai* are fast moving river (0.36 ms^{-1} -0.45 ms^{-1} and 0.46 ms^{-1} -0.54 ms^{-1} respectively) and due to heavy load of silt and sand a little growth of macrophytic vegetation was found. The river *Pandai* and upstream of the river *Harbora* showed alkaline pH in contrast to acidic pH on station II, III and IV of the river *Harbora*. The dissolved Oxygen values and free CO_2 values in sugar and distillery wastes were recorded nil, however a moderate amount of DO was recorded in the upstream of the river *Harbora* and river *Pandai*. The values of free CO_2 and chlorides were higher in the river *Harbora* in comparison to the river *Pandai*. Total dissolved solids of the effluents showed marked differences because of their variations in nutrient enrichment. Effluents possessed high temperature and BOD values. In the upstream of the river *Harbora* and *Pandai* sixty three species of fishes belonging to thirty six genera, nineteen families and five orders were reported including seven new records to the ichthyofauna of Bihar (*Mishra*, 1995). In the polluted zone of the river *Harbora* eleven species belonging to eight genera and eight families of tolerent fishes were found (Table-5). Straying fishes thrived only for short periods until the next flushing from the sugar factory and distillery killed them. During the course of investigation twenty four types of diseases were reported prevailing amongst the inhabitants of the area like *Malaria, Filaria, Eczema, Tuberculosis, Bronchitis, Jaundice, Amoeboic* and *Blood dysentry, Heart diseases, Worms diseases, Skin diseases, Indigestion, Ascariasis, Asthma, Constipation* etc.

Table 1

Atmospheric Temperature, humidity and Rainfall of Narkatiaganj during September 1994 to August 1995.

Months	Average atmospheric temperature (°C) Maximum	Humidity (%) Minimum	Max.	Rainfall (Inches)
September 1994	33.5	25.8	86.6	19.6
October	33.3	22.0	83.5	Nil
November	28.0	17.8	85.0	Nil
December	22.0	12.5	84.5	Nil
January 1995	22.1	12.5	88.7	0.14
February	25.7	15.4	85.6	0.71
March	31.1	19.8	84.4	0.35
April	35.2	24.3	79.7	0.25
May	39.3	28.7	75.0	1.05
June	34.8	29.2	81.0	10.6
July	33.5	27.7	85.3	12.3
August	34.4	26.7	83.9	11.7

Table 2

Physico-chemical characteristics of upstream and downstream water samples of the river Harbora.

Months 1994-1995	Station	Flow of water (ms^{-1})	Temperature (^0C)	pH	Dissolved Oxygen (mg/1)	Free CO_2 (mg/1)	BOD	Cholorides (mg/1)
September 1994	I	0.42	26	7.2	7.6	2.0	20	12.2
	IV	0.45	27.6	7.2	2.8	14.5	95	18.0
October	I	0.36	25.8	7.1	8.0	1.5	22	14.2
	IV	0.36	32.1	6.8	2.5	15.5	92	17.5
November	I	0.36	23.5	7.2	8.8	2.1	20	12.0
	IV	0.36	32.5	5.8	2.1	18.6	110	15.5
December	I	0.36	22.2	7.4	8.6	2.0	22	12.5
	IV	0.36	32.8	5.8	2.2	20.2	108	18.2
Jan. 1995	I	0.38	22.0	7.8	9.0	Nil	30	11.2
	IV	0.39	33.8	5.6	1.8	26.4	115	14.4
February	I	0.37	24.0	7.8	9.2	Nil	28	8.3
	IV	0.36	33.0	5.5	Nil	22.0	122	11.8
March	I	0.36	27.4	7.6	8.4	2.0	24	4.5
	IV	0.36	34.2	5.4	1.0	26.0	120	7.5
April	I	0.36	30.2	7.3	7.6	2.0	20	11.5
	IV	0.36	34.1	5.2	Nil	26.2	132	14.8
May	I	0.37	30.5	7.3	7.3	1.4	22	12.5
	IV	0.36	35.5	5.0	Nil	26.0	130	18.0
June	I	0.40	31.0	7.2	7.2	2.1	22	12.0
	IV	0.40	31.2	5.6	1.5	20.5	118	18.2
July	I	0.41	31.2	7.2	7.2	2.2	20	12.5
	IV	0.45	30.6	5.8	2.0	20.0	120	21.0
August	I	0.42	28.5	7.2	7.4	2.5	20	11.5
	IV	0.42	28.8	5.4	1.0	20.0	120	18.5

Table 3

Physico-chemical characteristics of sugar factory and distillery effluents.

Parameters 1994-95	Sugar factory effluent Station II	Distillery effluent Station III
Colour	Light Brown	Deep brown framy
Flow of effluent (ms^{-1})	0.27-0.26	0.40-0.36
Temperature (^{O}C)	40-43	50-57
pH	5.5-5.7	4.6-5.0
Dissolved Oxygen (mg/1)	Nil	Nil
Free CO_2 (mg/1)	Nil	Nil
Total solids (mg/1)	1200-1800	7000-8300
BOD (mg/1)	1200-1800	35,000-50,000

Table 4

Physico-chemical characteristics of water samples of the river Pandai.

Months 1994-1995	Flow of water (ms^{-1})	Temperature (^{0}C)	pH	Dissolved Oxygen (mg/1)	Free CO_2 (mg/1)	Chloride (mg/1)	BOD
September 1994	0.52	24.2	7.8	14.2	Nil	15.0	15
October	0.46	23.5	8.0	13.5	Nil	14.0	21.0
November	0.46	21.5	7.6	13.5	1.8	12.5	14.0
December	0.46	20.4	7.6	13.2	1.6	14.5	14.5
Jan. 1995	0.46	18.1	7.5	13.0	1.6	10.5	16.0
February	0.48	22.2	7.6	14.0	Nil	8.0	14.2
March	0.46	25.2	8.1	14.0	Nil	5.1	14.0
April	0.46	28.0	7.6	13.5	0.5	10.3	20.0
May	0.48	29.1	7.4	14.0	Nil	13.2	18.2
June	0.51	29.4	7.4	14.2	Nil	15.1	20.1
July	0.52	29.0	7.4	14.0	Nil	14.0	18.5
August	0.54	27.5	7.4	16.5	Nil	15.2	12.5

Table 5

List of tolerent species of fishes collected from the downstream of the river Harbora.

Family :	*Bagridae*		
	1)	*Mystus bleekeri*	(Day)
	2)	*Mystus cavasius*	(Ham.)
	3)	*Mystus tengra*	(Ham.)
Family :	*Belontidae*		
	4)	*Xenentodon cancila*	(Ham.)
Family :	*Gobiidae*		
	5)	*Glossogobius giuris*	(Him)
Family :	*Belontidae*		
	6)	*Colisa fasciatus*	(Schneider)
Family :	*Channidae*		
	7)	*Channa striatus*	(Bloch)
	8)	*Channa punctatus*	(Bloch)
Family :	*Mastacembelidae*		
	9)	*Mastacembelus armatus*	(Lacepede)
Family :	*Siluridae*		
	10)	*Wallago attu*	(Sehneider)
Family :	*Anguillidae*		
	11)	*Anguilla bengalensis bengalensis*	(Gray)

DISCUSSION

The analysis of the water and effluent samples shows that the condition of the downstream of river *Harbora* is very poor, since the parameters have achieved a very high magnitude. During crushing season sugar factory discharges thousand gallons of untreated effluent per day tales directly into the river *Harbora* in addition to the effluent of distillery through brick made channels. The physico-chemical data of station II and III show highly polluted nature of effluents. The absence of DO, free CO_2, high BOD, TS, temperature and low pH values, are the main indicators of its being heavily polluted. The polluted nature of station IV is further substantiated by great reduction in the diversity of fishes which are only represented by tolerent forms and heavy mortality of fishes. In the polluted zone straying fishes thrive only for short periods until the next flushing from the sugar factory and distillery killed them. The sugar factory effluent contains a high degree of organic pollution hence if effluents stagnates in an area for a few hours, biological action starts and septic condition gives H_2S gas imparting black colour to the effluent. Moreover the Oxygen is also exhausted cauring death of fishes and other aquatic lives. The distillery wastes are highly organic in nature and because of its high BOD demand quickly removes the Oxygen from the water producing pollution hazard in the aquatic ecosystem. Chloride content of the river *Harbora* at the confluences point and downstream is high. The higher concentration of Chlorides in the downstream of the river *Harbora* cause sufficient pollution due to sewage as well as industrial wastes. In addition to the effluents released by these industries these rivers are leading towards eutrophication due to man's influence too like cultural activities, open defecation, agriculture around the water bodies using insecticides, pesticides etc. *Kudesia* and *Verma* (1985) reported that the combined effluent of cane sugar, chemical industry, distillery and rubber industries is characterized by high load of zinc, cadmium, barium, sulphates, chlorides, suspended organic and inorganic matter, very high BOD and COD and low DO etc., a brown colour and offensive odour to water. This makes the water unfit for irrigation, fish catch, drinking of cattle and human beings. pH values of both the rivers were found between 5-8.1, which is not suitable for the production and propagation of fishes as pH 9 is considered most suitable for the fishes. It is further substantiated by the absence of many forms found elsewhere in the rivers of the district and extinction of many fish forms.

Human health and quality of life are strongly influenced by the environment. The most important environmental factor in the transmission of *infections disease* is water. The second major factor is the quality of urban and residential environment, and the third, the pattern of human movements and contracts (*Swarup, Mishra* and *Jauhari,* 1992). During course of investigation twenty four types of important diseases were reported prevailing amongst the inhabitants of area (*Mishra* and *Verma,* 1995), some of which are infections and may be considered to be transmitted by the polluted water of these rivers by various ways. Thus it may be concluded that the sugar factory and distillery of Narkatiaganj cause a pollution hazard in the aquatic ecosystem of the river *Harbora* and many important flora and fauna are on the verge of extinction unless the wastes are diluted and chemically treated before discharge. The water of these rivers may not be used for drinking by cattle and human beings.

Acknowledgement
The author express his gratefulness to the University Grants Commission, New Delhi for providing financial assistance to carryout this project.

❑ REFERENCES

APHA-AWWA-WPCF, 1976 Standard methods for the examination of water and wastewater. American Public Health Association, New York.

Agrawal D.K., Gaur S.D., Tewari I.C., Narayan Swami N. and *Magwah S.M.* 1976. Physico-chemical characteristics of Ganga water of Varanasi, *Indian J. Environ. Hlth.,* 18:201-206.

Arora H.C., Tapan R., Chattopadhya S.N. and *Sharma V.P.,* 1973. Survey of sugar mill waste disposal; Part I, S.K.G. Sugar factory Hathua (Bihar). *Indian J. Environ. Hlth.,* 15:28-30.

Bilgrami K.S. and *Datta Munshi J.S.,* 1966. Ecology of river Ganges. Impact of human activities and conservation of aquatic biota (Patna to Farakha). Tech. report submitted to D.S.T., Govt. of India, 97 pp.

1979. Limnological survey and impact of human activities on the river Ganges (Barauni to Farakha. M.A.B., Tech. report submitted to D.S.T., Govt. of India, 91pp.

David A. and *Ray,* P.1966. Studies on the pollution of the river Daha (N.Bihar) by sugar and distillery wastes. *Environ. Hlth.,* 8:6-35.

David A. 1956. Studies on pollution of Bhadra river fisheries and Bhadrawati, Mysore state with industrial effluents. *Proc. Nat. Sci. Acad. India.,* 22:132-145.

Datta N.C., Bandopadhya B.K., Majumdar A. and *Ahuja D.* 1988. Hydrobiological profile of Hooghly sector (Bally to Bandel) of river Ganga. (R.K. Trivedy edited). *Ecol. and Pollution of Indian River.* Vol. I, Ashish Publishing House, New Delhi, 173-129.

Day F. 1981. *The fishes of India; being a natural history of fishes known to inhabit the seas and freshwater of India, Burma and Ceylon* (Vol. I) Today and Tommorow's Book Agency, New Delhi, 778 pp.

Greenwood P.H., Rosen D.E., Weitzman S.H. and *Myers G.S.,* 1966. Phyletic studies of teleostean fishes with a provisional classification of living forms. *Bull. Amer. Mus. nat. Hist.,* 131:33-456.

Hakim M.Y.H., 1981, Impact of effluents of sugar mills on water quality and biota of Burhigandak near Samastipur town (N.Bihar). *Proc. Nat. Symp. Pollu. Envol. Environ.* : 29.

Jayaram K.C., 1981. The fresh water fishes of India, Pakistan, Bangladesh, Burma and Srilanka - A Handbook Pub. Director, Zoological Survey of India, Calcutta 475 pp.

Kudesia V.P. and *Verma S.P.* 1985. Physico-chemical studies on industrial pollution of Kali Nadi to combined effluents of cane sugar, chemical industry, distillery, and rubber industries in Meerut region. Paper presented in 2nd International

Conference on Environmental Education held at New Delhi from March 4-9, Abs. No. 50.

Mishra S.K. 1995. Fishes of Narkatiaganj, Bihar (India), *Flora* and *Fauna*, 1 (2) : 127-130.

Mishra S.K. and *Verma R.M.P.1995* Medicinal Plants of Champaran (Bihar) used in various diseases, *Flora* and *Fauna*, 1(1) : 75-76.

Shankar V. Sangu R.P.S. and *Joshi G.C.* 1986. Impact of distillery effluents on the water quality and river Reh in doon valley. *Poll. Resh.*, 5(3 & 4) :137-142.

Sreenivasan A. and *Sounder Raj R.* 1967. Effect of certain wastes on the water quality and fisheries of river Caum and Bhivani, *Environ. Hlth.* 9 : 13-21.

Swarup R., Mishra S.N. and *Jauhari V.P.* 1992. Encyclopaedia of Ecology, Environment and Polllution control, Envirionmental Health, Education and Public understanding, Mittal Publications, New Delhi, 374 pp.

Trivedy R.C. 1979. Pollution studies of Chambal river due to Nagda Industrial omplex. Ph.D. Thesis, Vikram University, Ujjain.

Trivedy R.K. and *Goel P.K.* 1986. *Chemical and Biological Methods for Water Pollution studies.*, Environ. Publi., Karad, India, 250 pp.

Trivedi R.K., 1990. *River Pollution in India*, Ashish Publishing House, New Delhi, 294 pp.

Talwar P.K. and *Jhingran A.G.* 1991. *Inland Fishes of India and Adjacent Countries* Oxford & IBH Publ. Co. Pvt. Ltd., New Delhi, Vols. I & II, 1158 pp.

Authors :

1. **Safal Kumar Mishra :** Department of Zoology T.P.V. College, Narkatiaganj-845 455. W. Champaran, Bihar-India.

■ ■

14

Effect of pollutants on the dissolved oxygen concentration of the river Ganga at Kanpur

❑ INTRODUCTION

Dissolved oxygen is one of the most important parameters in the assessment of water quality. It is used by almost all aquatic plants and animals for respiration. In the absence of dissolved oxygen aquatic life will be abolished. The main sources of dissolved oxygen in water are from the atmosphere and the photosynthetic process of the green plants. Oxygen from air is absorbed by direct diffusion and agitation of the surface waters by wind action and turbulences. The amount of oxygen dissoved in water depends on the surface area exposed, temperature and salinity. The amount of oxygen derived from green plants depends upon the density of plants, the duration and the intensity of effective light (Michael, 1990). Clean surface water is normally saturated with 7.60 mg/L of dissolved oxygen at 30ºC (Behera et. al., 1989). When it is below 3.0 mg/L due to the presence of pollutants, than the water is considered unfit to use for different purposes (Kudesia, 1990). Oxygen is gradually removed from the water by respiration of biota, decomposition of organic matter, rise in temperature, oxygen demanding wastes and inorganic reductants such as hydrogen sulphide, ammonia, nitrites, ferrous iron etc.

Study area

The Kanpur (26º 21' N latitude to 80º 21' E longitude) is located at an elevation of 410ft. above the mean sea level. The Ganga river is flowing on the Northern side of the city. This city uses near about 392.14 millilitres of water per day and releases approx. 274.50 millilitres of wastes like sullage water, sewage textile wastes, power plant wastes, sugar factory effluents and tannery effluents into the river Ganga at several points through 16 nalas (Pandey 1991).

Materials and methods

Water samples were collected from right bank, middle and left bank of the river near Suklaganj bridge of Kanpur (Fig -1) at 10 a.m. in every month from September 1993 to August - 1994 They were analysed for dissolved oxygen parameter immediately after collection by following the azide modification of winkler method described by Trivedy and Goel (1986)

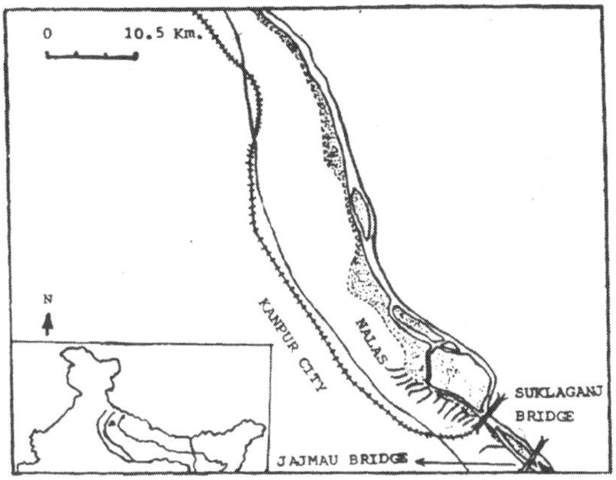

Figure 1. Location of sampling station and different nalas at the bank of the river Ganga near Kanpur.

Results and discussion

The dissolved oxygen of the Ganga river water recorded at Kanpur from September - 1993 to August -1994 are given in Table - 1

Table 1

Dissolved oxygen in mg/L during different months in 1993-94

Months	Right bank	Middle	Left bank
September	5.00	5.76	5.12
October	5.48	5.76	5.36
November	5.92	6.56	6.68
December	5.64	6.04	9.36
January	6.60	6.76	7.22
February	6.00	-	6.40
March	4.20	4.56	6.00
April	2.80	2.72	5.28
May	1.60	2.60	4.80
June	1.40	1.50	6.88
July	4.40	5.40	5.72
August	5.60	5.88	5.80

The dissolved oxygen varies from 1.40 to 9.36 mg/1. Lower amounts are found during April to June on the right bank and middle of the river which are below the limit (3 mg/`) (Kudesia 1990) prescribed for public water supply, bathing ghats, fish culture and irrigation. Maximum concentration 9.36 mg/L is occasional in the month of December on the left bank of the river. It is observed that the river shows lower values than normal i.e. 7.6 mg/L at 300 c, which may be due to oxygen demanding substances. The lowest and highest values are observed on the right and left banks, respectively. Seasonally, during winter (November to February) highest values are observed.

Conclusion

It is concluded that the river water shows low concentration of dissolved oxygen near Sukklaganj bridge of Kanpur. In the months of April, May and June the water in midstream and near right bank were unfit for public supply, bathing, fish culture and irrigation. Therefore, it is necessary to discharge the effluents and municipal wastes after proper treatment by the authorities of the industries as well as by the Kanpur municipality.

Acknowledgement

The authors are grateful to "The Ganga Project Directorate", Ministry of Environment and Forest, Government of India for financial support.

❑ REFERENCES

Behera R.K., Panda R.B., Sahu B.K., Sinha B.K. and Nayak A., 1989. Assessment of Brahmani river part - 1 : Water quality upstream and downstream of Rourkela Industrial Complex. *Poll Res. 8(4)* : 195-200.

Kudesia V.P. 1990. *Water Pollution*, Pragati Prakashan, Meerut.

Michael P. 1990. *Ecological method for field and laboratory investigations*, Tata McGraw-Hill publishing Company Limited, New Delhi.

Pandey U.K. 1991. Physico-chemical studies of the Ganga water at Kanpur (Kannauj to Shuklagang sector) In : *The Ganga, A scientific study*, (Ed.) Krishnamurti C.R., Bilgrami K.S., Das, T.M. and Mathur R.P., Ganga Project Directorate, Northern Book Centre, New Delhi.

Trivedy R.K. and Goel P.K. 1986. *Chemical and Biological methods for water pollution studies*, Environmental Publications, Karad

Authors :

1. **B.K.Sahu, R.J. Rao, S.K.Behera and R.K. Pandit :** School of Studies in Zoology, Jiwaji University, Gwalior-474 011. M.P. (India)

■

15

Arsenic in the Rivers of Safi and Damodar in Bachra Area of North Karanpura Coalfield of Hazaribagh District, Bihar, India

❑ INTRODUCTION

Safi and Damodar which originates from the Chotanagpur plateau forms the western and northern limits of the coal mines area. The water of both the rivers are used by the local resident of the coal mines area, especially for drinking purpose. Health problem like tuberculosis, stomach disorders, dehydration, skin disease and bronchoitis is present in that area.

Arsenic exists in the trivalent and pentavalent state and its compounds may be either organic or inorganic. Trivalent inorganic arsenicals are more toxic than the pentavalent forms both to mammals and aquatic species. It occurs naturally as sulfides in association with sulfides of silver, lead, copper, nickel, iron, cobalt and antimony. The environmental levels of arsenic are reported as total arsenic. Arsenic is released to the atmosphere from both natural sources and anthropogenic activities. Most of the arsenic compounds are soluble in water Sedimentation of arsenic in association with iron and aluminium may sometimes be considerable. Keeping in the view of the above fact the waters of both the rivers were investigated in detail.

The area under study is situated in Bachra area of North Karanpura coal field of Hazaribagh district, Bihar. The area is covered by toposheet no. 73 E/2 of the Survey of India. The area falls between latitudes 23° 41' 16" and 23° 42' 48"N and longitudes 83° 3' 45" and 83° 5' 41"E. Bachra colliery is connected to Khalari (14km.) by an all weather road and Khalari is connected to Ranchi via Bijupara by state highway nos. 47 and 48. The area is mostly hilly, interspersed with narrow valleys and small flood plains of numerous rivulets and is covered by thick forest. The highest elevation in 466. 34m.

The North Karanpura coalfield is characterized with rivers and nalas. The area falls in the drainage basin of the easterly flowing Damodar river which originates from famous watershed, about 10 kms. to the west of Chandwa village situated on the western fringes of the coalfields. The river enters the coalfield near Mahumilan railway station and exixt east of Bachra running in neighbouring South Karanpura coalfields. Among the numerous tributaries of Damodar river are the Haharo river, the Garhi or Barki nadi and the Safi nala which exposes good geological succession along their courses. The Garwa and Chundru nalas discharge into the Garhi nadi while the Badamahri and Ghaghara nalas in the northern and eastern part of the coalfields are the tributaries of the Haharo river. The Safi nala and Chatti nala in the southern part of the coalfield drain directly into the Damodar river. All these rivers and nalas are primarily fed by surface runoff and subsequently by sub-surface percolation of rain water. Some water courses, however, are fed by spring water and are perennial in nature.

The thick alluvium is noticed in the northern part of the Bachra block along the banks of Damodar river. No igneous intrusion has been found in the Bachra block. Barakar Formation consists of rocks of gritty sandstones, pebbly sandstones, fireclays shales and coals. there are symmetrical rhythmites present in that area.

The Karharbari Formation overlie the Talchir Formation and this has been classified on the palaeobotanical evidence. The sandstones are soft and gritty. Conglomerates have rounded, angular or subangular pebbles. Shales vary from bluish grey to dark grey. The conglomerates generally consists of pebbles of quartz, talchir sandstone and quartz mica schist.

Talchir Formation occurs as discountinuous strips along the northern as well as the southern side of the North Karanpura coalfield. It consists of fine to coarse grained sandstones, boulder beds, tilloids and varvites.

Metamorphic consists of limestone with chromite mineralisation and granitoids schists.

METHOD

Water samples have been collected from both the rivers by pressing a plastic container on the above surface of the water at five different locations in Safi river and ten different locations in Damodar river. Arsenic was analysed with the help of Atomic Absorption Spectrophotometer.

RESULTS AND DISCUSSION

Many arsenic compounds are water soluble and hence arsenic contamination of water can occur readily. Water is the major means of transport of arsenic in the environment compartments. Rivers and lakes generally contain less than 0.01 mg/L arsenic. The concentration in ground water depends on the arsenic content of the bed-rock.

Where ever arsenic is present in natural waters, it is most often found as the anion, either as arsenate or arsenite (Kapp and Kroner, 1967).

Table 1 and Table 2 shows arsenic concentration in both rivers. In the Safi the lowest concentration of arsenic is 0.001mg/L and the highest value is 0.003mg/L with mean value 0.002mg/L. In Damodar the lowest concentration is 0.001mg/L and the highest concentration is 0.004mg/L with mean value 0.002mg/L. Concentration is same in both the rivers with low dispersion. Arsenic has a tendency to become precipitated in the hydrolyzates. It is more concentrated near the surface than deep in the sediments. It is enriched in oxidate sediments, chiefly by adsorption on ferric ydroxide. The manganese rich oxidates are lower in arsenic than the iron-rich t, pe (Rankama and Sahama, 1950).

Table 1

Arsenic in the water of Safi river at Bachra.

Location	Arsenic
No. 1.	0.003 mg/L.
No. 2.	0.003 mg/L.
No. 3.	0.002 mg/L.
No. 4.	0.001 mg/L.
No. 5.	0.003 mg/L.

Table 2

Arsenic in the water of Damodar river at Bachara.

Location	Arsenic
No. 1.	0.003 mg/L.
No. 2.	0.003 mg/L.
No. 3.	0.002 mg/L.
No. 4.	0.002 mg/L.
No. 5.	0.003 mg/L.
No. 6.	0.004 mg/L.
No. 7.	0.001 mg/L.
No. 8.	0.004 mg/L.
No. 9.	0.003 mg/L.
No. 10.	0.002 mg/L.

Primary source of arsenic in both the rivers is due to erosion of coals and also due to the erosion and dissolution of mine spoil heaps by different rivulets present in that area, which finally drain into Safi and Damodar rivers. Average concentration of arsenic in the Permian coals of that area is 116 ppm. (Priyadarshi, 1995). Arsenic concentration in Bachra coals (40-200 ppm.) fall at the higher end of the estimated range for most world coals (0.5 to 80 ppm.). Overall we classify arsenic as being in "High" concentrations in Bachra coals.

Other minor sources are as follows.

1. Arsenic are released in the environment due to burning of coal by the local residents. Arsenic enter into the atmosphere through volatization and falls in the river through atmospheric precipitation or in the form of dry spray.
2. In the form of dry spray as Safi and Damodar river flows through the Bachra and Piparwar colliery, the source of dry spray is in the form of abundance of coal dust which results from mining processes of coal as well as transportation of coals.

Conclusion

1. Even though the concentration of arsenic in Safi and Damodar river is below tolerance limit, it may still cause problems to human well beings.
2. The main source of arsenic is due to erosion of permian coals and also due to the erosion and dissolution of mine spoil heaps.
3. Average concentration is same in both rivers with low dispersion, indicating uniform transportation from the source rock.
4. Samples were collected during the end of monsoon period when the concentration of trace elements in the water are generally low, because high discharge of water brings down the pollution by dilution alone. By considering above fact, it can be said that concentration of arsenic may rise during pre-monsoon period.

Acknowledgement

The Authors is thankful to Prof. G. Jha, Head of the Department of Geology, Ranchi University, for providing necessary guidance.

❑ REFERENCES

Kapp J.F. and Kroner R.C. 1967. Trace metels in waters of the United States. *A five year summary of trace metals in rivers and lakes of the United States.* U.S. Department of the Interior Federal Water Pollution Control Administration.

Priyadarshi N. 1995. Distribution and Behaviour of Trace Elements in Permian Coals of Bachra and their significance in Health and Environmental Problems. *Minetech.* 16:41-43.

Rankama K. and Sahama Th. G. 1950. *Geochemistry.* The University of Chicago Press, Chicago.

Authors :

1. **Nitish Priyadarshi** : Department of Geology, Ranchi University, RANCHI - 834 008. Bihar (India)

■

16

Occurrence of heavy metals in water and sediment of the River Satluj in Punjab

❑ INTRODUCTION

The river Satluj, one of the main rivers of Punjab, originates from 'Mansarovar Lake' in Tibet, enters Punjab near bhakra, flows through this state to finally join the river Beas at Harike. During its course, it gets heavily polluted by domestic sewage, agricultural run off and untreated or inadequately treated industrial effluents. The industrial effluents discharged generally in the form of colloids, suspensions and particulates, contain heavy metals which are ubiquitous and are cumulative poisons in both aquatic and terrestrial ecosystems (Midigosky et al, 1991). Studies on heavy metal pollution in aquatic systems indicate that these get incorporated in various components of ecosystems, especially in water, sediment and biota. (Veith et al 1972, Dixon and Sprague, 1981 and Whitton et, al, 1982). Of all the heavy metals, Zn, Pb, Hg, Cd and Cr are considered to be of principle interest in pollution biology. (Hellawell, 1986). Therefore, the concentrations of these heavy metals were determined in water and sediment of the river Satluj in Punjab.

Material and Methods

For the present studies, nine collection sites were selected on the river Satluj and its tributaries in Punjab. (Fig. 1) The first site was selected at Nangal reservoir where the water is without any polluting impregnation. The second site was selected at Nangal downstream which is about 4 km downstream from the first site. Prior to this collection site, the river receives effluents of the Panjab Nitrate Fertilizer Corporation (PNFC) and National Fertilizer Limited (NFL). The third site was selected about 14 km downstream of the second site at Garhshankar bridge. Here, the river is shallow and is without any known industrial pollution other than domestic sewage and agricultural run off. The fourth site was selected at Roper reservoir, a deep water reservoir formed due to the barraging of the river. The Ropar Thermal plant is located at about 5 km upstream. Here too, the industrial pollution is abesent. The fifth and sixth sites were selected on the Budha Nallah, a tributary of Satluj river, into which the effluents from industrial units of Ludhiana are poured in. The fifth site was located near the Central Jail before it enters the city and sixth was located near Hambran village after it crosses the city. The seventh site was selected on the river Satluj after it receives Budha Nallah. The eighth site was selected near the village 'Gidderpindi', at a place where east Bein (another tributary of river Satluj) meets the river Satluj. The East Bein brings industrial and domestic wastes from Jalandhar. The last site (Site IX) was selected at Harike reservoir. Here, the river Satluj confluences with river Beas.

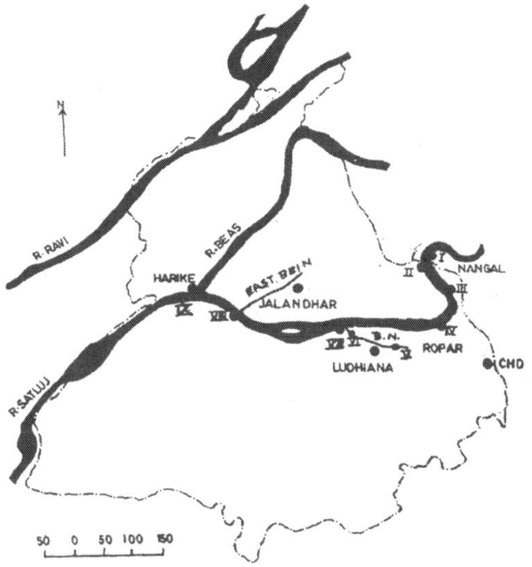

Figure 1. Map Showing the Sampling Sites.

For sample collection, Six trips were made during a period of one year (September, 1994 to August 1995). Each sample consisted of one litre of water to which 10 ml of 6N HNO_3 was added on the spot and brought to the laboratory for further analysis. 500 ml of each sample was further acidified with 5 ml of 6 N $HN0_3$, concentrated to 50 ml and processed for analysis of heavy metals as per APHA (1989) procedures. Sediment samples were dried at 40^0 C. ground and passed through a 100 mesh sieves. 10 gm was digested with 40 ml of the acid mixture (HNO_3 , H_2SO_4 : 3 : 1) for half an hour, cooled decented, Made upto 50 ml with distilled water and used for heavy metal analysis as per APHA (1989).

Results and Discussion

The average concentrations of Zn, Pb, Hg, Cd and Cr in water and sediment at all the sites are given in Tables 1 and 2.

TABLE : 1

Concentration of heavy metals in Water in river Satluj (mg./L)

S. No.	Parameters/ Sites	Site-I	Site-II	Site-III	Site-IV	Site-V	Site-VI	Site-VII	Site-VIII	Site-IX
1.	Zinc	NIL	12.0	NIL	NIL	NIL	20.15	10.12	0.25	0.06
2.	Lead	NIL	0.10	NIL	NIL	NIL	2.0	1.8	0.08	0.06
3.	Mercury	NIL	NIL	NIL	NIL	NIL	.005	.006	.002	.001
4.	Cadmium	NIL	NIL	NIL	NIL	NIL	NIL	NIL	NIL	NIL
5.	Chromium	NIL	NIL	NIL	NIL	NIL	NIL	NIL	NIL	NIL

TABLE : 2

Concentration of heavy metals in sediments of river Satluj (mg./kg.)

S. No.	Parameters/ Sites	Site-I	Site-II	Site-III	Site-IV	Site-V	Site-VI	Site-VII	Site-VIII	Site-IX
1.	Zinc	NIL	12.0	NIL	NIL	NIL	23.80	13.00	0.18	0.25
2.	Lead	NIL	0.12	NIL	NIL	NIL	2.00	2.00	0.06	0.06
3.	Mercury	NIL	NIL	NIL	NIL	NIL	.008	.007	.002	.001
4.	Cadmium	NIL	NIL	NIL	NIL	NIL	NIL	NIL	NIL	NIL
5.	Chromium	NIL	NIL	NIL	NIL	NIL	NIL	NIL	NIL	NIL

At site I, heavy metals were not at all detected both in water and sediment. Absence of heavy metals at site I is due to pollution free water in this reservoir. At site II, Pb and Zn were present while Hg, Cd and Cr were absent both in water and sediment. At this site effluents from NFL and PNFC are added. These results depict that effluents from fertilizer units add Zn and pb to water. The concentration of Zn was found to be within permissible limit but that of Pb was observed to cross the maximum limit as suggested by W.H.O. (Table I).

At site III all the heavy metals were found to be absent. Although sewage and agricultural run off are added along the course of water but pollution. of industrial origin at this site is none other than that added at the site II. This clearly shows that Pb and Zn added at site II. settle down there only and also that sewage and agricultural run off do not add heavy metals to the river water. Howeverit is well known that trace amounts of metals such as Cr, Zn and Pb are present in domestic sewage. Manivasakam 1984). Similarily, at sites IV and V no heavy metals were detected due to absence of any industrial effluent.

At site VI, Which receives effluents from industrial town of Ludhiana, Zn, Pb and Hg were detected while Cr and Cd were found to be absent. The concentration of all these three heavy metals crossed the maximum permissible limit. (W.H.O., 1984). Manivasakam (1984), Mishra (1990) and Kannan (1991) have also suggested that heavy metals are mostly found in effluents poured in from industrial units. At site VII, which is located after the confluence of Budha Nallah with Satluj river, Zn, Pb and Hg were present while Cd and Cr were absent both in water and sediment. Here the concentration of Zn decreased to lie within permissible limit. Concentration of Pb, too, declined slightly in water but still it remained above permissible limit. The concentration of Hg, however, increased in water but decreased in sediment and crossed the limits in both water and sediments.

At site VIII, where industrial effluents from Jalandhar city is brought in by the east bein, Zn, Pb and Hg were found to be present both in water and sediment. The concentration of both Zn and Pb were within permissible limits while the level of Hg was higher then permissible limits. At site IX also, Zn, Pb and Hg were present while Cd and Cr were absent both in water and sediment. The concentration of Zn and Pb remained within permissible limits while that of Hg crossed the limit.

In gennal terms, concentration of heavy metals was found to be considerably higher in sediments than in water. Similar observations were recorded by Kreuskopf (1955), Gurits (1966) and Singer (1977). Krauskof (1955) has suggested that the heavy metal concentration increases in the sediment due to the adsorption of cations by organic matter present in the sediment layers. Similarily Curits (1966) and Singer (1977) have also suggested that metals like Cu, Cr. interact with organic matter in aqueous phase and settle down resulting in high concentrations in sediments.

The high concentrations of the metallic components pose a threat to the survival of a vest majority of aquatic biota, as well as causing health hezards to the human beings. It is, therefore, suggested that the industrial effluents should be properly teated before it is allowed to enter and mix with the natural water resources.

Acknowledgement

The authors are grateful to P.S.C.S.T. and CSIR for financial assistance.

❑ REFERENCES

APHA (1989), *Standard Methods for the Examination of water and waste water.* 17th Ed. American Public Health Association, Washington, U.S.A.

Curits C.D . (1966). The incorporation of soluble organic matter into sediments and its effects on trace elements assemblages, In *Advances in Organic Geochemistry,* Hobson G.D. and Lours M.C., (Eds.) Oxford, Pergamon, pp 1-13.

Dixon D.G. and Sprague J.B. (1981). Copper bioaccumulation of hepatoprotein synthesis during acclimation to copper by juvenile rainbow trout. *Aquat. Toxicol.* 1 : 69-81.

Hellawell J.M. (1986). *Biological indicators of freshwater pollution and Environmental Management.* Elsevier applied Science London and New York.

Kannan K. (1991). *Fundamentals of Environmental Pollution* S. Chand and company Ltd. Ramnagar New Delhi.

Krauskopf K.B. (1955). Sedimentary deposits of rare metals. *Econ. Geol.* 50 : 411-463.

Manivasakam N. (1984) *Environmental Pollution.* National Book Trust, India. pp. 58

Midigosky S.R., Hernandez X.A. and Glass J. (1991). Lead, Cadmium and Aluminium accumulation in the red swamp crayfish *Procambarus clerkii* G. collected from roadside drainage ditches in Louisiana. *Arch.Environ. Contan. Toxicol.* 20 : 253-258.

Mishra P.C. (1990). *Fundamentals of Air and Water Pollution.* Ashish Publishing House, New Delhi.

Singer P.C. (1977). Influence of dissolved organics on the distribution, transport and fate of heavy metals in aquatic systems, *Fate of pollutants in the air and water environment, Part-I,* Suffet, I.H. (Ed.) New York, pp. 155-182.

Veith G.D., Defor D.L. and Bergastedt B.V. (1979). Measuring and estimating the bioconcentration factor of chemicals in fish. *J. Fish. Res. Bd. Can.* 36 : 1040-1048.

Whitlon B.A., Say P.J. and Jupp B.P. (1982). Accumulation of Zinc, Cadmium and Lead by the aquatic liverwort *Scapnia. Environ. Pollut. Ser. B.* 3 : 299-316.

World Health Organisation (1984). Guidelines for drinking water quality. W.H.O. Geneva.

Authors :

1. H.Kaur, S.S. Dhillon, G.Mander and K.S. Bath : Department of Zoology, Punjab University, PATIALA - 147 002. Punjab (India)

17

Some observations on Phytoplankton around Kakrapar Atomic Power Station

❑ INTRODUCTION

It is well documented in the literature that phytoplankton is a source of food in aquaculture industry. Harrison *et al.*, (1990) described the importance of phytoplankton study, as most groups of algae have great value in the food chain. It is known that natural aquatic bodies containing microphytic and macrophytic vegetations react to a variety of pollutants. Somashekhar (1988) has identified number of microphytic forms that can be used as biological indicators. Many varieties of algae have been successfully employed to detect europhication and assess the water quality. In addition, several workers (Besch *et al., 1972;* Bahl 1973; Schoeman, 1976; Venkateshwarlu, 1983) have notably used diatoms as indicators of water quality.

In the present piece of work three major groups of phytoplankton were studied qualitatively and quantitatively from seven selected spots from upstream to downstream of Tapi river near Kakrapar Atomic Power Station (KAPS) which is located about 80 km from Surat in South Gujarat. The aim of the study was to find out the qualitative and quantitative status of micro-vegetation in the area of KAPS which may be compared after a period of time to check if there is any threat to these organisms.

MATERIAL AND METHODS

a. Sample collection

Samples were collected from seven spots at different locations near KAPS from upstream to downstream of Tapi river (for details please refer the map) from October 1991 to April 1993. One hundred liters of sub surface composite samples were collected in the morning around 8 O'clock and filtered with plankton net of 75 mesh size. Samples were concentrated to 100 ml and kept in clear polythene bottles. Sampling was done in four seasons namely monsoon, autumn, winter and summer and preserved in 4% formalin with few drops of Lugol's solution added at the collection site.

b. Quantitative estimation

The quantitative estimation was done by Lackey's drop method (Lackey, 1938, Vollenweider, 1969) as described by Trivedy and Goel (1986).

c. Qualitative estimation

Identification of phytoplankton upto generic level was done using the reference of Bold and Wynne (1978), APHA (1976) and Fritsch (1965).

RESULTS AND DISCUSSION

a. Quntitative estimation

Quantitative estimation of phytoplankton included analysis of diatom (*Bacillariophyceae*), green algae (*Chlorophyceae*) and blue green algae (*Cyanophyceae*). Quantity of plankton was expressed as number per ml. (Table-1.)

Total number of phytoplankton (diatom, green and blue green algae) was found maximum during winter season from all the sports except spot no. 5. However, minimum number was recorded in summer season from spots. 1, 3, 4 and 5. Spots 4, 5 and 7 were more productive with higher counts of phytoplantkton throughout the year whereas spot 3 showed minimum number.

Table 1

Quantitative estimation of phytoplankton (Number/ml)

Spot No.	Seasons	Diatoms	Blue green Algae	Green Algae	Total
1.	M	24.5	4.1	0.6	29.2
	A	17.5	14.7	0.0	32.2
	W	36.4	11.2	0.7	48.3
	S	20.3	2.8	0.0	23.1
2.	M	4.5	9.0	0.0	13.5
	A	30.2	4.9	3.5	38.6
	W	32.9	18.9	2.1	53.9
	S	17.5	18.9	2.1	38.5
3.	M	4.0	6.2	2.8	13.0
	A	7.0	7.0	0.7	14.7
	W	21.0	3.5	0.0	24.5
	S	6.3	0.0	0.0	6.3

4.	M	36.8	12.6	13.0	62.4
	A	63.7	18.9	2.1	84.7
	W	107.5	11.9	15.4	134.8
	S	24.5	23.8	3.5	51.8
5.	M	52.4	29.4	20.5	102.3
	A	35.0	74.3	0.0	109.3
	W	40.6	25.9	12.6	79.1
	S	17.5	21.0	2.8	41.3
6.	M	11.6	1.0	0.6	13.2
	A	23.1	7.0	1.4	31.5
	W	48.3	12.6	0.0	60.9
	S	18.2	2.8	0.0	21.0
7.	M	41.8	2.8	4.9	49.2
	A	23.1	4.9	0.0	28.0
	W	45.5	29.4	29.4	104.3
	S	35.0	7.0	20.3	62.3

M = Monsoon, A = Autumn, W = Winter, S= Summer

In addition, diatoms were predominantly higher in all the spots during winter. However, the number of green algae was low as compared to diatoms and blue green algae in all the four seasons.

qualitative analysis

Total thirty-seven types of phytoplankton were identified upto generic level. The list of the phytoplankton is shown in Table-2. Three main categories of algae namely diatoms, green and blue green were found from all the spots.

There were fourteen genera among the diatoms, sixteen from green and seven from blue green algae. *Bacillaria, Navicula and Fragilaria* were the dominating genera among the diatoms. Similarly among blue green algae *oscillatoria* was common.

In the present investigation, the quantitative and qualitative analysis of phytoplankton was undertaken. We reported total thirty-seven genera of algae from *Bacillariophyceae, chlorophyceae* and *cyanophyceae* groups from seven spots around KAPS. All seven spots were located from upstream to downstream area of Tapi river. Dubey (1993) had carried out hydrobiological survey in Narmada basin. He reported thirty six species of phytoplankton of the same three groups, whereas Rao (1993) identified fifty seven species of *Bacillariophyta, chlorophyta* and *cyanophyta* from Himayat Sagar, Andhra Pradesh.

Table 2
List of Phytoplankton

No.	Name	Class
1.	*Bacillaria sp.*	Bacillariophyceae
2.	*Diatoma sp.*	Bacillariophyceae
3.	*Epithemia sp.*	Bacillariophyceae
4.	*Fragilaria sp.*	Bacillariophyceae
5.	*Gomphonema sp.*	Bacillariophyceae
6.	*Gyrosigma sp.*	Bacillariophyceae
7.	*Hantzschia sp.*	Bacillariophyceae
8.	*Melosira sp.*	Bacillariophyceae
9.	*Navicula sp.*	Bacillariophyceae
10.	*Nodularia sp.*	Bacillariophyceae
11.	*Nitzschia sp.*	Bacillariophyceae
12.	*Pinnularia sp.*	Bacillariophyceae
13.	*Pleurosigma sp.*	Bacillariophyceae
14.	*Synedra sp.*	Bacillariophyceae
15.	*Chara sp.*	Chlorophyceae
16.	*Chlorella sp.*	Chlorophyceae
17.	*Chlorogonium sp.*	Chlorophyceae
18.	*Chracium sp.*	Chlorophyceae
19.	*Cladophora sp.*	Chlorophyceae
20.	*Cosmarium sp.*	Chlorophyceae
21.	*Euglena sp.*	Chlorophyceae
22.	*Microspora sp.*	Chlorophyceae
23.	*Mougeotia sp.*	Chlorophyceae
24.	*Oedogonium sp.*	Chlorophyceae
25.	*Pediastrum sp.*	Chlorophyceae
26.	*Scenedesmus sp.*	Chlorophyceae
27.	*Sphaeroplea sp.*	Chlorophyceae
28.	*Spirogayra sp.*	Chlorophyceae
29.	*Spondylomorum sp.*	Chlorophyceae
30.	*Ulothrix sp.*	Chlorophyceae
31.	*Anabaena sp.*	Chlorophyceae
32.	*Gleotrichia sp.*	Chlorophyceae

33.	*Lyngbya sp.*	Chlorophyceae
34.	*Merismopedia*	Chlorophyceae
35.	*Nostoc sp.*	Chlorophyceae
36.	*Scytonema sp.*	Chlorophyceae
37.	*Oscillataria sp.*	Chlorophyceae

In the present study, distinct seasonal variation was noted in the number of diatoms. It was high during the winter season and could be correlated with high silicate content as reported by Patel (1995). It has already been established that silicate is an important factor which plays role in the growth of diatom. In an another study coducted by Saxena (1987) also confirmed that the number of diatoms was directly proportional to the silicate content.

In addition, present study revealed that spots 4 and 5 were more productive as compared to other spots with reference to phytoplankton. this could be due to the stagnant nature of the water from these spots. Sinha and Sinha (1993) made similar observations with stagnant water bodies and noted high growth of aquatic vegetation.

Acknowledgement
We acknowledge with thanks the financial assistance from Nuclear Power Corporation, Bombay. We also acknowledge the help rendered by Dr. G. Ragothaman, Head, Department of Aquatic Biology, South Gujarat University, Surat.

❑ REFERENCES

American Public Health Association. 1976. *Standard methods for examination of water and waste water*. Fourteenth Edition, Washington D.C.

Bahl L.L. 1973. Diatom community responses to primary waste water effluent. *J. Water Pollut. Control. Fed.* 45 : 134-144.

Besch W.K., Richard M. and Cautin R. 1972. Benthic diatoms as indicators of marine pollution in the North west Miramichi river system, New Brunswick, Canada. *Int. Rev. Gas Hydrobiol.* 57 : 39-74.

Bold H. C. and Wynne M.J. 1978. *Introduction to algae, Structure and reproduction.* Prentice Hall of India Pvt Ltd., New Delhi.

Dubey G.P. 1993. Status (Prior to the year 1980) of Hydrobiology, Fisheries and Socio-economic condition of fishermen in Narmada basin. In : *Recent advances in fresh water biology*. Vol-II (Rao K.S., 1993), Anmol Publication Pvt. Ltd. New Delhi-2. pp. 139-161.

Fritzch F.E. 1965. *The structure and reproduction of the algae* Vol. & II. Cambridge

University Press, London.

Harrison P.J., Thompson P.A. and Calderwood G.S. 1990. Effects of nutrient and light limitation on the biochemical composition of phytoplankton. *J. Appl. Physico.* 2 : 45-46.

Lackey J.B. 1938. The manipulation and counting of river plankton and changes in some organisms due to formalin preservation. *Pub. Health Report* 53:2080.

Patel P.B.1995. *Some observations on aquatic environment around Kakrapar Atomic Power Station* - Ph.D. Thesis submitted to the South Gujarat University, Surat, (India)

Rao M.B. 1993. Studies on the ecology and fish fauna of an oligotrophic lake Himayat Sagar, Hyderabad, Andhra Pradesh. In : *Recent advances in fresh water biology* Vol.-II (Rao K.S. 1993). Annual Publication Pvt. Ltd., New Delhi. pp.73-98.

Saxena M.M. 1987. *Environmental analysis, water soil and air.* Agro Botanical Publisher India. pp. 67.

Schoeman E.R. 1976. diatom indicator groups in the assessment of water quality in the Juksei-Crocodile river system. *J. Limm. Soc. S. Africa* 211 : 21-24.

Sinha K.K. and Sinha D.K. 1993. Observation on Macrophytes associated Benthic Macrofauna in some fresh water ponds of Munger (Bihar). *J. Ecobiol.* 5(2) : 89-93.

Somshekhar P.K. 1988. On the possible utilization of diatoms as indicators of water quality - A study of river cauvery In : *Ecology and Pollution of Indian river* Edited By Trivedy R.K. Ashish Publishing House, New Delhi. pp.375-382.

Trivedy R.K. and Goel P.K. 1986. *Chemical and Biological methods for water pollution studies.* Environmental Publication, Karad, India pp.1-34.

Venkateswarlu V. 1983. Taxonomy and ecology of algae in the river Moosi, Hyderabad, India. II, Bacillariphyceae. *Bibliotheca Physiologica* 66 : 1-42.

Vollenwider P.A., 1969. *A manual of methods for measuring primary production in Aquatic Environment* IBP Hand Book No. 12, Blackwell Scientific Publications.

Authors :

1. **M.Gadhia** : Department of Aquatic Biology, South Gujarat University, SURAT-395 007. Gujarat (India)

2. **P.B. Patel, S. Shaikh** : Department of Biosciences, South Gujarat University, SURAT- 395 007. Gujarat (India)

3. **P.K.Gadhia** : Department of Biosciences, South Gujarat University, SURAT-395 007. Gujarat (India)

■

18

Biomonitoring of Pollution in the Ganga River Uttar Pradesh

❑ INTRODUCTION

Pollution of Ganga River due to various human activities has become a matter of concern during early 80's (Krishnamurti *et al.*,1991). The Govt. of India has taken up an Action plan for prevention of pollution. Under the Action Plan considerable work has been done to control the pollution in the Ganga river. For proper management of aquatic ecosystem and water resources, the biological monitoring is essential. According to Cairns (1982), biomonitoring is often needed as a support to chemical monitoring because toxicity can not be tested without biota and biological response may be elicited by chemicals below analytical detection limits. The biota reacts to the pollution by translating chemical dosage to well defined biological response. Thus biological assessment techniques lead to information on the quality, condition and quantity of the streams. Since chemical analysis of water or sediments alone cannot give a quantitative idea of the amount, history, consequences of pollution biomonitoring is all the more important (Chawla and Vishwanatham, 1989). To assess the river quality studies have been initiated in the Ganga river in a stretch between Rishikesh and Kanpur to identify the status of higher vertebrates in the river, which will enable to indicate the quality and the biological restoration of Ganga river.

Study Area

Present study has been carried out in a stretch between Rishikesh and Kanpur in Uttar Pradesh (Fig. 1). The total length of the river under study is approx. 645 km. For monitoring of the river quality 8 sampling station have been selected (Table1). They are Rishikesh, Haridwar, Bijnor, Garmukteswar, Narora, Kachlaghat, Farrukabad, Suklaganj at Kanpur. River water from these stations was analyzed for Physico-chemical parameters. Planktons, and benthic fauna samples were also collected from these stations. Data on higher vertebrates were collected from these stations also from different suitable habitats in the study area.

Table 1

Details of sampling stations in the study area.

Sampling Station	District	Distance (km)	Sampling Point	Activities of Concern
Rishikesh	Dehradun	0	Downstream Muni ki Reti	CB, CR, DM, WR
Haridwar	Haridwar	29	Upstream barrage	CB, CR, WR.
Bijnor	Bijnor	129	Upstream Barrage	CR, AG, WR
Brijghat	Gaziabad	213	Below Road bridge	CB, CR, DM, AG
Narora	Bulundsahar	295	Upstream barrage	CB, CR, DM, AG, WR
Kachlaghat	Badam	362	Below Rail/Road bridge	AG, CB, CR, DM, ID
Ghatia ghat	Farrukhabad	485	Below Road bridge	CB, CR, DM
Kanpur	Kanpur	645	Below Sukhlaganj Rail bridge	AG, CB, CR, DM, ID

AG = Agricultural runoff sewage, DM = Point / Non-point domestic
CB = Community bathing effluents, ID = Point / Non-point Industrial
CR = Cremation & Post cremation Activity, WR = Water regulation

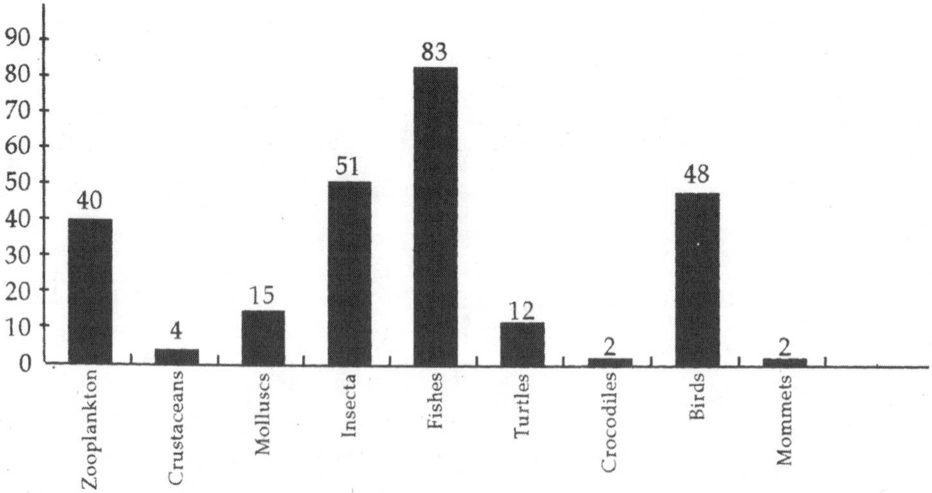

Figure 1. Biodiversity in the Ganga river between Rishikesh and Kanpur.

RESULTS AND DISCUSSION

Pollution Studies

The data on physico-chemical parameters at different sampling stations are shown in Table 2. Analysis of the data revealed that pollution load is still much higher in the lower Ganga near Kanpur than in the Upper Ganga. There are no major industries releasing any effluents into the river in this stretch except at Rishikesh and Haridwar, which release treated effluents. Disposal of human dead bodies into the river is still continuing at Kanpur. Due to the diversion of water from 4 barrages at Rishikesh, Haridwar, Bijnor and Narora quantity of water in the main river is very much reduced, which may not be sufficient for self purification of the river.

Biodiversity Studies

Data on Biodiversity in the study stretch of the Ganga river is represented in figure 2. The stretch of the Ganga river between Rishikesh and Kanpur is a good habitat for large number of vertebrate species. Although reduced due to human activities, breeding grounds of Mahseer fish are available in the upper stretches of the river near Haridwar. Migration / dispersal of Mahseer up to Narora is recorded. this fish is frequently available at Narora.

The turtle species diversity in this stretch is very rich. A total of 12 species of freshwater turtles are present. The *Kachuga* sp. are dominated with 5 species (*Kachuga smithii, K. tecta, K. tentoria, K. dhongoka* and *K. kachuga)* followed by two species of *Aspideretes* (*A. gangeticus* and *A. hurum)* and one species of *Chitra indica, Lissemys Punctata, Hardella thurjii, Geoclemys hamiltoni* and *Melanochelys trijuga.* Among them larger species are highly used by locals as food and smaller species are facing problems of drowning death in fishing nets. The present study has created greater awareness of the turtle conservation among the local people.

Two species of crocodiles *Gavialis gangeticus* and *Crocodylus palustris* occur in this stretch of the Ganga river. However, their population are very much reduced due to killing and habitat distraction. One adult female Gharial has been killed at Anupsahar. The Uttar Pradesh Forest Department released a total of 225 captive reared Gharial in the upstream of Bijnor. The area is protected under the Hastinapur Wildlife Sanctuary. During the monitoring studies 2 gharial and 2 mugger were sighted. Although the river stretch is most suitable for both the species of crocodiles, gharial will be thrive more successfully as this species never comes in conflict with human.

During the surveys data on avian species along the study area was collected. The river is a good habitat for a variety of migratory birds (more than 50 species) and resident birds, which feed on fishes and aquatic vegetation. Presence of such rich avifauna indicates good productivity of the river.

Table 2

Summary of some important physico-chemical parameters to show pollution status at different sampling stations in the Upper Ganga river.

Stations/ Parameters	Pollution Domestic/ Industrial	pH	DO (mg/1)	BOD (mg/1)	TDS (mg/1)	COD (mg/1)
Rishikesh	Domestic	7.48 (7.04-8.01) SD 0.27	8.36 (6.52-10.26) SD 1.23	2.34 (0.64-4.50) SD 1.49	84.7 (46.3-140.0) SD 27.3	13.9 (4.6-25.6) SD 7.0
Haridwar	Domestic/ Industrial	7.86 (7.05-8.52) SD 0.44	8.19 (6.40-10.32) SD 1.11	2.75 (1.76-5.0) SD 1.08	75.8 (52.5-93.5) SD 14.6	14.5 (5.6-31.6) SD 8.6
Bijnor	-	7.7 (7.05-8.65) SD 0.40	7.12 (5.50-8.60) SD 0.82	2.27 (0.32-5.55) SD 1.35	111.5 (61.0-169.5) SD 38.5	21.6 (6.0-58.8) SD 18.1
Brij ghat	Domestic	7.86 (7.10-8.70) SD 1.69	7.04 (4.85-9.56) SD 1.66	1.98 (0.32-3.60) SD 1.07	85.8 (65.0-110.0) SD 13.2	19.1 (6.4-43.2) SD 11.7
Narora	Domestic	7.66 (7.00-8.30) SD 1.22	6.80 (4.60-9.09) SD 1.53	2.45 (0.68-6.00) SD 1.64	90.0 (67.0-111.3) SD 33.5	18.7 (6.4-31.8) SD 8.2
Kachla ghat	Domestic/ Industrial	7.73 (6.90-8.20) SD 0.41	6.83 (4.77-8.52) SD 1.29	2.70 0.60-9.60) SD 2.41	100.0 (75.0-131.3) SD 33.3	27.0 (4.8-66.2) SD 20.0
Ghatia ghat	Domestic/ Industiral	7.66 (6.36-8.63) SD 3.31	6.83 (5.04-8.46) SD 1.04	1.99 (1.00-3.50) SD 0.85	114.8 (80.6-154.3) SD 24.1	20.7 (4.0-38.8) SD 11.6
Kanpur	Domestic/ Industrial	7.66 (6.86-8.50) SD 0.68	5.24 (3.00-7.01) SD 1.35	6.36 (2.32-18.5) SD 5.48	188.0 (101.0-357.6) SD 76.1	36.0 (8.0-61.8) SD 15.7

The tolerance limits of pollution discharge :

WHO	6.5-9.2	-	6.0	500	10
ISI standards	6.9	3.0	3.0	-	-

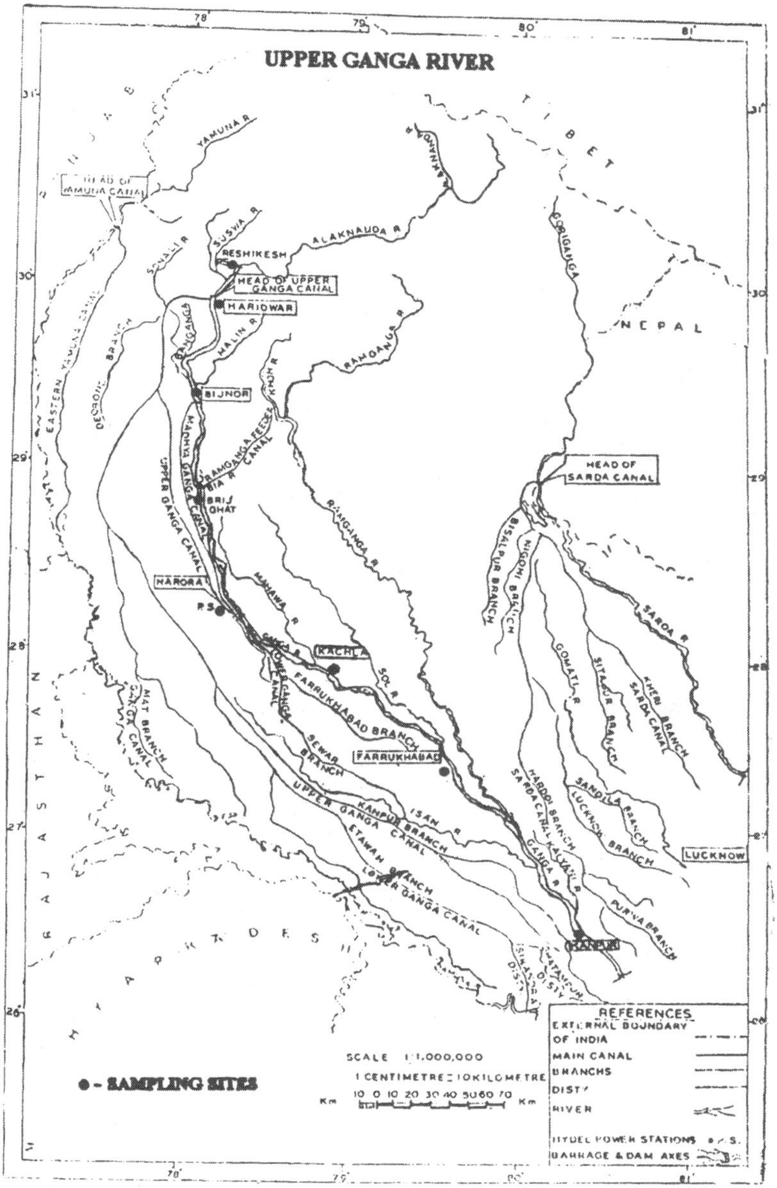

Figure 2. Upper Ganga River

Of the two species of freshwater mammals, dolphin *Platansta gangetica* is present in this stretch. This species was thought to be extinct from this stretch. Although the population of dolphin is not very large, the species is available from the downstream of Bijnor barrage. Brijghat at Garmukteswar is one of the good dolphin areas in this stretch. This species is also facing problems by large scale fishing operations. There are reports of mortality of 3 dolphins during the study period. One dead dolphin was collected at Narora. The skull is deposited in the Museum of School of Studies in Zoology, Jiwaji University, Gwalior.

Although local people along the study area reported the occurrence of Otters in the river, no otter was sighted during the study period.

Conservation Strategies

In the Ganga river due to increase in the human population their need for exploitation of the natural resources is also very much increasing. The continued exploitation of fish with untraditional capturing methods, turtles and migratory birds and degradation of habitat is alarming. If anything is to be done to the protection of different species and their habitat in the Upper Ganga river, the following steps need to be taken with priority.

1. Different sites along the Ganga like Haridwar, Balawali, Bijnor, Garmukteswar, Narora, Bittor and Sukhlaganj are worth their potentiality to support large Biodiversity. These areas should be declared as protected areas. In future at least one or two sanctuaries should be created in this stretch, preferably at Narora and Bittor.

2. Indiscriminate fishing practices should be stopped. The fishing methods should be modified to avoid drowning deaths of large vertebrate species such as turtles, crocodiles, and dolphins and also to allow growth of fish population.

3. Habitat degradation problems like water development programs, pollution from both domestic and industrial discharges should be strictly controlled.

4. There is a need for the Government and NGO's to take up widespread conservation and education programs in this area to highlight the problem of pollution and decrease of populations of different species of aquatic animals.

Acknowledgement

We are thankful to the Ganga Project Directorate for financial assistance.

❑ REFERENCES

Cairns J. 1982. *Biological monitoring of water pollution*. Pergamon Press.

Chawala G. and Viswanatham P.N. 1989. Role of predictive ecotoxicology and biological monitoring in the Management of aquatic ecosystems. Ed. Agrawal V.P., *et. al.*, Publ. society of Bioscience, Muzaffarnagar.

Krishnamurti C.R., Bilgrami K.S., Das T.M. and Mathur R.P., 1991. The Ganga A scientific study. Publ. Northern Book Centre, New Delhi.

Authors :

1. **R.J. Rao, B.K. Sahu, S.K. Behera and R.K. Pandit** : School of studies in Jiwaji University, GWALIOR- 474 011. M.P.(India)

19

Physico-Chemical and Ecological Study of River Ganga Near Barauni Industrial Area

❑ INTRODUCTION

Ganga being the lifeline of the region, it's quality and ecology has been attracting the attention of masses and is a matter of great concern. As the main drainage channel of the region,Ganga receives pollutional loads of domestic and industrial wastewaters as also agricultural run-off and return waters. Among the several industrial complexes / Congregations, that have developed along the Ganga, Barauni is one of the largest and the most polluting among those located in the state of Bihar. Most of the industries operating at the Barauni industrial area discharge their effluents in to river Ganga. The main industries which emptyout their effluents in to Ganga at Barauni include Barauni Refinery, Hindustan Fertilizer Corporation, Barauni Thermal power plant, Bata Industries Ltd. and McDowell Distillery etc. It was apprehended that the water quality and the biological productivity of Ganga may be significantly affected by these discharges and hence a study was carried out to assess the impact of effluents generated from the Barauni industrial discharges on Ganga water quality and ecology.

Methodology

Selection of sampling stations

To assess the quality of Ganga 5 stations covering a 20 km stretch were selected. Station 1 was selected 5 km upstream of Rajendra bridge, near Dharampura diara. No significant industrial or urban effluents are being discharged at this station or for a long distance upstream of it, so it was considered as background or reference station. Station 2 was selected near Rajendra bridge. This was slightly down stream of McDowell and Bata effluents discharge points the two major polluters located on right bank of Ganga. Station 3 was selected 4.5 km downstream of Rajendra bridge which was 0.5 km downstream of Hindustan Fertilizer Corporation outfalls. Station 4 was selected 5.2 kms downstream of Rajendra bridge which was 0.2 km downstream of the Barauni Refinery efluent out fall, and is expected to indicate the immediate impact of refinery effluent. Station 5 was selected at 15 km downstream of Rajendra bridge near Sihma diara, by which distance the river might have fully recovered from the effects, if any, of above mentioned pollution inflows. The location of industries and sampling stations are given in Figure 1.

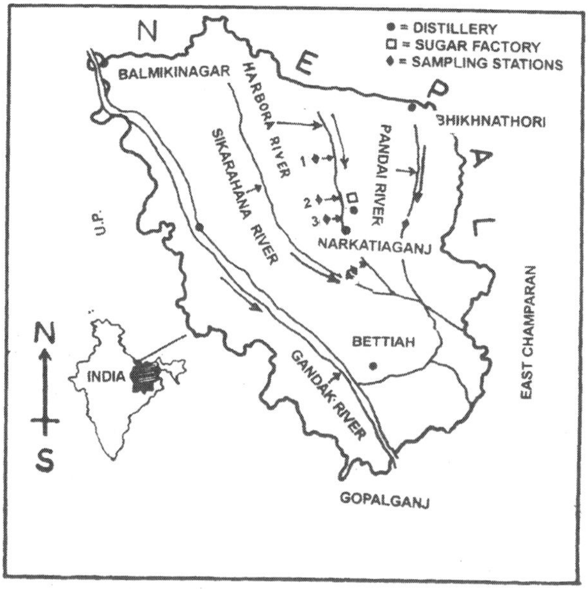

Figure 1. Location of Industries and sampling stations.

Analytical Techniques

Physical and Chemical parameters

To assess river quality, sampling was done at earmarked stations during three major seasons viz. pre-monsoon, post-monsoon and winter. Water samples were collected in polythene jerri-cans and analysed as per APHA Standard Methods. Parameters like pH, temperature, alkalinity, Hardness, transparency and dissolved oxygen were analysed on the spot.

Bio-productivity

Since the photosynthetic rate as measured by production of oxygen by the phytoplanktons reasonably measures the primary biological productivity of a water-body and is a better parameter for the real biological productivity than any measurements of bio-mass, this (viz oxygen production) was adopted as the indicator for bio-producitivity in this study. Thus bio-productivity at the different stations and in different seasons was estimated on the basis of dissolved oxygen produced following Light and Dark bottle techniques.

Table 1

Physico-chemical characteristics of river Ganga near Barauni Industrial area.

Sl No.	Parameters	Pre-Monsoon					Post-Monsoon					Winter				
		S1	S2	S3	S4	S5	S1	S2	S3	S4	S5	S1	S2	S3	S4	S5
1.	Air temperature, ^0C	30.0	35.0	37.0	38.0	37.0	33.5	29.0	30.5	30.5	30.0	24.0	22.0	24.0	23.0	23.0
2.	Water temperature, ^0C	29.0	30.0	31.0	31.0	31.0	29.0	27.5	28.0	30.5	29.5	19.0	20.0	21.0	22.0	22.0
3.	Transparency cm	15.0	15.0	20.0	17.0	18.0	11.0	11.0	13.0	15.0	16.0	59.0	41.0	32.0	43.0	53.0
4.	pH	8.8	8.7	8.6	8.6	8.8	8.4	8.5	8.5	8.5	8.7	8.6	8.7	8.8	8.8	8.8
5.	Electrical Conductivity, μmhos/cm	243	242	259	270	247	294	263	258	474	259	285	295	296	410	301
6.	Chlorides, mg/1	13.6	13.6	13.4	13.3	15.0	12.8	11.9	12.2	16.4	14.2	21.6	21.3	16.8	13.3	20.2
7.	Sulphates, mg/1	24.3	28.8	27.5	22.5	23.0	13.0	15.8	18.5	28.5	20.0	13.0	14.5	20.0	21.5	13.3
8.	DO, mg/1	7.9	7.3	7.4	7.0	7.5	6.4	6.8	6.8	6.8	7.2	7.6	8.3	9.2	10.4	8.8
9.	BOD, mg/1	2.4	6.0	5.9	7.8	3.8	0.3	0.3	0.2	0.6	0.7	1.2	1.8	2.0	3.8	1.3
10.	COD, mg/1	8.0	17.6	16.6	29.6	12.6	4.0	16.0	4.0	8.0	4.0	7.2	3.6	3.6	7.2	3.6

Results and Discussion

The analysed seasonwise results for various physicochemical parameters are given in Table 1. The results of productivity is given in Table 2. From these tables following inferences can be drawn.

Table 2

Primary-productivity (mg C/m^2/hr) at different selected stations of river Ganga

Seasons	Parameters	S1	S2	S3	S4	S5
Pre-monson	Gross primary Productivity	101.2	100.9	98.2	95.3	101.1
	Net primary productivity	70.5	70.1	61.3	62.7	70.7
	Respiration	30.7	30.8	36.9	32.6	30.4
Post-monsoon	Gross primary productivity	23.2	22.2	19.9	22.1	22.9
	Net primary productivity	15.2	14.9	10.9	13.5	15.1
	Respiration	8.0	7.3	9.4	8.6	7.8
Winter	Gross primary productivity	63.2	61.2	58.0	55.1	62.5
	Net primary productivity	38.5	37.8	35.2	31.5	37.1
	Respiration	24.7	23.4	23.8	23.6	25.4

I. Dissolved oxygen was found to be satisfactory at all stations through all seasons (all values were close to or higher than 7 mg/1). During winter and premonsoon, when the water level and flow were low and hence could become critical for the dilution of industrial effluents. the DO and pH were still found to be satisfactory for the health of river ecosystem.

II. During post-monsoon and winter the sulphate concentration increased from S1 to S4 and decreased between S4 and S5. This increase appeared to have been due to industrial discharges. The increase between S1 and S2 was probably due to the discharges of McDowell and Bata industries and around 4.5 mg/l in premansoon and winter. The increase between S2 and S4 and particularly between S3 and S4 was probably due to seepages from BTPS ash ponds. This was as high as 10 mg/l in postmonsoon when seepages would be high.

III. The main impact observed was on BOD and COD concentrations which during the low -flow pre-monsoon season steeply rose from S1 to S2, reduced marginally by S3, but again rose slightly at S4 due to discharge of Barauni refinery effluents. Finally bio-degradation reduced the BOD and COD so as to be only slightly higher at S5 than those at S1.

IV. During post-monsoon due to turbid nature and less transparency and high flow of water, the productivity was relatively low at all stations. and ranged between 19.9 to 23.2 mg C/m^2/hr of gross primary productivity. Compared to the post-monsoon season, the winter productivity were about 3 times higher and the pre-monsoon productivity were about 5 times higher. In all seasons, productivity gradually declined as one went from stations S1 to S4 and were 5 to 10% lower at S4 than at S1. However by SS the productivity more or less recovered to the S1 levels.

Conclusion

Ganga water was in a pretty healthy and clean state in this stretch and was not significantly affected by the pollutants loads entering Ganga from Barauni industries as evidenced by the consistent value of pH (around 8.0) and DO (all values were close to or higher than 7 mg/L) through all seasons. Since the DO in all seasons remained above 6.9 mgh/L, BOD and COD should not cause great concern. The bio-productivity was also affected only slightly and that too over a small stretch of 10-15 Km. Thus the industrial discharges entering in to Ganga in this reach have only a mild impact on the stream.

❑ REFERENCES

APHA, (1980) *Standard Methods for the Examination of Water and Waste water*, American Public Health Association, New York.

Chattopadhya S.N., Routh T., Sharma V.P., Arora H.C. and Gupta R.K. (1984) A short term study of the pollution status of river Ganga in Kanpur region, *Ind. Jr. of Env. Hllh., 16(3)* : 244-257.

Pande P.K. and Pande G.N. (1980) Physico-chemical characteristics of river Ganga at Kanpur, *Jr. Inst. Engr.*, 61 : 28-30.

Singh J.P., Yadav P.V. and Singh L., (1988) Pollution status on Sangam and its adjoining rivers before Kumbha Mela at Allahabad, *Ind. Jr.Env.Pro.,8(11)*: 839-842.

Sinha A.K., Srivastava R.K., Pandey D.P. and Modak D.P., (1989) Water quality characteristics of Ganga river from Karemanickpur to Phaphamau-Acase study, *Ind. Jr. of Env. Pro. 9(11)* : 845-852.

Upadhyay R., Dubey A.P. and Pandey G.N. (1982) Monitoring of Pollution of at Sangam during Ardha Kumbha, *Poll. Res., 1* : 1-2.

Authors :

1. **G.Kamalak Kannan** : Institute of Environmental Science, Mahatma Gandhi Gramodaya Vishwavidyalay, CHITRAKUT- 485 331. District - Satna, M.P.(India)

■

20

Impact of Physico-Chemical Characteristics on the distribution of zooplanktonic species in the Rivers of Chambal command area, Madhya Pradesh

❑ INTRODUCTION

River pollution in India has now reached a point of crisis due to unplanned urbanization and rapid growth of industrialization. Due to pollution in water the entire arry of life in water is affected. Zooplankton, the primary consumers, Which play an important role in fish production, are being adversely affected. These organisms are highly variable in nature from one water bodies to the other and acts as bio-indicators of pollution (*Arora, 1961, 1966, Sampath et al., 1979, Sharma, 1986 and Saksena, 1987*). Chambal command area lies in northern madhya pradesh and has several rivers. The problem of water quality deterioration is mainly due to human activities e. g., disposal of dead bodies, discharging of sewage wastes and agricultural runoff. Sharma et al. (1989) and Kaushik et al. (1992) have attempted to study these rivers and tried to assess their water quality and phytoplanktonic community, and hence an effort was made to assess the impact of physico chemical characteristics on the distribution of zooplanktonic species in four rivers namely Saank, Asaun, Kuari and Chambal in this area.

Material and Methods

The water and planktonic samples were collected from the selected points during winter season. The selection of sampling sites was done in relation to residential localities along these rivers. The sampling stations were established near Noorabad, Chhonda, Deori and Rajghat villages on Saank, Asaun, Kuari and Chambal rivers respectively. The surface water samples were collected and analysed as per the standard method of APHA et al. (1975). The zooplanktonic samples were collected with the help of Birge conical nylone plankton net (No.

20) and were preserved immediately by adding few drops of 5% formalin. They were studied with the help of sedgwick-rafter plankton counting cell. Identification was done with the help of standard books and research publications.

Results and Discussion

Normally, the physical properties of water are governed by the fundamental geographics and chemical properties are influenced by cyclical pattern of nutrients involving both abiotic and biotic components of aquatic ecosystem. The physico chemical characteristics have been interrelated to each other and have been found to exert influence on the biological communities and biological production in waterbodies and thus it becomes essential to understand this interrelationship. It has been reported earlier that all the river systems namely Saank, Asaun, Kuari and Chambal are polluted to some extent and among them river Saank is more polluted comparatively. The degree of pollution follows the order Saank > Chambal > Asam > Kuari (Kaushik *et al.*, 1992).

The zooplanktonic composition of any waterbody is always due to the history of area and present ecological conditions. The running water rheozooplankton contribute significantly to their secondary biological productivity. In all fifty seven species of zooplankton were identified (Table-1) As for the quantitative abundance the rotifers had twenty six species followed by Cladocera with ten species. Protozoa with nine species, Ostracoda with seven species and Copepoda with five species. Four species viz., *Arcella discoides* (Protozoa), *Moina Brachiata* (Cladocera), *Mesocyclops* sp. (Copepoda) and *Cypris* sp. (Ostracoda) were commonly present in all the rivers and thus, indicating preference of wide range of physico-chemical characteristics, whereas *Lecane (Lecane) Curvicornis, Lepadella ovalis, Cephalodella* sp., *C. catellina, C. forficula, C. intuta, Testudinella patina* (Rotifera), *Simocephalus exspinosus, Ceriodaphnia reticulata* (Cladocera) and *Spicodiaptomus* sp. *Heliociaptomus viduus* (Copepoda in Saank river, *Mytilina ventralis* (Rotifera), and *Centrocypris horrida* (Ostracoda) in Asaun river, *Difflugia acuminata, D. Lobostoma* (Protozoa), *Dadaya macrops* (Cladocera) and *Cyprinotus gunningi* (Ostracoda) in Kuari river and *Bullinularia indica, Centropyxis stellata, Difflugia* sp. (Protozoa), *lecane (lecane) Venusta, Brachionus forficula*, B. *angularis, Anureopsis fissa, Lepedella triptera, Polyarthra vulgaris, Asplanchna brightwelli*, and *Notholca* sp. (Rotifera), and *Cyclocypris* sp. (Ostracoda) in Chmbal river were present exclusively with a little narrow range of tolerance for changing physico-chemical characteristics.

TABLE : 1

Showing zooplanktonic distribution in Saank, Asaun, Kuari and Chambal river at Morena division.

Zooplanktonic Species	Rivers			
	Saank	Asaun	Kauri	Chambal
PROTOZOA				
Arcella discoidis	+	+	+	+
Bullinularia indica	-	-	-	+
Centropyxis ecornis	+	-	+	-
C. aculeata	+	+	-	+
C. stellata	-	-	-	+
Difflugia sp.	-	-	-	+
D. acuminata	-	-	+	-
D. lobostoma	-	-	+	-
D. muriformis	+	+	-	-
ROTIFERA				
Lecane (Monostyla) bulla	+	+	-	-
L. (M.) closterocerca	+	+	-	-
L. (M.) crenata	+	-	+	-
Lecane (Lecane) luna	-	+	+	-
L. (L.) curvicornis	+	-	-	-
L. (L.) ungulata	-	+	-	+
L. (L.) venusta	-	-	-	+
Brachionus calyciflorus	+	-	+	+
B. forficula	-	-	-	+
B. falcatus	-	-	+	+
B. patulas	-	-	+	+
B. angularis	-	-	-	+
Anureopsis fissa	-	-	-	+
Lepadella ovalis	+	-	-	-

L. triptera	-	-	-	+
Cephalodella sp.	+	-	-	-
C. catellina	+	-	-	-
C. forficula	+	-	-	-
C. intuta	+	-	-	-
Mytilina ventralis	-	+	-	-
Polyarthra vulgaris	-	-	-	+
Asplanchna brightwelli	-	-	-	+
Euchlanis dilatata	-	+	+	+
Trichotria tetraetis	+	-	-	+
Notholca sp.	-	-	-	+
Testudinella patina	+	-	-	-
CLADOCERA				
Moina branchiata	+	+	+	+
Simocephalus exspinosus	+	-	-	-
Alona intermedia	+	+	-	+
Bosmina longirostris	+	+	+	+
Scapholeberis kingi	-	+	+	+
Alonella sp.	-	-	+	+
Chydorus sphaericus	+	+	-	+
Macrothrix rosea	-	+	-	+
Ceriodaphnia reticulata	+	-	-	-
Dadaya macrops	-	-	+	-
COPEPODA				
Mesocyclops sp.	+	+	+	+
Thermocyclops sp.	+	+	+	-
Phyllodiaptomus blanci	+	-	-	+
Spicodiaptomus sp.	+	-	-	-
Heliodiaptomus viduus	+	-	-	-
OSTRACODA				

Cypris sp.	+	+	+	+
Hemicypris sp.	+	+	+	-
Centrocypris sp.	+	-	-	+
C. horrida	-	+	-	-
Cyclocypris sp.	-	-	-	+
Cypridopsis disper	-	-	+	+
Cyprinotus gunningi	-	-	+	+

Species richness is a good indication of the dynamic balance of the communities. The rotifers constitute from 33.33% to 53.84% in total zooplanktonic community in various Indian waters (Kaushik, 1992). Whereas in the present study, percentage of rotifers in total zooplankton varied from 28.56% to 43.75% (43.75% in Chambal river, 43.33% in Saank river, 33.33% in Kuari river and 28.56% in Asaun river). The difference in rotifer community structure in the river can be attributed to intense variability in physico-chemical and geographical condition. Other factors responsible can be nutrients, organic matter and phytoplanktonic growth. The species richness index has also reflected on the quality of water in these rivers. However, the water quality of Chambal and Saank river and Kuari and Asaun river are similar to each other when a quantitative comparison is made.

Pennak (1957) has pointed out that when more than one genera of the same group occur in any waterbody one genus is more abundant than the other. In the present study, *Brachionus* species were found to compete with *Lecane* numerically. As such, the number of species of *Lecane* were greater in Saank and Asaun river but less number of species in Kauri and Chambal river than that of *Brachionus*.

The species composition of zooplankton is directly or indirectly influenced by complexes of ecological factors and their seasonality. Among various physico-chemical characteristics, light, temperature, wind, dissolved oxygen and food (Welch, 1952) and temperature, pH, dissolved oxygen, salinity and other dissolved nutrients (Nayar, 1970) are known to control the production, composition and distribution of zooplankton. The tropical waters are said to be more productive on account of high temperature, however, reported that temperature alone is not a limiting factor for the growth of zooplankton though it plays an important role either direct or indirect in their growth and production. Water temperature between 10^0C and 29^0C are suitable for zooplanktonic development (Kaushik, 1992). In the present study, low water temperature is favourable for the development of rotifers and cladocerans.

Lakshminarayana (1965) and Chourasia and Adoni (1985) have found a direct relationship between pH and zooplankton production. It is quite evident from the present study that the moderate pH range of 7.5 to 7.8 favoured increased growth of Cladocera and Rotifera respectively. Nasar (1977) noted that pH between 7.6 and 7.7 is quite suitable for development of zooplankton. Apart from these factors, the population of phytoplankton and their production may also influence zooplanktonic population. Rao *et al.* (1981) and Kaushik (1992) have also observed that growth of zooplankton is directly proporationate to the growth of phytoplankton. Low level of nutrients act as limiting factor for zooplankton development despite good illumination (Sreenivasan, 1970). However, in nature no direct relationship can be traced between the quantity of zooplankton and nutrient content at any specific time as the limiting materials are completely utilized soon after their introduction and the optimal nutrient requirement vary greatly from one species to other (Chourasia and Adoni, 1985). It is more generally recognized that the higher biota is an index of rich nutrients in a water. The increase in zooplankton density with the increase in phosphate content and rich phytoplankton population has been noticed in the present study.

The zooplanktonic species have been recognised as pollution indicators by their usual absence or presence in a waterbody. Sampath *et al.* (1979), Verma *et al.* (1984), Sharma (1986) and Mishra and Saksena (1990) have commented on the basis of study of zooplankton communities that *Arcella discoides, Lesquereusia modesta, Difflugia corona, D. muriformes, Acanthocystis spinibera, Brachionus angularis, B. calyciflorus, B. flacatus, B. foficula, B. diversicornis, B. caudatus, B. rubens, Keratella lenzi, K. tropica, Moina sp., Daphnia (Daphnia) carinata, D. (D.) lumholtzi, Cypris sp., Cyprinotus gunningi, Spicodiaptomus chelospinus and Mesocyclops leuckarti* are present in mesotrophic waters. *Arcella* spp., *Difflugia* spp., *Centropyxis* spp. *Brachionus angularis, B. calyciflorus, B. rubens, B. urceolaris, B. bidentata, B. budapestinensis, B. caudatus, B. quadridentatus, B. petrodinoides, B. patulus, B. sessilis, B. forficula minor, Anuraeopsis fissa, K. tropica, K. cochlearis, Platyias quadricornis, Lecane cornuta, L. crenata, L. lunaris, L. decipiens, L. closterocerca, L. pyriformis, Colurella bicuspidata, Filinia longiseta, F. terminalis, pompholyx sulcata, Asplanchna* sp., *Rotaria neptunia, Cephalodella* sp., *Conochilus* sp., *Polyarthra sp., Daphnia parvula, Ceriodaphnia lacustris, Chydrorus sphaericus, Daphanosoma leuchtenbergianum, Moina* sp., *Bosmina longirostris, Mesocyclops edax* and *Diaptomus siciloides* well representative of eutrophic waters (David and Roy, 1966, Arora, 1961, 1966, Sharma, 1986, Bhatti and Rand, 1987, Cap, 1980, Janicki *et al.*, 1979 and Saksena, 1987). In the present study, *Arcella discoides, Centropyxis ecornis, C. aculeata, C. stellata, Difflugia acuminata, D. lobostoma, D. muriformis, Lecane (Monostyla) bulla, L. (M.) closterocerca, L. (M.) crenata, Brachionus calyciflorus, B. forficula, B. falcatus, B. patulas, B. angularis, Anureopsis fissa, Cephalodella catellina, C. forficula, C. intuta, Polyarthra vulgaris, Asplanchna brightwelli, Moina brachiata, Bosmina longirostris, Chydorus sphaericus, Ceriodaphnia reticulata, Mesocyclops* sp., *Spicodiaptomus* sp., *Cypris* sp. and *Cyprinotus gunningi* were present in rivers under study and they are suggestive of pollution.

Acknowledgement
We express our gratefulness to Late Prof. J. Bahadur and to Prof. R. Mathur, Head, School of Studies in Zoology, Jiwaji University, Gwalior for providing laboratory fecilities and encouragement during the course of this study.

❑ REFERENCES

APHA, AWWA, WPCF (1975). Standard methods for examination of and Waste-water. American Public Health Association, 14th ed.

Arora H.C. (1961). Rotifera as indicators of pollution. *CPHERI Bull.*, 3-4:24.

Arora H.C (1966). Rotifera as indicators of trophic nature of environments. *Hydrobiologia. 27* : 146-149.

Bhatti D.S. and Rana K.S. (1987). Zooplankton in relation to abiotic components in the Fort Moat of Bharatpur. *Proc. Nat. Acad. Sci. India,* 57 : 237-242.

Cap R.K. (1980). Comparison of zooplankton found in the eastern basin of lake Erie in 1928 and 1974. *Ohio J. Sci., 80 :* 114-118.

Chourasia S.K. and Adoni A.D. (1985). Zooplankton dynamics in a shallow eutrophic lake. In Proc. Nat. Symp. Pure and Appl. Limnology (Ed. A.D. Adoni). Bull. Bot. Soc. Sagar, 32 : 30-39.

David A and Ray P. (1966). Studies on the pollution of the river Daha, N. Bihar by sugar and distillery wastes. *Environ. Hlth., 8* : 6-35.

Janicki A.K., Decosta J. and Davis J. (1979). Mid summer crustacean planktonic communities of seven small compoundments. *Hydrobiologia. 64* :123-129.

Kaushik S. (1992). *Ecological studies of some water bodies at Gwalior, Madhya Pradesh.* Ph.D. Thesis, Jiwaji University, Gwalior.

Kaushik S., Agarker M.S. and Saksena D.N. (1992). Distribution of phytoplankton in riverine waters in Chambal command area, Madhya Pradesh. *Bionature, 12* :1-7.

Lakshminarayana J.S.S. (1965). Studies in the phytoplankton of river Ganges, Varanasi, India. I. Physico-chemical characteristics of river Ganges. II. The seasonal algae in the river Ganges. *Hydrobiologia. 25* : 119-165.

Mishra S.R. and Saksena D.N. (1990). Seasonal abundance of the zooplankton of waste water from the industrial complex at Birla Nagar (Gwalior), India. Acta *Hydrochem. hydrobiol., 18* : 215-220.

Nasar S.A.K. (1977). Investigations on the seasonal periodicity in the zooplankton in the fresh water pond in Bhagalpur. Acta. *Hydrochem. hydrobiol. 5* : 577-582.

Nayar C.K.G. (1970). Studies on the rotifers population of two ponds of Dilani Rajasthan. *J. Zool Soc. India,* 22 : 21-34.

Pennak R.V. (1957). Species compsotion of limnetic zooplankton communities. *Limnol. Oceanogr.* 2 : 222-232.

Rao V.N.R., Kannan V. and Duraisamy A. (1981). Relative concentration of Na, Ca and Mg for growth of some diatoms. *Proc. Indian Acad. Sci.* 50 : 169-175.

Saksena D.N. (1987). Rotifers as indicators of water quality. *Acta Hydrochem. Hydrobiol., 15* 169-175.

Sampath V., Sreenivasan A. and Ananthanarayanana R. (1979). Rotifers as biological indicators of water quantity in Cauvery river. In proceeding Symp. Environ. Bio., Muzaffarnagar, India (Eds. S.R. Verma, A.K. Tyagi and S.K. Bansal), pp. 441-452.

Sharma B.K. (1986). Assessment of pollution indicators in Indian Rotatoria. *J. Meghalaya Sci. Soc.,* 47-49

Sharma S., Kaushik S. and Mathur R. (1989). Pollution studies of Saank, Asaun and Kuari rivers in Madhya Pradesh. In proceeding of the Seminar on Management of Aquatic Ecosystems (Eds. V.P. Agarwal, B.N. Desai and S.A.H. Abidi). Society of Bioscience, Muzaffarnagar, p. 211-218.

Sreenivasan A. (1970). Limnology of tropical impondments. A comparatiove study of the major reservoirs in Madras State (India). *Hydrobiologia, 36* : 443-469.

Verma S.R., Sharma P., Tyagi A., Rani S., Gupta A.K. and Dalela R.C. (1984). Pollution and Saprobic status of eastern Kalinadi. *Limnologica (Berlin), 15* : 69-133.

Welch P.S. (1952). *Limnology*. McGraw, Hill Book Company, New York.

Authors :

1. **S.Kaushik, D.N. Saxsena and S.S.S. Verma** : School of Studies in Zoology Jiwaji University, GWALIOR - 474 011. M.P. (India)

21

Algal Indicators in a riverine Ecosystem Polluted By Glucose Factory In Tamilnadu, India

❑ INTRODUCTION

The river Amaravathi is tributory of the river Kaveri. It is a river originating in the W. Ghats in Kerala in the Moonar region. It flows through the range of hills called ANAIMALAI in the Coimbatore District in Tamilnadu. It flows into the Tiruchi District, where it merges with the main river Kaveri. The name Amaravathi indicates to an element of immortality. This is a very ironical situation, because the river suffers heavily on account of industrial pollution during its course of flow, but still survives to keep on flowing.

The Amaravathi river is impounded by a dam christened the Amaravathi dam in the Udumalai area 40 km. south east off Pollachi. From the dam, as stated earlier the river flows further down into the Tiruchi Dt. As the river so flows it is heavily polluted by wastes from a sugar factory, paper factory, chemical solvent factory and human tenements.

The algae, which are the primary producers form a part of the aquatic flora. Apart from their role as primary producers, many of them are known for their ability to tolerate and indicate pollution. This ability has been empolyed by Palmer (1969, 1980) and Nygaard to assess water quality (See Trivedy and Goel 1984) to calculate pollution indices based on the analysis of the algal composition (See also Hosmani & Bharathi 1982). The region of the tributary Amaravathi exposed to industrial pollution as above has been studied for a period of 26 months for its algal composition with particular reference to algal taxa in relation to pollution. The results pertaining to the spot of Amaravathi where the effluents from the glucose factory (KAMALA SUGAR) are discharged into the riverine system are presented and discussed here. The algal composition, froms the basis for the calculation of Palmer and Nygaard pollution indices. The observations and discussions pertain to this aspect only.

Materials and Methods

The methodology involved the collection of water samples from the affected spot in 5 L carboys every month. The samples were brought to the laboratory, filtered and allowed to sediment and centrifuged. By sedimentation and centrifugation a concentrated sample was obtained to which equal volume of 8% formalin was added. A few drops of idoine were added to the sample. The sample was then sedimented, washed and preserved.

The estimates included :

1. the identifikation of algal taxa (planctonic / otherwise) by using standard manuals/publications (Desikachary 1959, Iyengar and Vimala Bai 1941, Pandey & Pandey 1980, Philipose 1967, Ramnathan 1964, Randhawa 1962, Venkatraman 1939, West and West 1907)
2. Calculation of pollution index (generic). of Palmer (Palmer 1969, 1980) Vide Trivedy & Goel 1984). Nygaards modified componds quotient. (See Trivedy and Goel 1984, Hosmani & Bharathi 1982) for the polluted region for the whole period under investigation. (1990-92)

$$CQ = \frac{\text{Blue green + Diatoms + Chlorococcales + Euglenoids}}{\text{Desmids}}$$

Observations

The Palmer pollution index (generic) values and Nygaard compound quotient values for 26 months for the polluted region in Amaravathi are presented in figure 1. The high & low values in each of these are analysed here. High Palmer index values, nearing 20 (the eutrophication criterion sensu Palmer) is reached for this spot in May 1990, March 1991 and january 1992. The algal composition as revealed by Qualitative & Quantitative analysis includes the general *oscillatoria (O. Princeps, O. amoena, O. jasorvensis* and *O. annae* and *phormidilum (P. molles)* besides the diatom genera, *Gyrosigma, Pleurosigma* and *Mastogloia.*

In March 1991, the blue-green genera include *Oscillatoria (O. amoena), Lyngbya, Phormidium (P. molle).* The green algal genera include the filamentous *Spirogyra, Rhizoclonium,* the desmid genus *closterium* the chlorococcalean *Scenedesmus,* and the diatom genera *Synedra (S. ulna)* and *Nitzschia (N. vitrea and sp.)* In January 1992 the algal composition is as follows : The blue-green genera are *oscillatoria (O. princeps) (O. annae), Plectonema (P. Wollei)* the green algal genera, *Oeldogonium, spirogyra, Rhizoclonium, & Scenedesmus* and the diatom genera *pinnularia (P. fasciate, P. gibba), Navicula (N. pygmaea N. cuspidata) and cyclotella (C. menegheniana).* Of the genera mentioned above, *Mastogloea, Gyrosigm Pleurosigma, Oedogonium and Rhizoclorium* are not rated as tolerating pollution. All the other genera are among pollution tolerating taxa in terms of Palmer's rating.

Apart from the above mentioned months and years, the spot shows the presence of pollution indicators, but the pollution values are much less. For example in January & November 1990, February, March, April, July & October 1991 In January 1990. *Oscillatorina (O. princeps) Phormidicum, phormidicum, Lyngbya (P. molle) Fragilaria (F. intermedia) Naviculay (N. Cuspidata), Nitzschia (N. palea & N. Obtuse)*

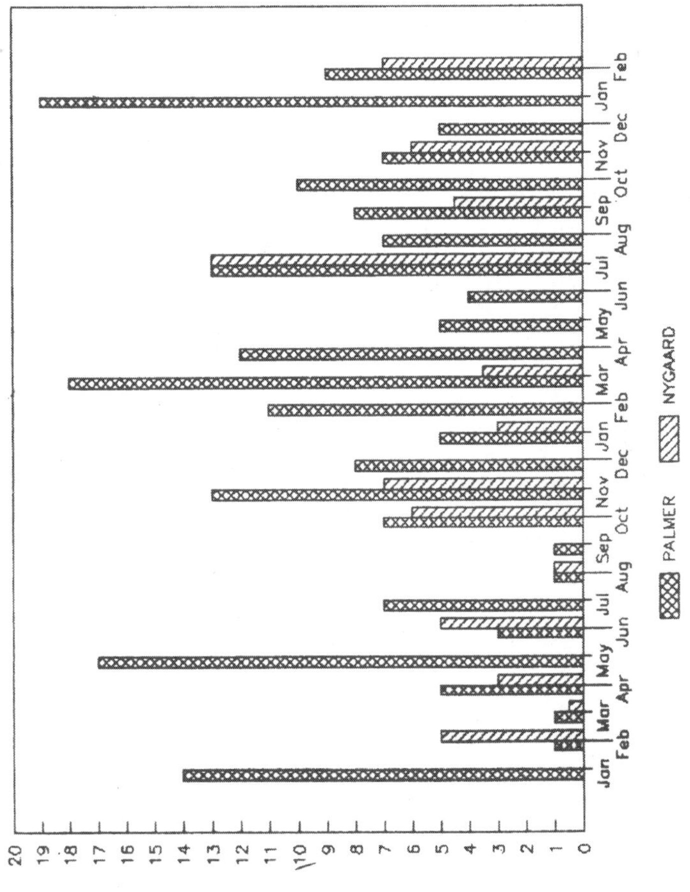

Figure 1. River Amaravathi Pollution Index - Sugar Factory Area.

constitute the algal flora. To give one more example of lower pollution index (Palmer), the composition of algae in October 1991 may be considered. Here the genera are Oedogonium, *Fragilaria (F. intermedia, F. brevistriata) Scenedesmus, Plectonema (P. Wollei), Spirogyra, Phormidium (P. molle), Phacus & Navidula (N. Cuspidata)*

The period when the Plamer values are lowest (1) are February, March, August, and September 1990; The algal composition for March 1990 is *Cosmarium Nostoc & Closterium.*

Examining pollution indices in terms of Nygaard's modified compound Quotient, as seen from text fig. 1. highest values (nearing 6, the eutrophication level) are seen during : February, June, October and November 1990, July & November 1991, February 1992. In July 1991, the Nygaard values are very high. The algal composition for this period is : Blue -greens *Phormidium pachydermaticum, plectonema wollei,* diatoms *Gomphonema lanceolatum, Achnanthes inflata, A. hadckiana, Nitzschia obtusa, Frgilaria intermedia F. brevistriata, Gomphonema parvulum, Amphora coffeaformis, Synedra ulna, Eunotia monodon,* the chlorococcalean alga *Scenedesmas* and the desmid *closterium.* When properly compounded the value is 13, which is much more than the eutrophication value of 6. The algal froms that are present at this time, but do not come into calculation as per formula are *Spirogyra, Rhizoclonium* and *Oedogonium.*

In August 1990, the Nygaard value is low (1). Here the diatom genera are *Fragilaria* and *cymbella* and the desmids are *cosmarium* and *closterium.* These are relevant for the calculations. Other algal forms include *spirogyra* and *Oedogonium.* When compounded the value is 1.

In January, May, July, September & December 1990, February, April, May, June, August, October & December 1991 and January 1992, the Nygaard values are 0. One such above case may be examined. The blue greens include *Oscillatoria princeps, Phormidium molle, Lyngbya* sp, and the diatoms include *Fragilaria intermedia, Navicula cuspidata, Nitzschia pale a N. obtusa.* Due to lack of any desmid genera, the compounded value is 0. Similarly in other seasons as indicated above, the Nygaard values are 0, due to absence of desmids in the composition.

The pollution tolerating or rated forms occuring commonly during the course of investigation at this spot are listed below. blue-greens *Oscillatoria amoena, O. princeps, phormidium molle,* The diatoms include *Achnanthes brevipes, A. hauckiana, Amphora Coffee formis, A. holsatica, Cymbella Cistula Nitzschia ritrea N. palea, Fragilaria brevistriata, F. intermedia, Gomphonema lanceolatum Hantzschia amphioxys, Navicula cuspidata, Synedra ulna, Rhopalodia gibra* the green algae, *Closterium, Cosmarium, Scenedesmus bijugatus,* and *Spirogyra.*

Discussion

The riverine system of Amaravathi in the said polluted spot shows diverse algal flora, which is far from uniform in composition from month to month. Quite a lot of importance has been assigned to a veriety of algal taxa by Palmer and Nygaard from the point of view of pollution. To reduce massive data presentation, only specified and typical examples are presented in the previous section. Palmers generic index used here is calculated by compounding the index values assigned to various genera. The values may be low or high, depending upon how far the particular genera tolerate extreme conditions of pollution. For instance the following forms are given high pollution index values.

Euglena, Scendesmes, Oscillatoria etc. As stated earlier, the Palmer index values are high, reaching almost 20, (the Palmer criterion for eutrophication) during May 1970; March, 1991, and January 1992.

The Nygaard, values are compounded by adding the blue-green diatmos, chlorococcales & Euglenoids and dividing these by desmids. The calculations are critically controlled by the presence of desmids. If there are no desmids, the CQ is only 0 as they are dividing factors. Highest pollution value as per this criterion is observed in July 1991. At this time, the palmer value is also fairly high, but definitely not highest.

Other periods when Nygaard values closely approximate the critical level are February, June, October, November 1990, July, November 1991 and February 1992. Of these, high palmer values are also seen for November 1990 and July 1991 only particularly, in May 1990, March 1991 and January 1992 when palmer values are high, Nygaard CQ is 0. Though the Palmer and Nygaard values are fluctuating, they hardly can be correlated with each other.

In terms of repeatebility, and intensity high pollution trends could be observed during the months of March, July, October and November. This observation is based on the Palmer and Nygaard values as generally considered and calculated from the observed algal composition for the region under consideration.

The dynamics of pollution cannot be predicted, though generalization as above may be made. There is a succession of pollution indicating and tolerating algal in this reverine system. The index become high or low depending upon the relative concentration of these algal populations. During the two year period of study as indicated earlier, these have been many occasions during which eutrophication trends could not be detected as judged from the algal indices employed. Only thrice, in terms of Palmer index and five times in terms of Nygaard index, the spots of the river shows eutrophication trends during the two year period of study. Though this number is negligible, it still indicates a degree of pollution, particularly in terms of Nygaard values. The riverine system is not clean throughout. Its cleanliness is interrupted to verying degree as revealed by text fig. 1.

This trend for eutrophication low or high is revealed throughout the period of study. The algal pollution indicators serve a very useful purpose in assessing the pollution load of the riverine system at the point of the sugar-factory.

Acknowledgement

The author wishes to thank the Madras Science foundation, Madras and the Management of N.G.M. College, Tamilnadu for Facilities.

❑ REFERENCES

Desikachary (1959). *Cyanophyta,* I.C.A.R., New Delhi 686 pp.

Hosmani S. P. and S.G., Bharati, (1982), Use of algae in classifying water bodies phykos, 21 : 48-51

Iyengar M.O.P., and B.Vimala Bai, (1941). Desmids from Kodaikanal, South India *J. Ind. Bot. Soc. XX (1 & 2) : 73-103.*

Palmer C.M. (1969). A composite rating of algae tolerationg organic pollution *J. Phycol.* 5 : 78-82.

Palmer C.M. (1980). *Algae and Water pollution* Castle House Publ. Ltd., U.S.A.

Pandey U.C. and Pandey D.C. (1980). Addition to the algal flora of Allahabad, V. *Desmids. Phykos 19* (2) : 161-170.

Philipose M.T. (1967). *Chlorococcales,* I.C.A.R., New Delhi. 365 pp.

Ramanathan K.R. (1964). Ulotricales, I.C.A.R., New Delhi.

Randhawa M.S. (1962). *Zygnemataceae,* I.C.A.R., New Delhi.

Trivedy R.K. and P.K. Goel (1984). *Chemical and Biological Methods of Water Pollution Studies.* Environmental Publications, Karad, (India).

Venkataraman G. (1939). A stystmatic account of some South Indian diatoms *Proc. Indian Acad. Sci.,* (B), (10) (b) : 293-368.

West W. and S. West (1907) Fresh Water algae from Burmainc a few from Bengal and Madras. *Annals of the Royal botanic gardens, Calcutta VI Part II, 260 pb.*

Authors :

1. **V. Sankaran** : Department of Botany, N.G.M. College, POLLACHI- 642 001. T.N. (India)

■

22

Algal Indicators in a riverine Ecosystem Polluted By Paper Mills Waste In Tamilnadu, India

❑ INTRODUCTION

As explained in an earlier paper in this vloume, the Amaravathi river is a tributary of the Kaveri river. It has its origin in Kerala and flows through the districts of Coimbatore and Trichy. During its course, the river flows near the vicinity of the famous pilgrim centre of Tamilnadu, viz., Palani. Before reaching this place, the Amaravathi river is polluted by wastes from an industry manufacturing papers and boards using mostly sugar cane waste as raw materials.

As mentioned earlier, this riverine system supports a fairly rich algal flora, which is highly variable in terms of both quality and quantity. The stretch of the river system near the paper factory has been analysed for algal flora for a period of more than two years viz., January 1990 to February 1992. Most of the algae collected and studied belong to different taxonimic groups and entities and have been used as indicators to assess the degree of pollution and eutrophication in fresh water ecosystems. (Palmer 1969, 80) and Nygaard (See Trivedy & Goel 1984, Hosmani & Bharathi 1982). Who have evoloved compounding formulae for calculation of the respective pollution indices. Based on similar terms, the algal composition of this polluted spot in the said riverine system has been used to calculate the Palmer and Nygaard pollution indices. The pollution dynamics, based on algal composition in ther region of the river affected by discharge of paper mill waste will be presented here. The analysis pertains to biological aspects only with particular reference to algae as biological pollution indicators, Comparison is also drawn between the area polluted by sugar factory waste and paper factory waste.

Materials and Methods

Standard procedures of samples collection and algal analysis were employed. Water samples were collected in 5 L carboys. From the affected spot. They were brought to the laboratory where they were filtered and sedimented. Sedimentation was followed by centrifugation to have concentrated samples. To the sample, equal volume of 8% formalin was added, followed by a few drops of iodine. The sample was subsequently washed and fixed in 4% formalin.

The algal taxa were identified by using standard manuals (Desikachary 1959, Iyengar and Vimala Bai 1941; Pandey and Pandey 1980, Philipose 1967, Ramanathan 1964, Randhawa 1962, Venkataraman 1939, West & West 1907).

Palmer generic index was calculated as per Palmers values (See Trivedy & Goel 1984).

Nygaard modified compound quotient was calculated

$$\text{by using the formula} = \frac{\text{Blue-green + Diatoms + Chlorococcales + Euglenoids}}{\text{Desmids}}$$

(Hosmani & Bharathi 1982)

The calculation were based on algal composition for each monthly collection for more than 2 years (1990-1992).

Observations

The Palmer generic index values and Nygaard compound quotient calculated for each of the 26 collections are given in Fig. 1. It is to be noted here that the Palmer values do not reach the criterion of eutrophication viz 20. The maximum reached here is 16 (March 1991). The algal population consisted of the following genera. (Species indicated within brackets). *Oscillatoria (O. princeps). Phormidium (P. molle), Frgilaria (F. intermedia), Navicula (N. cuspidata Amphora (A. Coffeaformis) Nitzschia (XN. Baka), Scenedesmus (S. acutus), Closterium and Spirogyra.* Lower palmer values (11) are indicated during January, June and September 1991 and February 1992.

The algal composition may be surveyed for some of the above Collections-January 1991 : *Oscillatoria (O. obscura), Phormidium (P. molle), Pleurosigma, Gyrosigma, Synedra (S. ulna) Fragilaria (F. brevistriata), Navicula (N. cuspidata), Amphora (A Coffeaformis) Oedogonium.*

Fabruary 1992 Collections reveal the presence of the following taxa. *Aphanothece Phormidium (P. molle), Risularia (R. braunii), Navicula (N. protracta, N. salinarum N. cuspidata) Cyclotella (C. meneghenian) Pleurosigma, Gyrosigma, Amphora (A. Coffeaformis_ Closterium, Rhizoclonium, Oedogonium.* The Palmer value is extremely low in May, August and December 1990. In these months, the algal composition is observed to be as follows, December 1990 - *Phormidium (P. pachydermaticum, Fragilaria (F. brevistriata), Pinnularia (P. fasciata), Achnanthes (A inflata) Amphora (A. coffeaformis) Oedogonium and Spirogyra.* The Palmer index touched 0 in July 1990 and April 1991. The algal composition in July 1990 is *Rhopalodia (R. gibba, R. gibberula) Fragilaria (F. brevistriata)* and *Merismopedia* The analysis for April 1991 revealed *Fragilaria (F intermedia) (F.brevistriata) Synedra (S. ulna) Pleurosigma, Gyrosigma, Hantzschia (H. amphioxys), Pithophora, Spirogyra.*

In terms of Nygaard's pollution index high values were recorded at this spot

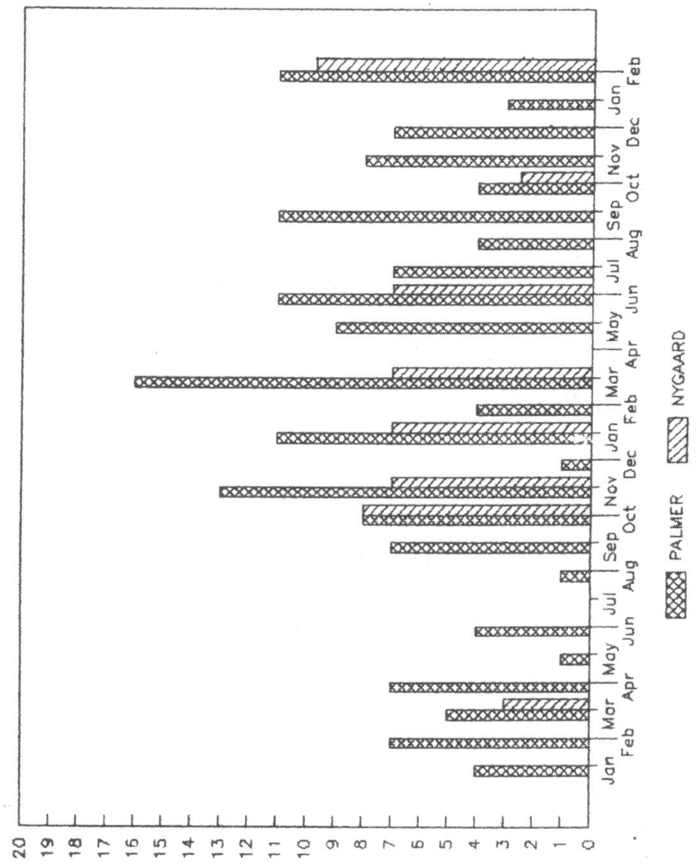

Figure 1. River Amaravathi Pollution Index - Paper Mill Area.

in October, November 1990, January, March & June 1991 and February 1992. In October 1990, the observational analysis reveals the following algal taxa. Blue-greens- *Phormidium (P. molle)*, diatoms *Rhopalodia (R. gibba), Navicula (N. Cuspidata), Nitzschia (N. palea) Fragilaria (F. intermedia), Amphora (A. Coffeaformis), Pinnalaria* (P. gibba), Gomphonema *(G. lanceolatum)* and the desmid genus *Cosmarium*. When compounded, the value is 8.0 the highest Nygaard value reached is 10 (February 1992). The algal composition is : blue - greens-*Phormidium, (P. molle) Aphanothece, diatom genera fragilaria (F. intermedia, pinnularia (P. braunii), Navicula, (N. protracta, N. salinarum, N. cuspidata), Cycfotela (C. mereghenìana), Plearosigma, Gyrosigma, Amphora (A. coffeaformis)* and the *desmid* genus *closterium*. On computation, the value is 10. The Nygaard value is low in March 1990, October 1991. The algal composition - blue-greens *Oscillatioria (O. obscura), Phormidium (P. molle, Spirulina* and the desmid *Cosmarium*. The diatoms were surprisingly absent.

The Nygaard value touches zero in January, February, April to September & December 1990, February, April, May, July to September, November, December 1991 and January 1992. In September 1990, the algal analysis reveals the following. blue-green *Phormidium molle*, diatoms *Gomphonema, Cancelatum, Amphora coffeaformis, fragilaria intermedia, Pleurosigma, Gyrosigma, Rhopalodia gibba, Synedra ulna, and Nitzschia closterium*. There is no record of any desmids. In January 1992 the algal taxa showing up are : blue-green, *Phormidium pachydermaticum, diatoms, Fragilaria bravistrita, Cyclotella mengheniana, Achnanthes hauckiana, Pinnularia interuppta, P. gibba, Gomphonema lanoerlatum,* Desmids are not recorded.

The common pollution tolerant and indicating algal taxa encountered at different times during the 26 month period study are blue-greens *Oscillatoria obscura, Phormidium molle*, the diatoms *Amphora coffea formis, Cyclotella mengheniana, Cymbella cistula, Fragilaria brevistriata, Fragilaria intermedia, Gomphonema lanceolatum, Navicula cuspidata, Nitzschia obtusa, N. Paea, Pinnularia barunii, P. gibba, Rhopalodia gibba, Synedra ulna* the desmids *closterium, cosmarium, the chlorococcalean Scened esmus.*

Discussion

The high or low Palmer or Nygaard values are due to the relative concentration of the pollution indicating tolerating algae at a given time. As indicated in the earlier paper, the method of compounding of the two types of pollution indices are basically different. The Nygaard calculations are critically controlled by the presence or absence of desmids. The absence of desmids will give 0 values even if other pollution tolerant groups are present in high number.

Based on the observations made, it may be stated that the trend of eutrophication is not so high in the paper mill waste discharge area as in the glucose factory discharge are. However the tendency towards eutrophication is present. In terms of Palmer, high trend is seen once (in March 1991) only. In terms of Nygaard, high trend is seen 6 times (See text Fig.1) The pollution dynamic reveals a very unpredictable picture as in the case of sugar factory effluent discharge spot.

The algae in the part of the riverine ecosystem carry out their function of primary production amidst adverse conditions promoted by the pollutants of the industrial unit. They are producers, tolerators and indicators. There are times when the riverine ecosystem is not eutrophic, as revealed by the absence/reduction of algal taxa concerned. Hence the role of algae in assessing water quality is highly significant.

Phytoplankton is a shifting and drifting community, especially in lotic riverine ecosystems. This community is also displaced from source of its attachment. Hence the investigator has to deal with benthic forms and a displaceable phytoplanktonic community. This has an important bearing with regard to the pollution indices that have to be compounded using at least two different formulae as in this case. Particularly, in calculating the Nygaard values, the desmids form the dividing factor. Irrespective of the presence of other pollution indicators, if desmids are absent due to natural reasons, the pollution value becomes zero. This is an ironical situation. However, even if desmids are absent, the presence of other polllution indicators ought to indicate eutrophication. In this case, Palmer's pollution index becomes more useful. However a judicious application of these two indices becomes all the more necessary in assessing water quality.

In the riverine ecosystem, or to be more general in any aquatic system, many algal populations constitute the primary producers. Many of them are known as pollution tolerators and have authentically assigned values in this capacity. There are also other forms which have not been assigned such values. The pollution indicators and non indicators coexist in the aquatic ecosystem the riverine ecosystem in this case. By, virtue of their ability to exist in a polluted environment the 'non-indicators' also must be recognised as tolerating pollution. Due attention and consideration becomes necessary to review these algal forms and revise our knowledge on algal pollution indicators. This will be highly useful in evolving a more comprehensive picture of pollution indicators and enable better formulation of pollution dynamic and assessment of water quality.

Acknowledgement
The author wishes to thank the Madras Science foundation, Madras and the Management of N.G.M. College, Pollachi, Tamilnadu.

❑ REFERENCES

Desikachary (1959). *Cyanophyta,* I.C.A.R., New Delhi 686 pp.

Hosmani S. Pand S.G., Bharati, (1982). Use of algae in classifying water bodies

Phykos. 21 : 48-51

Iyengar M.O.P., and B.Vimala Bai, (1941). Desmids from Kodaikanal, South India *J. Indian Bot. Soc. XX (1 & 2) : 73-103.*

Palmer C.M. (1969). A composite rating of algae tolerationg organic pollution *J. Phycol*. 5 : 78-82.

Palmer C.M. (1980). *Algae and Water pollution*, Castle House Publ. Ltd., U.S.A.

Pandey U.C. and Pandey D.C. (1980). Addition to the algal flora of Allahabad, V. *Desmids. Phykos 19* (2) : 161-170.

Philipose M.T. (1967). *Chlorococcales*, I.C.A.R., New Delhi. 365 pp.

Ramanathan K.R. (1964). Ulotricales, I.C.A.R., New Delhi.

Randhawa M.S. (1962). *Zygnemataceae*, I.C.A.R., New Delhi.

Trivedy R.K. and P.K. Goel (1984). *Chemical and Biological Methods of Water Pollution Studies*. Environmental Publications, Karad, (India).

Venkataraman G. (1939). A stystmatic account of some South Indian diatoms proc. Indian Acad. Sci., (B), (10) (b) : 293-368.

West W. and S. West (1907) Fresh Water algae from Burma Including a few from Bengal and Madras. *Annals of the Royal Botanic Gardens, Calcutta VI Part II, 260 pb.*

Authors :

1. **V. Sankaran** : Department of Botany, N.G.M. College, POLLACHI- 642 001. T.N. (India)

23

Phycological Approach for Assessing Pollution in River Pandu - A Tributary of the River Ganga

❑ INTRODUCTION

Algae constitute an integral part of the river ecosystem and occupy a unique position in aquatic ecosystem due to their significance as primary producers. These are of paramount importance in the economy of fresh waters as they from a link in the food chain and function as natural purifiers of the rivers due to their oxygen producing and nutrient Cleaning capabilities. But with growing industrialization and population pressure, the unabated use of rivers as waste depositories, has not only deteriorated their water quality but has also adversely affected the algal flora of the rivers eliminating some otherwise useful species leaving behind only a few pollution resistant forms. The use of algae in assessing the pollutional status of an aquatic habitat is well established (Palmer, 1963, 1969, Patrick, 1973, Hosmani and Bharati, 1982, Nandan and Patel, 1986, Trivedy, 1986). Palmer (1969) stated that the presence of pollution tolerant species of algae may be used as an indicator of pollution and compiled a list of pollution tolerant genera and species of algae. He has further developed an "Algal Pollutioon Index" for assessing the pollution of water bodies (Palmer, 1969). The wide application of algae in pollution assessment is probably due to the following obvious reasons :

1. They are the natural inhabitants of aquatic environment.
2. An inverse relationship generally exists between the pollution load and algal diversity.
3. Phycological investigations alone are sufficient to reveal the condition of a water body without undertaking the tedious process of physico-chemical estimations.

Prompted by these advantages of algal investigation over the physico-chemical ones, phycological investigations on river Pandu were made to know -

1. The algal flora of the river.
2. The algal distribution and community structure at different sites of the river.
3. The pollutional status of the river.

Material and Methods

The algal flora of the river Pandu, that flows through the outskirts of the Kanpur city and receives industrial and municipal wastes at a number of points, was studied from January 1993 to December 1994. Four sampling sites were selected along the stretch of the river between Panki Industrial Area and Hamirpur Road Crossing. Details regarding the sampling stations are given in Figure 1. Periodic sampling was done at these stations as per standard procedure. Taxonomic identification was done with the help of relevant literature (Desikachary, 1959, Tiffany and Britton, 1952). Most frequent twenty species have been considered in the present study for assessing the pollution of the river (Table -1). Each species was assigned a pollution index factor as per Palmer's index (Palmer, 1969) and the total score was arrived at by summing-up the pollution index factors at each station.

Results

Algae reported from the river and corresponding results obtained as per Palmer's index are given in Table 1.

Table 1

Algal flora of the river and corresponding pollution index. (Palmer, 1969)

	Algal group/ Species	Pollution Index No. Stations 1	2	3	4
I.	*Chlorophyceae*				
	1. * *Chlamydomonas sp.*	4	-	4	4
	2. * *Chlorella vulgaris*	-	2	-	2
	3. * *Scenedesmus quardicauda*	4	4	4	4
	4. *Scenedemus abundans*	+	-	-	-
	5. *Spirogyra sp.*	+	-	+	+
	6. *Hydrodictyon reticulatum*	+	-	+	-
	7. * *Stigeoclonium tenue*	3	3	3	3
	8. * *Closterium sp.*	1	-	1	1
	9. * *Pandorina morum*	3	3	3	3

II	*Cyanophyceae*				
	10. * *Oscillatoria tenuis*	-	4	4	4
	11. * *Oscillatoria limosa*	-	4	-	4
	12. * *Oscillatoria princeps*	1	-	1	-
	13. * *Phormidium sp.*	1	1	1	1
	14. *Nostoc sp.*	+	-	+	+
	15. *Merismopedia sp.*	+	-	-	-
III	*Euglenophyceae*				
	16. * *Euglena sp.*	-	5	-	5
	17. * *Euglena viridis*	-	6	-	6
	18. * *Phacus sp.*	2	-	2	2
IV	*Bacillarophyceae*				
	19. * *Nitzschia palea*	-	5	5	5
	20. * *Navicula sp.*	3	3	3	3
	Total Score	22	40	31	47

* Assigned by Palmer (1969). + Indicates presence. - Indicates absence.

Discussion

It is obvious from the data given in Table 1 that out of twenty species considered in the present study nine belong to Chlorophyceae, six to Cyanophyceae, four to Euglenophyceae and only two to Bacillariophyceae. Though only two species of Bacillariophyceae have been taken into account for assessing the pollution of the river-this group was most frequently represented throughout the study period. Maximum growth of Chlorophyceae was noticed from January to March and then again in December. At other times also this group was well represented among the algal flora of the river. Chlorophyceae dominated at station 1. Out of nine species of chlorophyceae, eight were reported from station 1. Cyanophyceae grew luxuriently from March to June. Out of six Cyanophyceae species, *Phormidium* alone was reported from all the stations whereas *Merismopedia sp.* was noticed only at station 1. Euglenoids, however, were abundant in May and June and were well represented in the algal samples from station 2 and 4. Diatoms were frequently present throughout the investigation period, but showed their abundance in January and February.

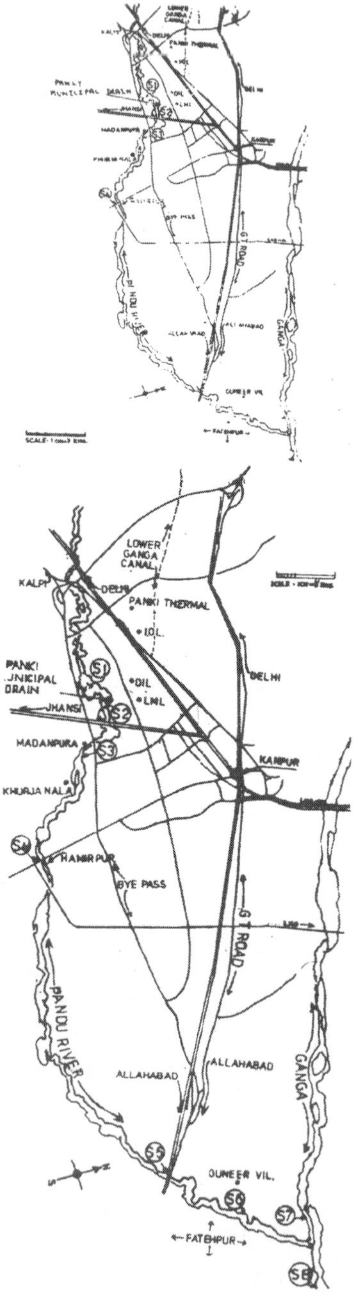

Figure 1. Stretch of river Pandu at Kanpur Showing Industrial and residential areas along with outlets points.

Except *Scenedesmus abundans, Spirogyra sp., Hydrodictyon reticulatum, Nostoc sp.* and *Merismopedia sp.*, all other species reported in the present study, have been cited as pollution tolerant species of algae (Palmer, 1969) Fourteen algal species were found at each stations 1 and 3, whereas stations 2 and 4 recorded eleven and sixteen algal species respectively. Though stations 1 and 3 recorded the same number of algal species; yet their taxonomic composition was different. Out of fourteen species, recorded at each of these stations, ten were common to both, indicating that somewhat similar environmental conditions were prevailing at these stations.

Pollution index score at different stations ranged from 22 to 47. It was highest at station 4 followed by stations 2, 3 and 1. All stations showed a score greater than twenty, indicating grossly polluted condition of the river. The finding is as per our expectations because this small river, with very little flow, receives huge amount of industrial and domestic wastes from Panki Industrial Area, Ratan Lal Nagar, Govind Nagar and other nearby areas through which it passes till it leaves the precincts of the Kanpur city. The wastes discharged into the river are much beyond its assimilating capacity, turning it into one of the most polluted tributaries of the river Ganga that joins it near Fatehpur with all its pollution load.

Acknowledgement
The authors gratefully acknowledge aid accorded by the Principal of the A. N. D. N. N. Mahila Mahavidyalaya, in providing the research facilities. S. Singh is thankful to C. S. I. R., New Delhi for providing a senior research fellowship.

❑ REFERENCES

Desikachary T.V. (1959), *Cyanophyta* Indian Council of Agricultural Research, New Delhi.

Hosmani S.P. and Bharati S.G. (1982). Use of algae in classifying water bodies. *Phykos 21* : 48-51.

Nandan S.N. and Patel R. J. (1986). Assessment of water quality of Vishwamitri river by algal analysis. *Ind. J. Environ. Hlth., 29* : 160-161.

Palmer C.M. (1963). The effect of pollution on river algae. *Ann. N. Y. Acad. Sci., 108.* : 389-395.

Palmer C.M., (1969). A compositing rating of algae tolerating organic pollution. *J. Phycol., 5* : 78-82.

Patrick R. (1973). Use of algae especially diatoms in assessment of water quality. ASTM. STP. 528 Amer. Soc. for testing of materials : 76-95.

Tiffany L.H. and Britton M.E. (1952) *The Algae of illionis*. The University of Chicago Press.

Trivedy R. K. (1986). Role of Algae in Biomonitoring of Water Pollution *Asian Environment, phillipines*.

Authors :

1. **D. Tiwari, P. Kaur and S. Singh** : Department of Botany, A. N. D. N. N. Mahila Mahavidyalaya, KANPUR. U.P. (India)

■

24

Sediment Quality of the River Ganga in the deltaic region and its utilisation

❑ INTRODUCTION

Estuarine sediments are sinks for many materials transported from the land. Many substances that occur naturally such as exchangeable ions (like Na^+, K^+ etc) and nutrients like N, P etc are expected to be mobilised in sediment as a result of natural processes and human activities.

It is well known that due to enormous soil erosion and unscientific discharge of wastes from sources manifold, annually the river Ganga transports more than 83 million tons of dissolved solids which is the highest for all Indian rivers. The sediment transport is about 328 million tons (Sinnarkar et al 1987). The Ganga-Brahmaputra system transports some 118 million tons of dissolved solids annually to the Bay of Bengal of which Baghirathi (lower region of the river Ganga) carries highest concentration of suspended solids. The sediments have caused heavy siltation and filling up of the river beds. The filled river beds are unable to contain the discharge of water during the monsoon resulting in heavy floods. This naturally causes extensive damage to the life and crops incurring recurring losses to the exchequer and untold miseries to the people at large. But the siltation causes enormous mobilisation of essential ingredients from the ecosystems. Obviously desiltation of the river beds should be carried out in right earnest. It is true not only for the river Ganga but for the other rivers as well. But the process of desiltation is difficult and apparently a costly process. However, the cost of desiltation may be effectively neutralised with the possible use of sediments for agronomy and other purposes. These considerations led us to undertake studies for the possible use of sediment particularly in agronomy. Samples of sediments along both the banks of the river Ganga were collected from Bandel to Diamond Harbour (in the lowere deltaic region of the river Ganga). The air-dried samples were analysed to for Na^+, K^+, Ca^{++} and Mg^{++}), nutrient concentrations (like P N), S (as Sulphate), C-content, moisture content and pH values. The values of the constituents can be converted to kg/ha by multiplying ppm values with 2.24 as one hectrere furrow slice soil weighs 2.24 X 10^6 kg. The results are described in this communication.

Material and Methods

Samples along the banks of the River Ganga were collected using common appliances like spade, kurpa etc. 10 grab samples from nearby places were collected and mixed together to give a representative sample for a particular place. Usually, three such replicate samples were taken from a particular place. The samples were collected in polyethylene bags. Bottom deposits were collected at several locations at different depths during slack water periods using an Ekman dredger.

The samples were soaked with filter paper to remove water and air-dried at room temperature.

Discussion

The results as given in table show wide variations in the estimated values which are quite natural for a continually changing river system. Another important reason for fluctuation is the variation of temperature for the air-dried samples. We have taken the mean of the results leaving very low or very high values. It is to be noted that a vast collection of data are needed for adequate correlation of the results but due to lack of funds and facilities, the data in the present case are limited.

Table 1

Average values of chemical parameters of air-dried sediments of the River Ganga from Bandel to Diamond-Harbour

(20 Stations from the banks on both sides of the river)

Year	K^+	Na^+	Ca^2	Mg^{2+}	C	N	P	So $So4^2$	pH
	(mg/g)		(mg /100g)		%				
1987	0.30-0.70	0.18-0.78	0.80-1.28	0.20-0.32	0.50-2.90	0.10-1.18	0.60-1.70	0.03-0.26	7.2-8.4
1988	0.35-0.60	0.35-0.90	0.60-1.05	0.25-0.36	0.35-3.10	0.15-1.32	0.72-1.90	0.04-0.22	7.2-8.6
1989	0.20-0.90	0.24-0.84	0.72-1.20	0.20-0.38	0.45-3.92	0.20-1.50	0.85-2.02	0.02-0.30	7.2-8.6
Mean	0.30	0.37	0.93	0.30	2.35	0.64	1.20	0.13	7.8

Studies on pollution can be regarded to have three different aspects :

i. Monitoring of pollution i.e. proper identification of pollutants and their sources.

ii. Prevention of pollution and preservation of ecosystem.

iii. Resource generation by recycling or reuse of pollutants.

Most of the studies on the pollution of the River Ganga are limited to monitoring the water quality of the river at different points and use different remedial measures like construction of electric creamatorium, the disposal of dead bodies, sewage treatment plants for the treatment of effluents to be discharged in the river. Though sediment quality can serve as a method of monitoring pollution but no systematic studies have been made to monitor the sediment quality of the River Ganga. Author and his co-workers have undertaken a systematic study of the sediment quality of the River Ganga. The study was further extended to derive useful conclusions.

The physico-chemical properties and nutrient contents of the sediments of the river Ganga give us valuable information regarding the feasibility of sediments for use in agronomy, i.e. resource generation and preservation of ecological balance.

The C and N contents of the Ganga sediment have been found tobe fairly good and amenable for agricultural purpose. The C/N ratio is usually much less than 10 in most cases. The condition of soil as regards to breakdown of organic materials is indicated by a C : N ratio which should be 10 : 1 under satisfactory conditions.

The amount of Phosphorus in sediments, in general, exceeds 12 kg/ha and usually around 18-20 kg/ha. Thus, P-content can be regarded to be medium to high, whereas K-content can be regarded to be very high as K-content in almost all cases exceeds 280 kg/ha. It is to be noted that the Ganga water is rich in K-content.

Inspite of fluctuations, it is apparent that the nutrient (like N and P) budget of the sediments are considerable and has been on the increase. This is reasonable in view of increasing drain-off of fertilizers, pesticides from the agricultural fields and discharge of effluents containing synthetic detergents and other organic matters. The P-budget of the river Ganga has been reported to be high. Naturally sediments should be enriched with phosphate deposits. The N-budget containing NH_3 (new pollution) and NO_3 had also been found to be high in the water of the river Ganga. Naturally, this may arise from anaerobic decomposition of organic matter in sediments. This is probably the reason for the usual low C/N values of the sediment.

Cation-exchange capacity of the soil is usually limited. But the cation exchange capabilities of the Ganga sediments have been found to be appreciable as observed from the estimates of Na^+, K^+, Ca^{2+} and Mg^{2+} suggesting appreciable clay minerals in the sediments. It is known that clay materials have significant ion-exchange capability, a factor responsible in holding nutrients like Na^+, K^+, Ca^{2+} and Mg^{2+} as well as NO_3 and PO^{3-}_4 Causing soil fertility. Moreover, the availability of P is very much dependent on these elements.

Sulphur is an essential nutrient and upto 70 kg ha^{-1} may be needed for optimum growth of plants with intensive agriculture. We have determined the amount of S as SO_4 and S-content is sufficient in sediments. Moreover, atmospheric input of S from rain occuring in most parts in sufficient for most of the crops.

Martin and Maybeck (1979) observed considerable depletion of Na, K. Ca and Mg as a result of chemical erosion which globally represent 25% of the total river load. Our studies (Mukherjee et. al 1993) with sediments show considerable depletion of these elements due to chemical erosion and soluble compounds are transported easily by the river water causing ecological imbalance.

The possible use of sediments for agronomy may help to arrest the depletion of these elements.

The pH of sediment range between 7.4 - 8.6 (usually around 7.8). The pH of soil influences the suitability of soil for crop production by ascertaining (i) the physical properties of soil like structure, permeability, availability of nutrients (ii) lime and gypsum requirements and (iii) microbial activity.

It is apparent that the pH of the sediments will be suitable for agricultural purposes and no treatment is usually necessary. The pHs are amenable for microbial activity. The pH-values and the cation exchange capabilities of the sediments show that the soil can absorb considerable acid rain.

It is to be noted that we have to depend on increased imported fertilizers to augment the soil fertility. However, the sediments contain enough P and other nutrients. The sediments alone or with very little a amount of fertilizer can be utilised for agricultural purpose resulting in reduced dependence on fertilizer. This may save foreign exchange. Desilting would reduce floods causing less damages to crops and properties.

The demand of fertilizers in soil increases by about 5-10% annually. The soil texture and the soil fertility decrease with years with increased use of inorganic fertilizers. Moreover, the phosphate fertilizer is usually fixed but the considerable exchange phosphate in soils resulting in the improvement of soil fertility and soil texture.

Moreove, sediments contains a number of phosphate solubilizing bacteria (Mukherjee et al. 1995) which under proper condition will enhance phosphate availability by solubilization.

Thus, the use of sediments means better soil management with resource generation and compensation of soil erosion leading to the restoration of ecological balance.

However, extensive work is needed and systematic analysis of more samples are needed to derive useful conclusions.

❑ REFERENCES

(Smt) S.N. Sinnarkar, S.K. Kesarwani and S. G. Bhat (1987) River Ganga - An overview of Environmental Research, NEERI, Nagpur. p 18.

F. J. Welcher, Standard Methods of Chemical Analysis 6th Edition, Vol 2 Part B D. Van Nostrand Company Inc Princeton New Jersey, 2399-2467.

C.A. Black, Methods of Soil Analysis Part 2 American Society of Agronomy Inc. Madison. Wisconsin U.S.A. pp. 1112.

M. L. Jackson, Soil Chemical Analysis Prentice Hall Inc. Englewood Cliffs NJ 2nd Printing pp 111-183.

(Mrs.) M. Chattopadhyay (nee Ray), D. Mukherjee, S.K. Bhattacharyya and S. C. Lahiri, The Environmentalist, (Communicated)

D. Mukherjee, (Mrs.) M. Chattopadhyay (nee Ray) and S.C. Lahiri *J. Ind. Chem. Soc., 1993, 70,* 56-58.

M. Chattopadhyay (nee Ray) and S.C. Lahiri *J. Ind. Chem. Soc., (in press)*

D. Mukherjee, M. Chattopadhyay and S.C. Lahiri, *The Environmentalist, 1993, 13,* 199-210.

C. Sreekumaran, K. T. Joseph and M. Parameswaran *Current Science, 1970,* 5, 105.

J. M. Martin and M. Meybeck, *Marine Chemistry 1979,* 7, 173.

D Mukherjee, J. Das, S.C. Lahiri and G.M. Sarkar, *J. Ind. Chem. Soc., 1995, 72,* 767-770.

Authors :

1. **M. Chattopadhyay (nee Ray) :** Department of Chemistry, Kalyani University, KALYANI - 741235. W.B.

2. **S.C. Lahiri :** Department of Chemistry, Kalyani University, KALYANI - 741235. W.B.

■

25

Physico-chemical Characteristics of Ganga River Water at Mokamah (Begusarai) Bihar

❑ INTRODUCTION

There are a total of 113 rivers in the country with a total length of 45,000 km. out of these rivers. 14 are major. 44 medium and 55 minor ones. Ganga River called "The Mother River of India" is one of the 14 major rivers in the country. No doubt its unique holiness has attracted a large number of Indian scientists to carry out investigations with respect to its physical, chemical & Bacterielogical quality at different locations in the country.

The Ganga about 2500 km. long enters Bihar State at Buxar & exits at Rajmahal covering a distance of nearly 560 Km. In Bihar four class-I cities four class-II cities & seven towns having a population of less than 0.5 lakh & a number of villages are located on the bank of Ganga River. But an investigation into its physico-chemical quality in & around Rajendra Bridge at Mokamah (Begusarai). Bihar has special significance because of the fact that Barauni - Mokamah Industrial complex is regarded as a centre of major Industries in North Bihar comprising H. F. C. Ltd. Barauni Thermal Power Station, I. O. C. Ltd., Barauni. Bata India Limited, Mokamah, Mc Dowell & Co. Ltd., Hathidah etc. In order to assess physico-chemical quality of Ganga River at Mokamah with reference to the possible polluting effect caused by these Industries, more than ten parameters have been analysed.

Bassed an analytical findings a conclusion has been drawn with respect to the suitability of Ganga River water for different purposes.

Material and Methods

Water samples of Ganga River were collected for a period of one year (1991) from different sampling points in & around Rajendra Bridge at Mokamah during winter, Summer & Rainy seasons throughtout the year. The analytical methods were adopted as per Standard Methods for Examination of Water and Waste Water. *(APHA-AWWA-WPCF 1989).*

Analytical Results

Table 1 indicates the physico-chemical quality of Ganga-River water at Mokamah Collected from several sampling points during summer. Winter and Rainy seasons in the year 1991.

TABLE : 1

Physico-chemical characteristics of Ganga River water at Mokamah, (Begusarai)

Sr. No.	Parameters	Winter Season		Summer Season		Rainy Season	
		Min.	Max.	Min.	Max.	Min.	Max.
1.	Temp. ^{0}C	19.6	22.6	26.8	31.6	27.0	30.2
2.	pH	7.2	7.5	7.3	7.6	7.0	7.3
3.	Electrical Conductivity μmhos/cm.	327	345	335	418.0	308	346
4.	DO mg/1	7.0	7.3	6.8	7.2	6.7	7.0
5.	* BOD 5 at 20^{0}C mg/1	0.9	2.1	1.0	2.7	1.8	3.2
6.	** COD mg/1	10.8	18.5	12.0	23.0	16.0	26.0
7.	Chloride mg/1	13.2	16.5	14.6	17.8	9.8	12.6
8.	Sulphate mg/1	6.2	7.5	7.8	9.4	6.4	7.2
9.	Total Hardness (as CaCO3) mg/1	146	165.4	138	152.5	145	158.0
10.	Total solids mg/1	258	384	242	358	252	324
11.	*** T.S.S. mg/1	34	42	27	36	24	32
12.	Turbidity (N.T.U.)	8.0	11.0	7.0	12.0	14.0	26.0

* Biochemical Oxygen Demand in mg/1.

** Chemical Oxygen Demand in mg/1.

*** Total suspended Solids in mg/1.

Results & Discussion

Most of the parameters analysed for assessing physico-chemical characteristics of Ganga water have been found to be within tolerance limits as preseribed by I. S. I. (I. S. 10500-1983). The overall values of pH, DO and Bio-chemical oxygen Demand indicate that the river can be used as a drinking water source without conventional treatment but with sufficient disinfection. The river water seems to be suitable for bathing, swimming & recreation purpose.

Low COD values clearly indicates the river at the particular site is not organically polluted. But all the industrial establishments in & around Rajendra Bridge at Mokamah must be cautioned against discharge of untreated industrial effluents into the river with a view to maintain its physico-chemical quality in future.

So far as its microbiological quality is concerned. Ganga water at the site may be polluted due to the disposal of residues of dead bodies. Although it has not been ascertained by Bacteriological investigations.

Conclusion

Due to the fast flow rate & abundance of water in the river, the values of pH, DO, BOD & other significant parameters undergo slight variation within the tolerance limits prescribed by I.S.I. & W. H. O. self purification capacity of the Ganga River is also an important factor for its satisfactory physico-chemical quality.

Acknowledgement

The authors are thankful to the Principal. *Samastipur college, Samastipur* for providing laboratory facilities as well as his encouragement during the present work.

❑ REFERENCES

A.P.H.A. (1980). *Standard Methods for Examination of Water and wastewater*. APHA, USA.

Rao S.N. et. al (1990) " Ganga water quality in Bihar", *Indian J. Environ Health Vol. 32*, No. 4. 393-400.

"River Ganga" *An overview of Environmental Research*" National Env. Engg. Research Institute, Nagpur (India)

Bilgrami K. S., Datta Munshi J. S. & Bhowmick B. N., Biomonitoring of River Ganga at pollnted sites in Bihar, *world science news 1985. 22(2)* 14-15.

Pandey N. C. Pollution of River Ganga in U. P. with specific Reference to Varanasi, civic Affairs, 1985. 32(11) 52-59.

Dey A. K.Environmental chemistry, wiley eastern Ltd. India (1987)

World health Organisation 1971. *International standards for drinking water 3rd edition, W. H. O.* geneva.

Trivedy R. K. and Goel P. K. "*Current pollution Researches in India*", Environmental Publications, Karad (India)

Trivedy V. N., Control of Industrial Discharge Effluent in River Ganga, *civic Affairs 1986. 33 (11)* 39-47.

Sandwar B. B., and Prasad J., "Quality of Ganga River water in Bihar W. R. T. Heavy metals pollution. 82nd Annual conference of Ind. Sc. Congress Association at Jadavpur University, Calcutta (3-8 Jan' 1995).

Authors :

1. **B.B. Sandwar** : Samastipur College, Samastipur. (Bihar) India.
2. **J. Prasad** : P. G. Deptt. of Chemistry, Samastipur College, Samastipur (Bihar) India.

■

26

Studies on Distributions and Speciations of Arsenic on Najafgarh Drain Entering River Yamuna in Delhi

❑ INTRODUCTION

Environmental pollution by toxic metals has assumed global significance in recent years. All forms of ecological systems are affected to varying extent by their toxicity, mobility, bio-availability and solubility in aquatic environment. They are most insidious pollutants because of their non-biodegradable nature and have the potential to cause adverse effects at certain levels of exposure and absorption. Speciation of toxic metals is defined as the identification and quantification of the individual physico-chemical forms of an element in a sample. Speciation studies are important because the toxicity and availability of an element ofen depends on its chemical form. (Sean X.C. Le et al 1994). In recent years attention has been paid to those elements that exhibit multiple oxidation states in water. In aquatic environment inorganic arsenic is known to exist in the +5, +3, 0 and -3 oxidation states. Arsenite As (III) and Arsine (AsH_3) are considered more toxic to human than the +5 oxidation state. (Seyler et al 1989, Wai-manMok et al. 1990, Aurillo et al 1994 and Chatterjee et al. 1995). Therefore, knowledge of Arsenic speciation in aquatic environment is of particular concern and interest.

Najafgarh drain is the largest open waste water drain which passes through most populous industrial and residential colonies creating very unsightly and unaesthetic appearances throughout it route to the River Yamuna. Najafgarh drain basin area within the National Capital territory is 832 sq. km. and it receives discharge from various small drains, which ultimately outfall into the river Yamuna at down stream of Wazirabad barriage. Najafgarh drain basin area is one of the twelve identified problem areas in the country from pollution point of view (CPCB Report 1990). In addition to domestic wastes, the drain carries industrial effluents from the industrial areas viz Wazirpur, Lawrence Road, Najafgarh Road, Kirti Nagar, Mayapuri and Anand Parbat. Industries located in these areas include electroplating, chemical, pesticides, fertilizers, dyeing, engineering, food processing and many others. Wastes from these industries have enough potential to increase the existing pollution of Najafgarh drain considerably due to toxic metals. Keeping in view of above, studies on distributions and speciations of Arsenic on the course of drain is significant for environmental interest.

Study Area

The sampling sites were selected on the consideration of maximum representativeness and approachability. Five sampling points were selected to cover the whole stretch of drain starting from the village Najafgarh down to its discharge point into river Yamuna at Wazirabad barriage, to assess the broad spectrum of the distribution of metallic pollutants. Two sampling stations were selected on River Yamuna viz. Up-stream and down-stream of the River, for assessment of possible impact on it. All these strategic points were designated as follows

S-1 : Up-stream of drain at Najafgarh village.

S-2 : Down-stream of drain after Kirti Nagar industrial area.

S-3 : Down-stream of drain after Lawerence Road Industrial area.

S-4 : Down-stream of drain after Wazirpur Industrial area.

S-5 : Tail end point of drain at Wazirapur barriage.

S-6 : Up-stream of river Yamuna.

S-7 : Down-stream of river Yamuna.

MATERIAL AND METHODS

Sampling programme

The objective of water sampling is to collect a portion of water, small enough in volume to be transported conveniently to the laboratory, while still accurately representing the water source being sampled. For the present study, sample from each strategic point were collected in four different seasons viz summer, monsoon, post-monsoon and winter of the year 1994-95. Collected samples were preserved as per the established guidelines and transported to laboratory within the maximum permissible recommended time (APHA 1992).

Experimental

Determinations of arsenic was performed with a VARIAN-A A 20 PLUS atomic absorption spectrophotometer equipped with varian hollow-cathode lamps. For determination of arsenic at ultra-trace level, an automated continuous flow vapor generation accessary VGA-76 was used, which gives convenience of an automated system in addition to the superior accuracy and precision associated with a continuous-flow system, providing steady-state atomic absorption signals. Silver Diethyldithio Carbmate method (APHA-1992) was used for determinations of in-

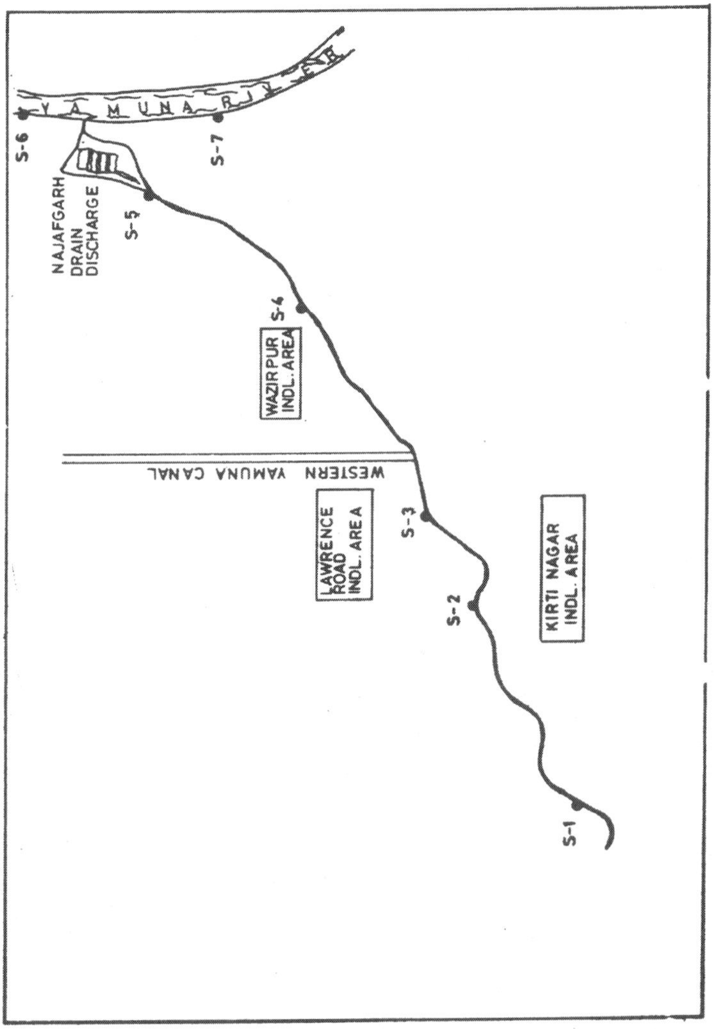

Figure 1. Map of Najafgarh Drain with Sampling sites.

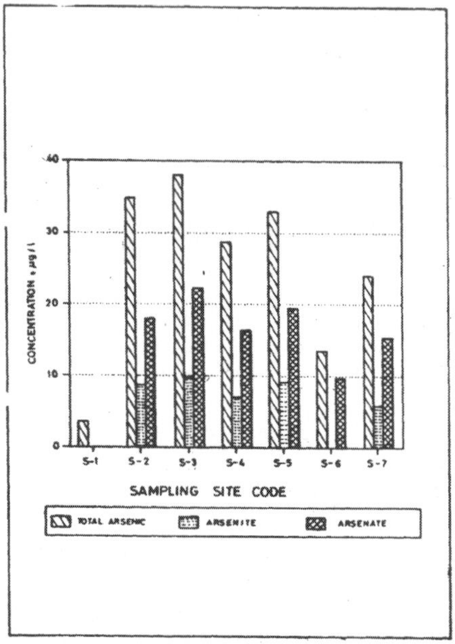

Figure 2. Speication of **Arsenic** on Najhafgarh Drain Entering River Yamuna in Delhi.

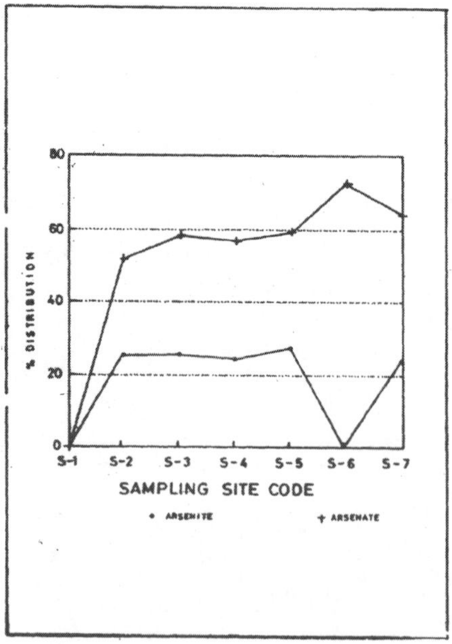

Figure 3. Distribution of Arsenite and Arsenate (% of total Arsenic)

organic speciation of arsenic, i.e. arsenate and arsenite at ultra-trace level.

All reagents used in experiments were of analytical-reagent grade or higher purity. De-ionised water from milli-Q system was used for reagent preparation and blank correction.

Results and Discussion

Results obtained with all the samples from each strategic point in four different seasons of the year are presented in table-1. As indicated by analytical data, higher values of pH are encountered at each point in the post-monsoon season. This reflects on the alkaline nature of the washout added during monsoon season. Concentration of suspended solids were also found higher at each sampling station during post-monsoon season. As is clear from the graphical representation of the analytical data, concentration of total arsenic, arsenite, & arsenate shows increasing trend at down-stream of the drain after Kirti Nagar industraial area (S-2_. Concentration of arsenate and arsenite at above site were found 25.2 and 51.6% respectively of the average concentration of total arsenic, which is 34.9mg/L.

Table 1

pH and total suspended solids at different sampling sites

Site Code	pH value				TSS, mg/L			
	Summer	Monsoon	Post-Monsoon	Winter	Summer	Monsoon	Post-Monsoon	Winter
S-1	7.8	7.7	8.0	7.7	35	32	48	37
S-2	7.2	7.3	8.0	7.8	90	75	120	84
S-3	7.1	7.2	8.2	7.9	82	68	98	77
S-4	7.2	7.4	8.0	7.5	68	45	90	55
S-5	7.3	7.2	8.1	7.7	60	48	80	48
S-6	7.6	7.7	7.9.	7.6	44	30	70	32
S-7	7.4	7.5	7.7	7.5	55	38	80	45

Down-stream of drain after Lawerence Road Industrial ares (S-3) exhibits the average concentration of total arsenic 38.1 mg/L, while arsenite and arsenate at above site were found 25.7 and 58.5% respectively of total arsenic. Increasing concentration of each form of arsenic at above point may attributes to the intense industrial activites in above areas. Decreasing concentration of each form of arsenic is noticed on the drain at the down-stream of the drain after Wazirpur industrial area as evident by the dilution of the drain by western Yamuna Canal before the above point. Upstream of river Yamuna was monitored to get the baseline data for different forms of arsenic and investigations carried out at above point shows that arsenite is found below detection limit in all the four season of the year, while average concentration of arsenate is observed 9.8 mg/L, which is 72.6% of total arsenic. Investigations carried out at down stream of river Yamuna after the tail end point of Najafgarh drain at Wazirabad Barriage, reveals that the average concentration of total arsenic and arsenate at down-stream is increased by 77.8 and 57.1% respectively, as compared to the analytical data of the same observed at up-stream of the river. Average concentration of arsenite and arsenate at the down-stream of the river are found 24.2 and 46.2% respectively of the average concentration of total arsenic, which is observed as 24.0 mg/L.

Table 2

Concentrations of Total Arsenic, Arsenite & Arsenate at different sampling sites (μg/L)

Site	Summer			Monsoon		
Code	As	As (III)	As (V)	As	As (III)	As (V)
S-1	4.0	ND	ND	2.0	ND	ND
S-2	35.0	8.0	18.0	26.0	5.0	14.0
S-3	42.0	10.0	26.0	27.0	5.0	16.0
S-4	28.6	6.0	14.0	20.0	4.0	12.0
S-5	30.5	7.5	20.0	25.0	5.0	15.0
S-6	12.5	ND	8.0	8.5	ND	8.0
S-7	22.0	5.0	15.0	14.0	ND	10.0

Site	Post-monsoon			Winter		
Code	As	As (III)	As (V)	As	As (III)	As (V)
S-1	4.5	ND	ND	3.5	ND	ND
S-2	40.5	12.0	22.0	38.0	10.0	18.0
S-3	38.0	9.0	21.0	45.5	15.0	25.0
S-4	32.0	8.0	20.0	34.0	10.0	19.0
S-5	36.0	10.0	20.5	40.0	14.0	22.0
S-6	15.0	ND	12.0	18.0	ND	14.0
S-7	28.0	8.0	18.0	32.0	10.0	18.5

* ND = Not Detected

Conclusion

The investigations presented herein clearly indicate that there is significant increase in the concentration of total arsenic and arsenate in river Yamuna by Najafgarh drain carrying industrial effluents from various industries. Eventhough concetration of total arsenic as indicated by our study, has increased at the downstream of the river, still concentration observed is within the prescribed norms of arsenic on surface waters. Therefore, further investigations are required to be done, on the sediments of the river for the same parameters.

Table 3

Distribution of Total Arsenic, Arsenite & Arsenate at different sampling sites

Site	Total Arsenic (As)	Arsenite As (III)		Arsenate As (V)	
Code	Average	Average	% of Total	Average	
S-1	3.5	-	-	-	-
S-2	34.9	8.8	25.2	18.0	51.6
S-3	38.1	9.8	25.7	22.3	58.5
S-4	28.7	7.0	24.4	16.3	56.8
S-5	32.9	9.1	27.6	19.5	59.3
S-6	13.5	-	-	9.8	72.6
S-7	24.0	5.8	24.2	15.4	64.2

Acknowledgement

Authors are highly grateful to Dr. D.A. Dabholkar, Director Shriram Institute for Industrial Research, Delhi for his ever encouragement and support to such kind of studies. Valuable suggestions on the subject from Mr. M. S. Dhingra and Mr. V. G. K. Nair are duly acknowledged thankfully.

❑ REFERENCES

APHA, AWWA, WPCF (1992). *Standard Methods for Examinations of Water and Wastewater*. APHA, U.S.A.

Aurillo, Robert P. Maron, and Harold F. Hemond (1994). Speciation and fate of arsenic in three lakes of the Aberjona Watershed. *Environ. Sci. Technol.* 28 (4) : 577-585.

Amit Chatterjee, Dipankar Das, Badal K. Mandal, Tarit Ray Chowdhury, Gautam Samanta and Dipankar Chakraborti (1995). Arsenic in ground water in six district of West Bengal, Arsenic species in drinking water and urine of affected people. *Analyst*. 120 (3) : 643-650.

Central Pollution Control Board Report - 1990.

Patrick H. Masscheteyn, Ronald D. Delanna, and William H. Patrick, Jr. (1991). Effect of redox potential and pH on arsenic speciation and solubity in a contaminated soil. *Environ. Sci. Technol.* 25 (8) : 1414-1418.

Patrick, Seyler and Jenn-Marie Martin. (1989). Biogeochemical processes affecting arsenic species distribution in a permanently stratified lake. *Environ. Sci. Technol.* 23 (10 1258-1263.

Sean X.C., William R. Cullen, and Kenneth J. Relmer. (1994). Speciation of arsenic compounds in some marine organism. *Environ. Sci. Technol.* 28(9) : 1598-1604.

Wai-Manmak and Chian M. Wal (1990). Distribution and mobilization of arsenic and antimony species in the Coeur D'Alene river, Idaho. *Environ. Sci. Technol.* 24(1) 102-108.

Authors :

1. **Jagdish Kumar, D. P. Singh, R. K. Banerji, C.L. Jain** : Shriram Institute for Industrial Research, 19, University Road, DELHI 110 007.

27

Water Quality of Bhadra River in Karnataka Due to the Effluent Discharge From Paper and Steel Industries

❑ INTRODUCTION

It is an established fact that the water quality of rivers is getting deteriorated due to the disposal of waste from townships and industries. Due to urbanisation and industrial growth and also due to the unscientific utilization of water resources, the pollution level of almost all the rivers has increased.

In this paper, an attempt has been made to findout the pollution level in Bhadra river, as the two major industries Mysore Paper Mill (MPM) and Vishweswaraya Iron & Steel Ltd., (VISL) are discharging their effluents to this river. In the present study, only two stations have been selected, one at upstream (U/S) and another at downstream (D/S) from the point of effluent discharge. The physico-chemical parameters are analysed at monthly interval for a period of one year.

Many investigators like Sinha (1988) Shukla et al (1988), Trivedy (1989), Singh and Singh (1992), Kapoor (1993) have reported work on Indian rivers But enough data on Bhadra River is not available. Bhadra river orginates from western Ghats before joining Tunga River which later becomes Tunga Bhadra River.

MATERIALS AND METHODS

River water samples were collected once in a month from January to December 1995. The samples were collected on 5th of every month between 9 AM to 11 AM taking care to get the representative samplings. Water analysis was carried out as per APHA (1992) Trivedy and Goel (1986) and NEERI (1980). The water temperature, pH & DO were determined at the site while the other parameters were determined in the laboratory.

RESULTS AND DISCUSSION

It is observed that DO values are quite low for all the months in D/S compared to

U/S water (Table No. 1). The seasonal variation in DO are also observed. The minimum DO was found to be zero and maximum 2.8 mg/lit for D/S, whereas for U/S the DO found to be in the range of 6.3-8.0 mg/lit. It is also observed that the DO of U/S water on an average remains constant for all the months except for the month of July and August. (Table 1 & 2)

Table 1

Variations in Dissolved oxygen in Bhadra River (Year 1995, Unit : mg/lit)

Month	U/S	D/S
January	6.7	0.7
February	7.3	1.2
March	7.4	Nil
April	8.0	2.8
May	7.5	2.5
June	7.8	Nil
July	6.3	Nil
August	6.5	2.6
September	7.0	2.5
October	7.0	2.5
November	7.3	0.4
December	7.5	0.7

The low value of oxygen content could be due to the release of oxygen demanding wastes. In the rainy season, in an unpolluted river, the D.O. value usually remain high but in the present study the D.O. value are much lower than the standard value even in the rainy season.

The BOD values of U/S were water found to be in the range of 1.0-1.5 mg/lit and for D/S water values are in the range of 13.4-222 mg/lit. (Table No.2). A comparison of D.O. & BOD values with its standard limits is shown in Table No. 3.

Table 2

Variation in BOD of Bhadra River (Year 1995, Unit : mg/lit)

Month	U/S	D/S
January	1.3	120
February	1.3	110
March	1.2	170
April	1.1	220
May	1.0	222
June	1.1	120
July	1.2	170
August	1.0	124
September	1.2	134
October	1.0	160
November	1.2	180
December	1.5	150

From the values obtained, it is evident that the effluent added high concentration of organic matter responsible for the degradation of the river water quality.

Table 3

Indian Standards Limits for Bathing and Drinking Water

Parameter	Bathing limits	Drinking limits
pH	7.0 - 8.5	7.0-8.5
DO mg/lit	5.0	6.0
BOD mg/lit (5d, 20^0 c)	3.0	Nil

The chloride concentration varied between 7.1 to 12.8 mg/lit in U/S and between 7.1 to 68.7 mg/lit in the D/S. Similarly hardness varied from 22-80 mg/lit in the U/S and 72-160 mg/lit in the D/S. Total dissolved solids varied from a minimum of 16 mg/lit in the U/S to a maximum of 512 mg/lit in the D/S and similarly the average values for sulphates, sodium and calcium are more in the D/S compared to U/S water. (Table No. 4).

Table 4

Variations in pH, Conductivity, Temp., Hardness, Sodium, Calcium, TDS, Chlorides and Sulphate in Bhadra River (Year 1995).

Month	pH		Conductivity in Micromhos/cm		Temperature		Hardness mg/lit		Sodium mg/lit		Calcium mg/lit		TDS mg/lit		Chloride mg/lit		Sulphate mg/lit	
	U/S	D/S	U/S	D/S	U/S	D/S	U/S	D/S	U/S	D/S	U/S	D/S	U/S	D/S	U/S	D/S	U/S	D/S
Jan	6.9	5.5	150	450	25	28	64	160	19	36	36	112	16	216	10.6	42.5	20	38
Feb	7.7	7.4	90	160	24	25	32	100	24	49	20	30	24	200	7.1	12.05	15	32
Mar	7.0	7.4	180	400	25	26	80	120	22	47	32	58	56	512	9.21	34.1	20	34
Apr.	7.0	8.4	70	110	29	29	42	80	20	35	16	22	18	156	12.1	17.1	5	40
May	7.0	8.2	280	390	28	28	54	72	15	27	24	42	38	130	9.2	22.7	15	50
Jun	6.9	7.5	140	520	30	30	56	140	22	38	22	60	36	140	12.1	68.7	20	58
Jul	7.1	7.8	110	370	26	27	56	124	18	30	28	80	62	152	12.8	47.1	02	58
Aug	7.9	8.5	110	210	27	27	62	86	14	28	30	48	82	158	12	24	18	45
Sep	7.5	7.6	70	82	25	25	22	72	05	10	10	12	44	152	7.1	8.5	5	25
Oct	7.8	7.8	130	280	26	27	48	86	10	30	22	54	82	174	9.9	33.3	2	20
Nov	7.5	7.8	100	230	27	28	44	80	15	40	28	42	65	150	8.5	24.8	04	35
Dec	7.7	7.8	130	350	25	26	48	120	25	60	26	100	82	218	9.9	31.9	02	20

The conductivity values are indicative of solute content in water. In the present study the conductivity values are increaed from U/S to D/S in the average range of 70-200 micro mhos/cm, indicating water pollution.

On the basis of physico-chemical factors it can be concluded that the fluctuations observed in various parameters could be due to the inflow of industrial effluents and also due to the quantity of water released from Bhadra reservoir into the river. The overall results show the pollution in the D/S of Bhadra river is due to the discharge of industrial effluent specially from Paper and Steel industries.

ACKNOWLEDGEMENT

The authors wish to thank Principal J.N.N. College of Engineering, Shimoga for providing necessary laboratory facilities and also to All India Council for Technical Education, (AICTE) and Karnataka Pollution Control Board, Bangalore, for providing financial support to carry out this research work.

❑ REFERENCES

American Public Health Assoc. (1992). *Standard Methods for examination of water and waste water.*, APHA, 18th Edition, U.S.A.

Bharathi S.G. and Krishnamurthy S.R. (1990). Effect of Industrial effluents on River Kali around Dandeli, Karnataka I, Physico-chemical complexes. *Indian J. Environ. Hlth.* (2) : 167-171.

Desai P.V. (1995) Water quality of Dudhsagar River at Dudhsagar (Goa), *Poll. Res.* 14 (4) : 377-382.

De A.K. (1987). *Environmental Chemistry*, Willey Eastern Limited, India.

Israili A.W. and Ahmed M.S. (1993). Chemical characteristics of river Yamuna from Dehradun to Agra *Indian J. Environ. Hlth* 35 (3) 199-204.

Sawyer C.N. and Mcarty R.C. (1978). *Chemistry for Environmental Engineers* 3rd Edition.

Singh D.K. and Singh C.P. (1990). Pollution studies on River Subarnarekha around Industrial belt of Ranchi (Bihar). *Indian J. Environ. Hlth* 32 26-37.

Sinha M.P. (1988). Effect of waste disposal on water quality of river Damodar (Bihar) Physico-chemical characteristics. In : R. K. Trivedy Ed. *Ecology and Pollution in Indian Rivers,* Ashish Publishing House, New Delhi.

Trivedy R.K. and Goel P.K. (1984). *Chemical and Biological methods* for water pollution studies, Environ. Publ. Karad. (Maharashtra)

Wilcox L.V. and Hatcher I.T. (1950). *Methods of analysis used in the Rubidoux laboratory Riverside, California,* U.S. Dept. of Agriculture, 6th Edition Revised.

Authors :

1. **S. K. Raikar (Ph.D.)** : Prof. & Head, Department of Applied Chemistry, Jawaharlal Nehru National College of Engineering, Shimoga - 577 201. Karnataka.
2. **Sri. Ravindranath Sabhahita** : Department of Civil Engineering, Jawaharlal Nehru National College of Engineering, Shimoga - 577 201. Karnataka.

■

28

A Quantitative Study of the pollution and Physico-chemical Conditions of the River Mayurakshi in Santhal Pargana, Bihar

❑ INTRODUCTION

The Mayurakshi is one of the major rivers in Bihar and most important one in Santhal Parganas (lies between 23^0 40' and 25^0 18'NL and between 86^0 28' and 87^0 57' EL). It is a rainfed perennial river taking its origin from Trikuti hills (Deoghar dist.) and meandering over 1860 km in the state of Bihar and West Bengal ultimately joins the Bay of bengal via the river Ganges. It swells up during rainy season but gets reduced to narrow stream in the summer. Due to Canada dam at Masanjore, the depth of water gets sometimes changed by controlled flow of water.

The Mayurakshi is the main source of irrigation and drinking water for the people of Santhal parganas. Not only this, the municipality of Dumka (a divisional town) pumps its water into the municipal area for domestic affairs, whereas the nearby tribal people use its water dircetly from the river for drinking and other purposes of life. Due to uncontrolled discharge of municipal and domestic sewage, the ecology of the river has been assaulted. The tribals of this area get effected by varios diseases, such as, enteric diseases, skin disorders and eye-defects on large scale. In recent years, by consuming such contaminated water there has been a serious out breaks of blood dysenter, diarrhoea, gastroentritis, leprosy and eczema in different parts of Santhal Parganas (Bihar), in the vicinity of the Mayurakshi river. It is, therefore, necessary that for preserving the ecological balance of this river in the most backward and tribal dominated area of Bihar, a continuous monitoring of the quality of river water should be carried out. Practically, till date, this type of investigation is still lacking in this tribal zone of Bihar except the work of Kumar (1994 a and b), who studied the nature of water in only lentic freshwater ecosystems of Santhal Parganas.

MATERIALS AND METHODS

Monitoring of the water quality was carried out at five sampling stations along the streteh of 60 km. of the Mayurakshi from Maharo (Upstream) to Masanjore (down stream) from March, 1991 to February, 1994. The selected five sampling stations were near Maharo bridge (station-1.) about 60 km. south-west to masanjore dam, Khetha (station-2), about 5 km. north-east to Maharo bridge, Hizla (station-3), 2 km. north-west to Dumka town, Masanjore reservoir (station-4), where the river water is stored (Gross storage capacity - 616.74 million cubic metre) and Masanjore village (station-5), 1/2 km. away from Canada dam where water comes out through a big canal. At station 1, 2 and 5 there was no sewage outfall, but at station-2, domestic, sewage drains, dead bodies, carecases, ashes etc. from a burning ghat were added and station-3 receives a big drain which adds a big amount of municipal sewage to the river. The water samples collected twice a month between 8.00 to 10.30 am. using glass stoppered bottle for spot analysis and in the polythene bags for laboratory analysis. All the physico-chemical parameters were determined following the standard methods (APHA 1985 Trivedy and Goel, 1986, NEERI, 1986.)

RESULTS AND DISCUSSION

The minimum-maximum and mean value of various physico-chemical parameters at all five sampling stations from March, 1991 to February, 1992 and their permissible limits have been depicted in table -1.

Temperature, pH and Electrical conuctance (EC)

The temperature ranged from low to high durning December to February and May to October, respectively at all the stations.

The water was found slightly alkaline as the pH fluctuated between 7.1 to 8.5. It is obvious that pH 7 is considered to be the most standard for various reactions as well as for human beings. The pH limit was found within the acceptable range for various purposes (IS:1982).

The electrical conductance was always within the tolerance limit (100μ S) (Ellis (1937) Its higher value at station -2 and -3 in comparision to rest of the stations indicated eutrophication (Radwan, 1976).

Total dissolved solid (TDS) and Total suspended solid (TSS)

TDS and TSS were within the tolerance limit at all the stations and were found suitable for drinking and othere domestic purpose after conventional treatment following disinfection. The water having high residue is not fit for the consumer because it may produce undesirable change in the normal physiology of the body. Therefore, the direct consumption of water at station-2 and -3 for domestic as well as irrigational purposes is not suggested and requires filtration and additional treatment to cope with the impurities existed therein.

Dissolved oxygen (DO) and Biochemical Oxygen Demand (BOD)

The high value of DO was recorded during winter months when temperature was low and its value was found low during summer when the temperature was high and river flow was slow. BOD was found inversely correlated with DO and near the sewage disposal sites (station -2 and -3), it was not only high but it was under stress, specially at station 2. It may be due to higher activities of human beings and reduced quantity of the river water. To remove this problem, the treatment and disinfection of water is suggested.

Free carbondioxide (F CO_2) and Alkalinity

The CO_2 was found absent during early summer and winter due to increased photosynthetic activity of bottom vegetation (Pahwa and Mehorotra, 1966) and deoxygenation (Talling, 1957) while it was present during monsoon which indicates its influx through rain in the form of carbonic acid (Chakraborty et. al., 1959). Its higher value at station -2 and -3 was due to increased decomposition of organic matters at bottom.

The total alkalinity index was only due to carbonate and bicarbonate ions. The relative carbonate concentration was recorded low at site 1, 2 and 5. This high concentration of carbonate ions at site -2 and 3 was due to high concentration of sewage (Robert, 1977).

Calcium and Magnesium

The river water analysis revealed the dominance of calcium over magnesium (Sharma, and Ghose, 1987). The concentration of these metallic ions was much lower than common recommended level (IS : 1982) for agriculture and domestic purposes.

Chloride and Nitrate

The data indicate a clearly defined increase in the concentration of chloride at both stations viz. 2 and 3 from the major source of pollution in the river. (Table-1) This increasing trend of chloride concentration was due to entirely sewage being carried into the river. Thresh *et. al* (1944) pointed out that higher chloride concentrations are indicators of large amount of organic matters in the water. Sharma and Ghose (1987) also noticed that the higher value of chloride in water occurs mainly due to the municipal sewage. Klein (1957) found a direct correlation between chloride concentration and pollution load. But in general, higher chloride value were during the dry season at all the stations and this increase is due to reduced flow and relatively large amount of sewage waste and human activities in the river.

Table 1

Minimum, maximum and average physico-chemical characteristics of River Mayurakshi at different sampling sites. (All values are in mg/*l*, except Temp. (^{0}C), pH, Transparency (cm), and Conductivity (*u*S/cm)

Parameters		Site-1	Site-2	Site-3	Site-4	Site-5	Tolerance Limit Class-C
Water Temp.	Range-	20.8-30.5	20.5-31.5	20.2-30.4	20.4-32.4	18.2-30.4	
	Mean -	24.5	24.22	24.50	23.88	24.48	
pH	Range-	7.1-8.5	7.2-8.4	7.1-8.3	7.1-8.0	7.3-8.5	6.5-8.5
	Mean -	7.85	7.75	7.77	7.52	7.59	
Conductivity	Range-	105-280	110-310	135-350	191-212	102-250	
	Mean -	150	190	280	170	155	
F CO_2	Range-	0.0-5.5	1.2-12.4	2.2-13.3	0.0-8.8	0.5-7.2	
	Mean -	1.8	4.2	6.0	2.2	1.7	
DO	Range-	6.8-10.2	5.1-9.5	4.5-9.2	6.6-12.6	6.00-12.9	4mg/L
	Mean -	8.88	6.5	5.5	8.5	7.4	
Alkalinity	Range-	80-115	125-210	180-299	100-145	85-105	
	Mean -	98.5	150.3	201.5	110.4	95.4	
BOD	Range-	2.1-4.5	3.5-12.5	4.1-15.6	2.5-5.2	2.0-4.1	3mg/L
	Mean -	3.5	7.1	12.5	3.8	3.4	
TDS	Range-	60-110	80-130	85-150	65-105	58-108	1500
	Mean -	70	90	110	71	65	
TSS	Range-	180-305	210-410	208-425	198-360	185-310	
	Mean -	250	348	390	260	255	
Chloride	Range-	7.5-15.4	15.6-30.4	12.5-50.5	10.5-24.6	10.6-20.4	
	Mean -	10.5	25.4	40.0	18.8	12.3	
Nitrate	Range-	0.001-0.022	0.003-0.031	0.012-0.041	0.002-0.021	0.001-0.020	50
	Mean -	0.009	0.012	0.022	0.010	0.005	
Sulphate	Range-	5.1-10.5	12.5-20.8	14.3-35.2	6.0-11.4	5.2-9.8	400
	Mean -	8.1	16.2	22.5	8.8	7.8	
Phosphate	Range-	0.005-0.135	0.041-0.268	0.048-0.325	0.007-0.150	0.005-0.132	
	Mean -	0.038	0.112	0.128	0.042	0.036	

Although the nitrate anions were found within the permissible limit, the nitrate concentration was recorded comparatively higher at station -2 and -3. This is attributed to the high load of organic wastes (Shrama *et. al.* 1982).

Sulphates and Phosphates

The sulphate ions were found within the permissible limit at all the sampling stations. However, it was found higher at station -2 and 3 than the remeining stations due to the municipal sewage (Raina *et. al.*, 1984).

The higher amount of phosphate was also recorded due to high load of organic wastes (Atkins, 1923) in sewage discharge. Thus, a control over the sewage discharge is suggested so that the ecobalance of the river is not distarbed.

Therefore, the careful analysis of the pollution parameters reveal that the station -1, 4, and 5 are least polluted but the site -2 and 3 (near the sewage outlet points)are considerably polluted. The municipal sewage which contributed the maximum organic pollution resulting in an increased BOD and nutrient levels, is playing a key role as a main source of pollution in the water of Mayurakshi. The high concentration of BOD, TDS, TSS, metallic ions and dissolved anions at station -2 and 3 are definitely undesirable and unfit for domestic purposes. To keep the river free from pollution, discharge of unterated municipal sewage and waste disposal must be controlled and a continuous monitoring of the river water for various physico-chemical parameters should be done to preserve the ecobalance of the Mayurakshi.

Acknowledgements

The author is grateful to professor J.S. Dattamunshi, FNA, FNASC, Emeritus Scientist (CSIR) and Prof. S.P. Roy of Bhagalpur University for encouragement.

❑ REFERENCES

APHA 1985. *Standard Methods for the Examination of Water and Wastewater*. 16 ed. American Public Health Association, New York.

Atkins, W.R.G. 1923. The phosphate contents of fresh and salt waters in relation to the growth of the algae plankton. *J. Mar. Biol. Ass. U.K.* 13 : 119-150.

Chakraborty R.D., Roy R.D. and Singh S.B. 1959. A quantitative study of of the pollution and the physico-chemical conditions of the river Jamuna at Allahabad in 1954 - 1955. *Indian J. Fish.* 6 : 186-203.

Ellis M.M. 1937. Detection and management of stream pollution. In *Biology of water pollution* (D.F. Kuep *et. al.* ed.), USDI, FWPCA, Cincinnati.

I. S. I. 1982. Standard Tolerance limits for Inland surface waters subject to pollution, second revised edition, *Bureau of Indian Standard*, New Delhi.

Klein L. 1957. *Aspects of river pollution.* ButtorWorths Scientific Publication, London.

Kumar A. 1994 a. Seasonal trends in biological and physico-chemical properties of a fish pond of Dumka (Bihar). *Acta Ecol.* 16(1) : 50-57.

Kumar A. 1994 b. Role of species diversity of aquatic insects in the assessment of pollution in wetlands of Santhal Pargana (Bihar). *J. Environ. Poll.* 1 (3 & 4) : 117-120.

NEERI 1986. *Manual on Water and Wastewater Analysis.* National Environmental Engineering Research Institute, Nagpur, India.

Pahwa D. V. and Mehrotra S. N. 1966. Observations in the abundance of plankton in relation to certain hydrobiological conditions of river Ganga. Proc. Nat. Sei. Aca. India. 36 B (2) : 157-189.

Radwan S. 1976. Planktonic rotifers as indicator of lake eutrophy. *Ann. Univ. Mariae, Curie, Skiodowake. Sec. C..Biol.* 31 : 227-236.

Raina V., Shah A.R. and Shakti R.A. 1984. Pollution studies on river Jhelum -1; An assessment of water quality. *Indian J. Env. Hlth.* 26 (3) : 187-201.

Robert S.A. 1977. Quality of water for irrigation and drainage. *Deviation.* 103, I.R. 2 : 125-154.

Sharma C.B., Sinha S.N. and Ghose N.C. 1982. Pollution study of the river Gánga at Patna. *Proc, Poll, Control and wat. Conserv.,* Inst. Engg., Patna. 111-117.

Sharma C.B. and Ghose N.C. 1987. Pollution of river Ganga by municipal waste : a case study from Patna. *J. Geo. Soc. India.* 30 (5) : 369-385.

Talling J.F. 1957. Diurnal changes of stratification and photosynthesis in some tropical African waters. *Proc. Roy. Soc.,* 147 (B) : 57-83.

Thresh J.C., Suckling E.V. and Beale J.F. 1944. Chemical and zooplankton studies of lentic habitas in North Eastern New South Wales. Aust. *J. Freshwat Res.* 21 : 11-33.

Trivedy R.K. and Goel P.K. 1986. *Chemical and Biological Methods for Water Pollution Studies.* Environmental Publications, Karad.

Authors :

1. **Arvind Kumar :** Environmental Biology Research Laboratory Post Graduate Department of Zoology, S.K. University, DUMKA - 814 101. Bihar India.

■

29

Pollution Status of the River Complex Sabarmati At Kheda Region of Gujarat - II Phycological Characters

❑ INTRODUCTION

River Sabarmati rises from Aravalli hills, flows through Sabarkantha, Ahmedabad and Kheda districts of Gujarat. The river before discharging in the Bay of Khambhat, confluences with river Khari, Meshwo, Shedhy and other minor rivers near village Vauttha. The river Sabarmati brings enormous discharges of Sewage and industrial effluents particularly from Ahmedabad city and its industrial areas. The pollution status in and around Ahmedabad was carried out by Venkateshwarlu and Jayanthi (1968) and Shaji and Patel (1991), however, no attempts have been made so far to study the downstream of the river where it forms a complex of many rivers at village Vauttha.

METERIALS AND METHODS

In the present study, six sampling sites were selected to collect the water and algal samples. Three sites were fixed along the river Sabarmati over a stretch of about 10 km. The first site was near village Rasikpura before the confluence of Khari, while the second was near village Vauttha after the confluence of river Khari, and third site little away from Vauttha after mixing of river Vatrak. Sampling sites for Khari, Vatrak and Shedhy were also fixed in the Kheda region before they mix with Sabarmati (Fig. 1) Algal samples were collected both from water and nearby soil enriched with river water from all the sites. Water samples were also collected to analyse the physicochemical parameters. The algal samples were brought to the laboratory and were preserved in 4% formaldhyde. The camera lucida sketches of all possible taxa were made and they were identified using standard monographs and published literature (Desikachary 1959, Pressocot 1962, Philopose 1967, Iyengar and Desikachary 1981 and Patel 1989). Palmer's (1969) pollution index bases on genera and species were used for rating river waters for high or low organic pollution.

RESULTS

Algal taxa of four groups viz., Cyanophyceae, Chlorophyceae, Bacillariophyceae and Euglenophyceae were encountered from the different sites. The list of algae is given in Table I. In all 83 taxa belonging to 35 genera, 53, taxa belonging to 31 genera, 68 taxa belonging to 37 genera were recorded from Sabarmati, Vatrak and Shedhy respectively. In river Khari algal flora was almost absent. In Sabarmati, 37 taxa were common to all the three sites. The Cyanophyceae was represented by 28 taxa belonging to 10 genera of which *Chroococcus minor, C. minutus, Merismopedia minima, Oscillatoria acuminata, O. acuta, O. schultzii, O. tennuis and Arthrospira jenaeri* were more common. In river Vatrak, it was represented by 15 taxa of which *Merismopedia minima, Arthrospira jenneri, Oscillatoria schultzii* and *Anabaena* sp. were frequently found. River Shedhy was represented by 19 taxa consisting mainly *Chroococcus minutus, Merismopedia minima, M. tennuissima, Arthrospira jenneri, Spirulina major, Oscillatoria chalybea* and *Anabaena* sp.

Table 1

Algal flora of different study sites

Sr.	Name of the species	S1	S2	S3	V	SH
CYANOPHYCEAE						
1.	*Microcystis flos-aque*	+	+	+	+	+
2.	*Chroococcus minor*	+	+	-	-	-
3.	*C. minutus*	+	+	-	+	+
4.	*Synechococcus aeruginosus*	-	-	+	+	+
5.	*Merismopedia minima*	+	+	+	+	+
6.	*M. tenuissima*	+	+	-	-	+
7.	*Arthrospira jenneri*	+	+	+	+	+
8.	*A. platensis*	-	+	-	-	-
9.	*Spirulina major*	+	+	+	+	+
10.	*S. subsalsa*	+	+	+	+	+
11.	*Oscillatoria acuminata*	+	+	-	+	+
12.	*O. acuta*	+	-	+	-	-
13.	*O. boryana*	+	+	-	-	+
14.	*O. chalybea*	-	-	-	-	+
15.	*O. chlorina*	+	+	+	-	-
16.	*O. foreaui*	+	-	+	-	-
17.	*O. formosa*	+	+	+	-	-
18.	*O. laetevirens*	+	+	-	-	-

19.	*O. lemmermanni*	+	+	+	-	+
20.	*O. limosa*	+	+	+	+	+
21.	*O. obscura*	-	+	+	-	-
22.	*O. proteus*	+	+	+	-	-
23.	*Oscillatoria schultzii*	+	+	+	+	+
24.	*O. subbrevis*	+	+	-	-	-
25.	*O. subtilissima*	+	+	+	-	+
26.	*O. tenuis*	+	-	+	-	-
27.	*Phormidium sp.*	+	+	+	+	+
28.	*Lyngbya sp.*	-	-	+	+	-
29.	*Anabaena sp.*	+	+	+	+	+
30.	*Anabaenopsis sp.*	+	+	-	-	+
CHLOROPHYCEAE						
31.	*Pediastrum duplex* var. *reticulatum*	-	-	-	+	+
32.	*P. duplex* var. *clathratum*	-	-	-	+	+
33.	*P. Simplex*	-	-	+	+	+
34.	*P. simplex* var. *duodenarium*	-	-	-	+	+
35.	*P. tetras* var. *exisum*	-	-	-	-	+
36.	*P. tetras* var. *tetraedron*	-	-	-	-	+
37.	*Pandorina morum*	-	+	-	+	+
38.	*Chlorella vulgaris*	+	-	-	+	+
39.	*Ankistrodesmus convolutus*	+	-	-	-	-
40.	*A. falcatus*	+	-	-	-	-
41.	*Coelastum microporum*	-	+	+	-	-
42.	*Scenedesmus armatus*	-	-	+	-	-
43.	*S. benardii*	+	-	+	-	+
44.	*S. bijugatus* var. *linearis*	-	-	-	+	+
45.	*S. bijugatus*	-	-	-	-	+
46.	*Scenedesmus bijugatus* var. *irregularis*	-	-	-	-	+
47.	*S. denticulatus*	-	-	-	-	+
48.	*S. dimorphus* var. *tortus*	+	+	+	-	+
49.	*S. dimorphus*	+	+	+	+	+
50.	*S. obliquus* var. *alternans*	+	-	-	-	-
51.	*Scenedesmus protuberans*	+	+	-	-	+

52.	*S. protuberans* var. *minor*	+	-	+	-	+
53.	*S. quadricauda* var. *bicaudatus*	+	+	-	+	+
54.	*S. quadricauda* var. *intermedius*	+	-	+	-	+
55.	*S. quadricauda* var. *longispina*	-	-	-	+	+
56.	*S. quadricauda* var. *maximum*	-	+	+	-	+
57.	*S. quadricauda* var. *microspina*	+	+	-	-	+
58.	*S. quadricauda* var. *quadrispina*	-	+	+	-	+
59.	*S. quadricauda* var. *westii*	-	-	-	-	+
60.	*Scenedesmus sp.*	+	+	+	+	+
61.	*Closterium sp.*	+	-	-	-	+
62.	*Closteridium lunuta*	-	-	-	-	+
63.	*Cosmarium bicardia*	-	-	-	+	+
64.	*C. laeve*	-	-	-	+	+
65.	*C. lundellii* var. *ellipticum*	-	-	-	+	+
66.	*C. subcostatum*	-	-	-	+	+
67.	*Hydrodictyon sp.*	-	-	-	+	+
68.	*Spirogyra sp.*	-	-	-	+	+
69.	*Oedogonium sp.*	-	-	-	-	+
EUGLENOPHYCEAE						
70.	*Euglena agilis*	+	+	+	+	+
71.	*Euglena acus*	+	+	+	-	-
72.	*E. caudatus*	+	+	+	-	-
73.	*E. gracilis*	+	+	+	-	-
74.	*E. pascheri*	+	+	+	-	-
75.	*E. viridis*	+	-	+	-	-
76.	*Euglena sp.*	+	+	+	-	-
77.	*Phacus Oribicularis*	+	+	+	-	-
78.	*Phacus sp.*	+	+	-	-	-
79.	*Trachelomonas hispida*	-	-	+	+	+
BACILLARIOPHYCEAE						
80.	*Cyclotella menenghiniana*	+	+	+	+	+
81.	*Melosira sp.*	+	-	-	+	+
82.	*Fragillaria gracillima*	-	-	-	-	+
83.	*Fragillaria sp.*	-	-	+	+	+

84.	*Synedra tabulata*	+	-	-	-	-
85.	*Synedra ulna* var. *subequalis*	-	-	-	+	-
86.	*Synedra sp.*	+	+	+	-	-
87.	*Achnanthes karyophila*	-	-	-	-	+
88.	*Achnanthes exigua*	+	+	-	-	-
89.	*Cocconeis placentula*	+	+	+	+	+
90.	*Diploneis sp.*	+	+	+	-	-
91.	*Anomoeoneis sphaerophora*	+	+	+	+	+
92.	*Navicula anglica*	+	+	+	-	+
93.	*N. cuspidata*	+	+	+	+	-
94.	*N. thienemanniodes*					
95.	*N. pupula*	-	+	+	-	-
96.	*N. Salinarum*	+	+	+	-	-
97.	*Pleurosigma itium*	-	-	+	-	+
98.	*Gyrosigma sp.*	-	-	+	+	+
99.	*Gomphonema sp.*	+	+	+	-	+
100.	*Amphora acutiuscula*	+	+	+	-	-
101.	*A. angusta*	-	+	-	-	+
102.	*A. granulata*	-	+	+	-	-
103.	*Amphora sp.*	+	+	+	-	+
104.	*Nitzschia amphibia*	-	+	+	-	+
105.	*N. filiformis*	+	+	+	-	-
106.	*N. intermediata*	-	-	-	+	-
107.	*N. macilanta*	-	-	-	-	+
108.	*N. palea*	+	+	+	-	-
109.	*N. pygmea*	-	+	-	-	-
110.	*N. reversa*	-	-	+	+	-
111.	*N. sigma*	-	-	-	+	+
112.	*N. subtilis*	+	+	+	-	-
113.	*N. thermalis*	+	+	-	-	-
114.	*Liemophora sp.*	-	-	+	-	+
115.	*Surirella capronioides*	-	-	+	+	-

S1= Sabarmati site-I, S2, Sabarmati site - II, S3 - Sabarmati Site-III, Sh - Shedhy river Site - VI, V - Vatrak site - V.

No algae were present in Khari river + = Present - = Absent

Chlorophyceae was represented by 13 taxa belonging to 6 genera in river Sabarmati of which *Scenedesmus dimorphus, S. quadricauda, Pediastrum simplex* and *P. protuberans* were more frequent, while species like *Ankistrodesmus* and *Coelastrum* appeared only in few collections. River Vatrak was represented by 20 taxa belonging to 8 genera of which *Pediastrum duplex, P. simplex, Scenedesmus armatus, S. quadricauda, Clostridium lunata, Cosmarium* Sp. and *Spirogyra* were frequent, while in Shedhy it was represented by 32 taxa belonging to 10 genera. Here the common taxa were *Pediastrum simplex, P. duplex and Scenedesmus bijugatus.*

Bacillariophyceae was represented by 30 taxa belonging to 16 genera, among which *Nitzschia thermalis, N. subtilis, N. palea, N. filiformis, N. amphibia, Navicula anglica, N. cuspidata, Anomoenoeis spheaerophora* and *Synedra* sp. were the most frequent in river Sabarmati, while 17 taxa belonging to 12 genera in Vatrak and 15 taxa belonging to 14 genera in Shedhy were recorded. In river Vatrak *fragilaria* sp., *Anomoenoeis sphaerophara, Nitzschii reversa* were more frequent, while in Shedy *Cyclotella menenghiniana, Nitzschii amphibia, Pleurosigma* and *Anoemoenoeis sphaerophora* were found.

Euglenophyceae was represented by only 8 taxa belonging to 3 genera among which *Euglena pascheri, Phacus orbicularis* were more common in Sabarmati, in Vatrak *Euglena pascheri* and *Trachelomonos* were recorded, while sparse growth of *Euglena agitis* and *Trachelomonas* were found in Shedhy.

Table 2

Palmer's Pollution Index at different sampling sites

Site		Palmer's pollution index	
		Genus	Species
Sabarmati	I	33	34
Sabarmati	II	28	26
Sabarmati	III	29	29
Khari	IV	--	--
Vatrak	IV	30	17
Shedhy	VI	30	17

Out of the 20 most pollution tolerant genera and species reported by Palmer (1969) 14, 12 and 13 genera, and 10, 8 and 6 species were recorded from the three sites of river Sabarmati. Palmer's pollution index were scored for algal genera and species. Total score was 33, 28 and 29 for generic pollution index and 34, 26 and 29 for species pollution index for the three sites respectively. In Vatrak and Shedhy rivers the values were comparitively low for species index. For Vatrak the generic index value was 30 while species index value was 16. In Shedhy it was 30 and 17. The values are shown in Table 2.

DISCUSSION

Out of the different algal groups Cyanophyceae was a dominant group in all the rivers. The wide occurrance of Cyanophycean members in these rivers is due to their pollution tolerant nature as also recorded by many authors (Palmer 1969, Kumar et al. 1974, Palria and Rana 1985, Rana and Palria 1988, Shaji and Patel 1991). The presence of high level of pollutants as revealed by the chemical evaluation also supports this point. Seenayya (1971), Palria and Rana (1985) have recorded good Cyanophycean growth in effluents of alkaline pH.

Chlorophyceae, which was the dominant group in Vatrak and Shedhy, was the sub-dominant group in Sabarmati. Dominance of this grooup in less polluted habitats of many Indian rivers have been reported (Venkateshwarlu 1986, Somashekhar 1988), however *Scenedsmus* has wide occurence (Govindan and Sundaresan 1979). Chlorococcales were abundant in Shedhy, but they were less in Sabarmati. Desmids were absent in Sabarmati, while only few species were recorded from Shedhy. Many workers have pointed out the low diversity of desmids in polluted waters (Round 1973, Venkateshwarlu 1986, Shaji and Patel 1991).

Diatoms were well represented in all the three sites of Sabarmati, while it was poor in Vatrak and Shedhy. Wider occurence of diatoms in polluted habitats is well-established (Rai and Kumar 1977, Sengar and Sharma 1987). Sengar and Sharma (1987) are of the view that more hardness, sulphate, phosphate and nitrate are favourable for the growth of diatoms support the present data. *Nitzschia* appeared to be the dominant genus in the present study as found by Govindan and Sundarsan (1979) and Prasad and Singh (1982).

The presence of Euglenoids in our collection of polluted habitats have also been recorded by many workers (Venkateswarlu 1981). Both genus and species score of Palmer's pollution index above 20 indicate confirmed organic pollution, is revealed in the three sites of Sabarmati which scored very high pollution index. Similarly, Vatrak and Shedhy rivers score also indicate its polluted nature.

Considering the physico-chemical, phycological and Palmer's pollution index values of different study sites, different degree of pollution occurs in these rivers and phycological evaluation is a helpful parameter in the pollution studies.

Acknowledgement

The authors are thankful to the University Grants Commission, New Delhi for financial assistance.

❑ REFERENCES

Desikachary T.V. 1959., *Cyanophyta,* I.C.A.R., New Delhi.

Govindan V.S. and Sundaresan B.B. 1979. Seasonal succession of algal flora in polluted region of Adyar river *Indian J. Environ. Hlth.* 21 : 131-42.

Iyengar, M.O.P. and Desikachary T.V. 1981. *Volvocales,* I.C.A.R., New Delhi.

Kumar H.D., Bisaria G.P., Bhandari L.M., Rana B.C. and Sharma V. 1974. Ecological studies on algae isolated from the effluent of an oil refinery, a fertilizer factory and brewary. *Indian J. Environ. Hlth.* 16 : 247-255.

Palmer C.M. 1969. A composite rating of algae tolerating organic pollution *J. Phycol.* 5 : 78-82.

Palria S. and Rana B.C. 1985. Ecological studies of a river polluted by dyeing and textile factory waste. *Comp. Physiol. Ecol.* 10 : 235-37.

Patel R.J. 1989. *'Algae Novo'* Avichal Science Foundation, Vallabh Vidyanagar.

Philipose M.T. 1962. *Chlorococcales,* I.C.A.R., New Delhi.

Prasad B.N. and Singh Y. 1982. On diatoms as indicators of water pollution. *J. Indian Bot. Soc.* 6 : 326-336.

Prescott G.W. 1962. *Algae of the Western Great lake area.* W. M. C. Brown Company Publishers, Dubuger, Iowa.

Rai L.C. and Kumar H.D. 1977. Studies on the seasonal variation in algal communities of ponds polluted with fertilizer factory effluent. *Indian J. Ecol.,* 4 : 124-31.

Rana B.C. and Palria S. 1988. Phycological and physico-chemical evaluation of the river Ayad, Udaipur, *Phykos* 27 : 211-217.

Round F.E. 1973. *The Biology of Algae,* Edward Arnold (Publishers) Limited, 25, Hill Street, London.

Seenayya G. 1971. Ecological studies in the plankton of certain freshwater ponds of Hydrabad, India, Part II. Phytoplankton. *Hydrobiologia* 37 : 55-68.

Sengar R.M.S. and Sharma K.D. 1987. Role of algae in the assessment of pollution in river Yamuna, *J. Indian Bot. Soc.* 66 : 325-334.

Shaji C. and Patel R.J. 1991. Chemical and biological evaluation of pollution in river Sabarmati at Ahmedabad, Gujarat. *Phykos* 30 : 91-100.

Somashekar R. K. 1988. Ecological studies on two major rivers of Karnataka. In : *Ecology and Pollution of Indian Rivers* (Ed. by R.K. Trivedy) Ashish Publishing House, New Delhi.

Somashekar R.K. and Ramaswamy S.N. 1984. Biological assessment of water pollution : A study of river Kapila. *Inter. J. Environ. Studies* 23 : 261-267.

Venkateshwarlu V. 1981. Algae as indicators of river water quality and pollution, p. 93-100. In : *Biological indicators and indices of pollution,* WHO Workshop.

Venkateshwarlu V. 1986. Ecological studies on the rivers of Andhra Pradesh with special reference to water quality and pollution. *Proc. Indian Sci. Acad.* 96 : 495-508.

Venkateshwarlu V. and Jayanthi T.V. 1968. Hydrobiological studies of river Sabarmati to evaluate water quality. *Hydrobiologia* 31 : 442-448.

Authors :

1. **J. Jameson and B.C. Rana * :** Department of Bio science, Sardar Patel University, VALLABH VIDYANAGAR- 388 120. Gujarat (India)

* Since deceased

■

30

Studies on the Physico-Chemical Parameters of Shankha River At Bareilly

❑ INTRODUCTION

Since time immemorial, Ram Ganga is regarded as sacred river and people of Garhwal & Rohilkhand region worship it with supreme reverence. Millions of people take holy dip in the river Ram Ganga and have a strong faith that its water can never be impure. But this contention no longer holds good because the prestine beauty of Ram Ganga has been adversely affected due to enormous discharge of urban, agricultural and industrial wastes. River Shankha swallows bulk of pollution load and dumps the same in river Ram Ganga which carries it up to Farrukhabad in U.P. and finally merges in river Ganga at Kannauj.

Large scale industrialization at Bareilly since last decade has led to mushroom growth of several indutries viz; Synthetics and Chemicals Ltd., Indian Terpentine and Rosins Co. Ltd, Wimco Ltd., Camphor and allied products Ltd., Moracco Tannery Ltd., Aqua victuals Ltd., Froosh Valley product Ltd., Sugar industry and distillery. All these indutries release their toxic effluent in the rivers flowing in their vicinity through open drains ignoring the norms laid down by the DOE. The water of river Shankha is neither potable nor of irrigational use and posing a serious threat to the survival of aquatic life. People have fallen victim of dreaded water born diseases after consuming effluent mixed water of Ram Ganga. Biodegradable pollutants have caused depletion of dissolved oxygen in water thus enhancing its BOD. The non biodegradable pollutants have undergone biomagnification at higher trophic levels and causing mortality in fishes and other animals. Decline in water level in the river during summer is a step further to increase concentration of inorganic and organic pollutants. Studies pertaining to industrial water river pollution have attracted the attention of a large number of investigators in recent years (Reddy and Venkateshwaralu 1986, Bahadur and Sharma 1988, Iqbal et al., 1989, Trivedy 1989, Bahadur & Sharma 1990, Bannerjee and Singh 1993); Gupta and Singh (1995). Present study, therefore, is aimed at investigating the chief causative factors involved in the contamination of water in river Shankha.

MATERIALS AND METHODS

River Shankha is located at a distance of 1.8 km. towards North east of Rampur road near Fatheganj. One site was fixed for the study near Roop Pur village.

Water samples of river Shankha were collected at monthly intervals (Table 1) from April 1994 to March 1995. Standard methods (AOAC, 1965, Odum 1971, Waldron 1980, APHA 1989) were employed during collection, preservation and analysis of water samples. Temperature, colour, pH, EC, T.D.S., T.S.S., and D.O. by portable Kit CT-105, sodium, potassium, calcium, magnesium by Flame Photometer, Total hardness, Total alkalinity. BOD, COD, chloride, sulphate, phosphate, dissolved silica, oil & grease were analysed according to APHA. (1989)

RESULTS AND DISCUSSION

Physico-chemical parameters of river Shankha revealed (Table 1 & 2) that its water is unfit for drinking, recreational, fishery, agricultural and other uses. The contamination is due to effluent discharge from synthetics and chemicals industry as well as faecal matter of human origin and cattle excreta.

Colour of the river Shankha was yellow. Colour of the water appears to be due to presence of suspended matter, metallic ions, humic acid and other industrial wastes as has also been reported by Sharma & Habib (1994) for Shankha river.

Higher temperature of er water during summer appears to be due to greater heating and insolent heat from the sun. Onset of monsoon results in fall of temperature of water bodies. Further, clouds act as barrier in the process of heating. This behaviour finds support by the observations of Parvateesam and Mathur (1992). The pH of river water at the study site varied from 7.1 - 8.8. It was maximum in November 1994. However, minimum pH was observed in August 1994. This fluctuation seems to be due to dilution of water in rainy season. pH has been reported to be influenced by the carbonates, bicarbonates and total alkalinity (Gupta 1992) as has been observed in the present study.

Maximum conductivity was observed in June 1994 and minimum in March 1995. Higher conductivity is due to its high ionic concentration. This is substantiated by the findings of Bharti and Krishnamurthy (1990) for Kali river.

Dissolved solids were highest (860 mg/1) in February 1995 and Minimum (45 mg/1) in October 1994. The pollution has direct relationship with the solids may be suspended or dissolved. This is in confirmity with the findings reported by Prasad and Saxena (1980).

Table 2

Physico-chemical characteristics of the River Shankha at During April 1994 to March 1995.

Months	Temp. °C	Colour	pH	E.C.	T.D.S.	S.S.	D.O.	BOD	COD	Chloride	Sulphate	PO4	Na	K	Ca	Mg	Dissolved Silica	Total Hardness	Total Alkalinit
April-1994	29.50 ±0.00	Brown	7.47 ±0.05	79.33 ±1.50	606.00 ±2.64	68.00 ±1.00	2.60 ±0.00	23.47 ±0.01	150.40 ±0.95	80.66 ±1.52	356.0 ±2.88	2.20 ±0.00	127.0 ±1.32	30.0 ±0.50	25.40 ±0.60	23.50 ±0.49	3.50 ±0.00	124.0 ±1.12	76.0 ±1.00
May	29.50 ±0.00	Light Brown	7.48 ±0.04	77.33 ±0.30	561.00 ±1.60	118.00 ±0.57	2.40 ±0.05	13.46 ±0.01	134.20 ±0.00	63.73 ±3.06	642.0 ±1.00	1.26 ±0.05	153.0 ±1.00	26.0 ±0.50	24.40 ±0.20	20.86 ±0.51	3.20 ±0.10	112.0 ±2.08	82.0 ±0.05
June	32.00 ±0.00	Light Brown	7.52 ±0.03	83.33 ±0.41	675.00 ±2.51	121.00 ±0.50	2.65 ±0.02	14.60 ±0.05	146.00 ±0.00	78.26 ±0.50	155.0 ±0.86	1.30 ±0.10	176.0 ±0.86	33.0 ±0.50	28.33 ±0.15	21.52 ±0.21	3.40 ±0.20	118.0 ±1.00	86.0 ±0.10
July	31.50 ±0.00	Light Brown	7.21 ±0.07	66.40 ±0.10	564.00 ±2.00	120.00 ±0.57	1.10 ±0.05	19.56 ±0.20	172.00 ±1.00	79.67 ±0.15	161.0 ±0.53	1.20 ±0.04	18.0 ±0.50	7.0 ±0.50	35.28 ±0.12	17.17 ±0.15	2.26 ±0.10	109.0 ±0.57	36.20 ±0.50
August	33.20 ±0.00	Dark Brown	7.19 ±0.05	60.53 ±0.02	525.00 ±0.88	124.00 ±0.57	1.90 ±0.05	36.00 ±0.02	200.00 ±0.00	42.71 ±0.12	144.0 ±0.86	2.85 ±0.07	22.0 ±1.00	16.0 ±0.50	17.63 ±0.02	13.68 ±0.14	2.65 ±0.02	74.0 ±0.57	31.00 ±1.00
September	34.50 ±0.00	Dark Yellow	7.36 ±0.00	60.20 ±0.00	557.00 ±1.50	143.00 ±1.50	2.60 ±0.05	9.33 ±0.00	41.00 ±0.50	68.19 ±0.13	122.0 ±1.00	3.53 ±0.02	118.0 ±1.00	15.0 ±0.50	51.46 ±0.15	25.56 ±0.20	2.73 ±0.05	158.0 ±1.00	58.0 ±1.00
October	34.00 ±0.00	Light Brown	7.61 ±0.00	54.20 ±0.00	451.00 ±2.08	116.00 ±2.00	6.60 ±0.10	15.00 ±0.00	61.00 ±0.50	31.00 ±0.50	202.0 ±1.00	1.74 ±0.01	131.0 ±0.50	21.0 ±0.50	94.00 ±0.50	16.08 ±0.00	2.46 ±0.05	161.0 ±0.50	60.0 ±0.50
November	22.00 ±0.00	Light Brown	8.80 ±0.00	58.60 ±0.10	480.00 ±4.35	148.00 ±1.00	4.80 ±0.10	12.20 ±0.50	85.00 ±0.50	28.00 ±0.10	122.0 ±1.00	1.78 ±0.01	62.0 ±1.00	14.0 ±1.00	33.80 ±0.10	25.72 ±0.13	2.40 ±0.10	141.0 ±0.50	88.0 ±1.00
December	18.00 ±0.00	Clear	8.32 ±0.00	79.73 ±0.32	495.00 ±2.56	148.00 ±1.00	5.10 ±0.05	17.00 ±0.50	88.00 ±1.00	98.00 ±1.00	260.0 ±0.57	1.82 ±0.01	67.0 ±0.50	20.0 ±1.00	55.33 ±0.57	40.00 ±0.27	5.20 ±1.00	222.0 ±1.00	106.0 ±1.00
January-1995	16.00 ±0.00	Clear	8.40 ±0.01	76.70 ±0.05	770.00 ±2.00	98.00 ±2.00	8.00 ±0.00	6.20 ±0.10	52.00 ±1.00	116.00 ±0.50	202.0 ±1.00	1.83 ±0.91	27.0 ±0.50	24.0 ±1.00	38.80 ±1.03	39.40 ±0.57	5.00 ±0.00	203.0 ±1.32	110.0 ±1.00
February	22.00 ±0.00	Clear	8.62 ±0.05	78.64 ±0.01	860.00 ±2.00	118.00 ±2.00	8.00 ±0.00	6.80 ±0.10	39.00 ±1.00	120.00 ±1.00	180.0 ±0.00	2.00 ±0.10	30.0 ±0.50	12.0 ±0.50	38.86 ±0.25	38.66 ±0.84	6.00 ±0.50	200.0 ±1.00	82.0 ±1.00
March	23.00 ±0.00	Clear	8.60 ±0.01	40.00 ±1.20	528.00 ±2.00	113.00 ±1.00	8.00 ±0.05	4.60 ±0.05	68.00 ±1.00	40.00 ±0.50	140.0 ±1.00	3.20 ±0.10	21.0 ±0.50	18.0 ±0.50	36.56 ±0.32	28.67 ±0.30	5.20 ±1.00	156.0 ±1.00	89.0 ±1.00
F	-	-	-	399.04	45429.00	247.97	779.83	1186.36	4612.58	625.25	1174.89	163.11	3859.25	98.91	1345.54	637.10	44.99	1195.72	626.71
C.D. AT 5%	-	-	-	1.99	8.57	4.10	0.28	0.74	2.37	3.62	4.00	0.17	2.75	2.23	1.60	1.05	0.57	3.72	2.84
+SEM	-	-	-	0.20	0.85	0.41	0.03	0.07	0.23	0.36	0.40	0.02	0.27	0.22	0.16	0.10	0.06	0.37	0.28

All values expressed in mg/1 except E.C. (dsm/m) & pH

Table 2
Physico-chemical characteristics of the River Shankha at Bareilly in down stream studied at one site from April 1994 to March 1995.

S.No.	Parameters	Max.	Mini.	DOE(1992)	I.C.M.R.
1.	Temperature	34.5 (Sept.)	16.0 (Jan.)	-	-
2.	Colour	Dark Brownish (July)	Clear Water (Feb)	-	-
3.	pH	8.8 (Nov.)	7.19 (Aug.)	6.5-8.5	7.0-8.5
4.	E.C.	83.33 (June)	40.0 (March)	-	40(ds/m^2)
5.	T.D.S.	860.0 (Feb.)	451.33 (Oct.)	-	-
6.	T.S.S.	148.0 (Nov.)	98.0 (Jan.)	10.0	-
7.	D.O.	8.0 (Jan.95)	1.1 (July)	6.0	5.0
8.	B.O.D.	36.0 (Aug.)	6.2 (Jan.)	0.2	5.0
9.	C.O.D.	200.0 (Aug.)	39.0 (Feb.95)	4.0	-
10.	Chloride	120.0 (Feb.)	28.2 (Nov.)	-	2.50
11.	Sulphate	642.0 (May)	44.0 (Aug.)	150.0	-
12.	Phosphate	3.53 (Sept.)	1.17 (July)	-	-
13.	Sodium	176:0 (May)	18.0 (July)	-	-
14.	Potassium	33.0 (June)	7.0 (July)	-	-
15.	Calcium	94.0 (Oct.)	17.63 (Aug.)	75.0	75.0

16.	Magnesium	40 (Dec.)	13.68 (Aug.)	30.0	30.0
17.	Dissolved Silica	6.0 (Feb.)	2.26 (July)	-	-
18.	Total Hardness	222.0 (Dec.)	74.66 (Aug.)	-	120
19.	Total Alkalinity	110 (Feb.)	31.0 (Aug.)	-	-
20.	Oil and Grease	Nil	Nil	-	-

Maximum D O (8.0 mg/1) was observed in January 1995 & minimum in July 1994. In winter, dissolved oxygen was higher (8.0 mg/1) due to lower temperature, high retaintion capacity of oxygen. Poor DO indicates increased pollution load due to entry of material, such as fertilizers, insecticides, pesticides from the top surface of the soil during rainy season.

COD values varied from 39-200 (mg/1) and have been found to be much higher than BOD (4.0-36.0 mg/1). High COD values indicate the presence of chemically oxidisable carbonaceous matter as well as the inorganic matter such as nitrates, sulphides and reduced metal ions.

Maximum concentration of chloride was obtained (120 mg/1) in February 1995 and minimum 31.0 (mg/1) in October 1994. Presence of higher chlorides considered is as pollution indicator (Manikay Reddy and Venkatesh Warlu, 1987). Sulphates varied from 122 - 642 (mg/1). They have been reported as pollution indicators (Krishnamurthy and Bharti 1994). Phosphate ranged from 1.2 - 3.53 indicators (Krishnamurthy and Bharti 1994). Phosphate ranged from 1.2 - 3.53 (mg/1). High concentration of phosphate is due to industrial discharge in river. Presence of phosphate is associated with Pollution (Prakash, 1980).

Concentration of sodium, potassium and calcium ranged from 18 - 176 mg/1, 7 - 33 mg/1 and 17.6 - 94 (mg/1). Calcium content was higher than the tolerance limit. Magnesium varied from 13.68 - 40 (mg/1) which was also higher than the I.S.I. limit Hardness ranged from 75-222 mg/1. Total alkalinity varied from 31-110 mg/1. Higher values of total alkalinity in water indicates pollution of organic and inorganic nature as has been reported by Phillip, (1960).

CONCLUSION

All the parameters studied exhibited variations during the study period. The electric conductance (dsm-1), BOD, COD, DO, TSS, Calcium, magnesium, pH were found to be higher, which indicated greater pollution and rejects the utility of river water for potable use, recreational and fishery. The parameters viz. pH, E.C., BOD, COD, alkalinity, calcium, magnesium, sodium values exceeded the tolerance limits recommended by I.C.M.R. and DOE.

❑ REFERENCES

A.O.A.C. 1965. Association of official agriculture chemists. *Official methods of analysis.* 8th Ed. Washington D.C.

APHA. 1989. *Standard methods for the examination of water and waste water.* 17th ed., APHA Washington D.C.

Bahadur B. and Sharma B.K. 1988. Effect of industrial effluent on various growth attributes of *Hordeum vulgare* L. (Var. Jyoti). *Vegetos,* 1:45-50.

Bahadur B. and Sharma B.K. 1990. Effect of industrial effluent on seed germination and early seeding growth of *Triticum aestivum* var. U.P. 115. *Acta Botanica Indica,* 18 : 80-83.

Gupta M.C. 1992. *Nutrient dynamics, plankton and productivity of reservoir Amarchand, Southern Rajasthan in relation to fisheries development.* Ph.D. thesis Rajau, Bikaner Campus.

Gupta, Ravi Kumar and Gurdeep Singh 1995. Water Pollution profile of coal washeries. *Poll Res.* 14(2) : 203-213.

Iqbal Habib, Pandey U.C. and Mazhar M. 1989. Certain physico-chemical factors and phytoplanktons of Ram Ganga at Bareilly District (U.P.) *Him. J. Env. Zool.,* 3 : 251-254.

Krishnamurthy S.R. and S. G. Bharti 1994. Studies of River Kali around Dandeli (North Karnataka District) *Poll Res.* 13 (3) : 249-251.

Laxmaiah Chandra, S.K. Mahmood and Narendra Rao 1994. Nutrient status and biological characteristcs of Nacharam lake *Poll. Res.* 13(3) : 253-258.

Odum E.P. 1971. *Fundamentals of Ecology.* 3rd ed. W.B. Saunders Company, Philadelphia.

Pravateesam M. and Sudha Gupta. 1994. Physico Chemical characteristics of a lake receiving effluent from textile mills in Rajasthan. *Poll Res.* 13(4) : 317-321.

Phillips G.L. 1960. The mineral nutrient levels in three norfolk broads differing in Trophic status and on annual mineral content buget for one of them. *J.Ecol.* 65 : 447-474.

Prakash C. 1980. *Limnological study of river Yamuna with reference to pollution.* Ph.D. thesis. Agra University, Agra, India.

Prasad B.N. and Saxena M. 1980. Ecological study of blue green algae in river

Gomti. *Indian. J. Environ. Hlth.* 22 : 151-158.

Srivastava S.K., Kumar R. and Srivastava A.K. 1994. Effect of Textile Industry Effluent on the biology of river Tons at Mau (U.P.) Physico-chemical characteristics. *Poll Res.* 13 (4) : 369-373.

Venkatesh Warlu V. 1986. Ecological studies on rivers of Andhra Pradesh with special reference to water quality and Pollution. *Proc. Indian Sci Acad.* 96(6) : 495-508.

Waldron H.A. 1980. *Metals in the environment.* Acad. press, London.

Authors :

1. **Rakesh Chandra, Y. Bahadur :** Department of Chemistry, Bareilly College, BAREILLY - 243 001. U.P. (India)
2. **B.K. Sharma :** Department of Botany, Bareilly College, BAREILLY - 243 001. U.P. (India)

■

31

Heavy Metals Content of the Waste drains and the River Panchaganga At Ichalkaranji (Maharashtra)

❑ INTRODUCTION

Industrial, agricultural and domestic waste pollute the water bodies with heavy-metals which affects the aquatic-ecosystem in adverse manner. An attempt has been made to quantitatively assess the various heavy metals present (Cu, Mn, Cd, Ni, Zn, Fe, Co, Pb) in Panchaganga river water and waste drains meeting the River, during the different seasons and the result are presented in this paper. Selected sampling station are shown in fig.1.

After entering in Kolhapur city, the river passes through various regions of city and gets grossly polluted by various activeties. Before entering in Ichalkaranji it is already polluted due to various activities.

The load of pollution increases when the river flows thorugh the city Inchalkaranji since the city is grossly crowded by textile mills, dying and printing mills and sugar mill. The waste is partially treated which meets the river Panchganga when the river enters the city and the drain which is meeting is known as 'Kabnoor drain'. (Drain--I).

Before the entry of the drain the Chandur village is situated on the banks of the river, which is station I (site--I). Aftr this drain, the river bank is fully covered by agricultural activities and site-II is 3km far away from the drain, the area is known as "Ghat region'. On this region culutral pollution in the form of mass-bathing, cloth and automobile washing, catttle wading and religious activities are evident. The maximum pressure of human activities is found at this site below this river. The river is crossed by two bridges. Site --III is before meeting the drain and site-IV is after meeting the drain. Municipal drain is taken as drain-II. Which brings domestic waste with sewage waste from North region of Inchlkaranji township mixed with the effluents from the cottage dye industries which are also partially treated or untreated.

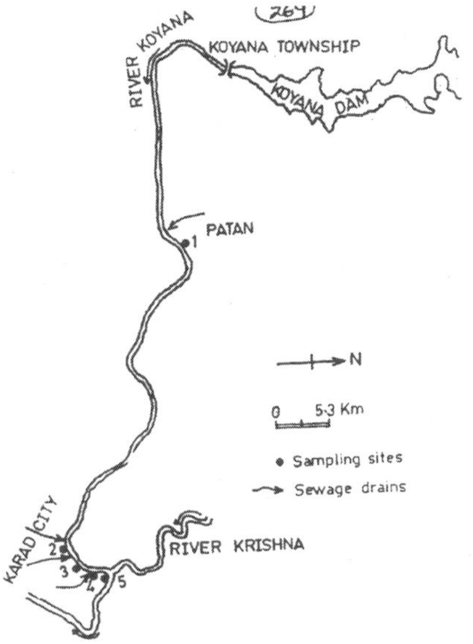

Figure -1.

METERIALS AND METHODS

The sampling is done bimonthly from above sites. The suitable container, time and chemicals of preservation were selected as per APHA (1986). Well mixed samples of each site were taken and acidified with conc. HNO_3. Estimations of heavy metals were done with the help of Atomic absorption spectrophotometer.

RESULTS

Table 1 shows the data on heavy metal concentration in the river. In all nine metals were studied.

Copper was present in concentration varying from 10 μg/L to 50 μg/L. The manganese concentration varied from 20 μg/L to 2160μg/L. At site 1, site 2 and site 3, the concentration was very low. However, concentration was very high at site 4, which suggests a source of manganese between site 3 and site 4. Cadmium was present in extremely low quantity and was detected only at site 2 and site 4. Nickel concentration ranged from 10-60 μg/L in the river. Higher concentration was noted at site 4. Zinc concentration at site 1 was very low (30-40 μg/L) but the concentration was found immediately going up at site 2 (110-370 μg/L). The concentration was again increasing till site 3 (300-900 μg/L). A heavy decline was noted at site 4. Iron concentration was fluctuating between 160-230 μg/L at site1. A slight decline was noted at site 2. However, the values increased by 5 to 75 times during different months at site 3. The values further increase considerably at site 4 and were never belwo 1010 μg/L and went upto 7110 μg/L. Cobalt concerntration ranged from 10-50μg/L at various sites. Lead concentration varied from 10-118 μg/L; values were higher at site 3 and site 4.

Table 1

Heavy metal concentrations in the Panchaganga river water

Months	Site1	Site 2	Site 3	Site 4
COPPER				
March 1988	20	20	30	10
May 1988	30	30	40	20
July 1988	10	10	20	10
September 1988	10	--	10	10
November 1988	20	--	10	20
January 1989	20	20	10	50
MANGANESE				
March 1988	50	60	80	2160
May 1988	30	40	51	2130
July 1988	30	30	40	90
September 1988	40	130	40	1340
November 1988	30	40	20	1120
January 1989	30	30	20	2160
CADMIUM				
March 1988	--	2	--	4
May 1988	--	3	--	3
July 1988	--	--	--	--
September 1988	--	--	--	--
November 1988	--	5	--	2
January 1989	--	3	--	1
NICKEL				
March 1988	20	--	30	40
May 1988	30	--	40	60
July 1988	10	--	30	20
September 1988	30	30	30	20
November 1988	10	--	20	30
January 1989	30	--	40	52

ZINC				
March 1988	30	210	530	40
May 1988	40	300	400	30
July 1988	30	110	300	30
September 1988	30	250	900	60
November 1988	40	370	630	58
January 1989	30	280	520	41
IRON				
March 1988	210	30	2120	6320
May 1988	230	130	1180	7110
July 1988	160	160	1110	5740
September 1988	190	160	1040	3230
November 1988	230	190	1230	2120
January 1989	190	190	920	1010
COBALT				
March 1988	10	30	50	40
May 1988	10	20	20	30
July 1988	10	10	10	20
September 1988	10	10	10	20
November 1988	30	--	30	50
January 1989	20	20	30	50
LEAD				
March 1988	40	30	20	30
May 1988	30	20	10	30
July 1988	30	60	60	60
September 1988	20	30	100	180
November 1988	60	50	40	50
January 1989	30	10	30	40

All values are in µg/L. -- = absent.

Copper concentration in waste-drain ranged from 10-750 µg/L. In Kabnoor drain this metal was present only in the months of September (20 µg/L). While in municipal drain it went upto 150 µg/L. The manganese concentration in the Kabnoor drain ranged from 100µg/L to 1120 µg/L and in the municipal drain from 140 mg/L to 5160 µg/L. The concentration was very low during July. Cadmium concentration varied from 5-60 µg/L. In Kabnoor drain the concentration of Cadmium was higher than the municipal drain. Nickel concentration was found to range from 10-90 $\mu g/L$ in the waste drains. Higher concentration were noted in municipal drain. Zinc concentration in the Kabnoor drain was very high (210-1320 $\mu g/L$). In Municipal drain however it ranged between 40-130 $\mu g/L$. The values of Iron (Fe) at Kabnoor drain were below 1000 $\mu g/L$ and ranges from 290-860 $\mu g/L$. In the municipal drain all values are above 3000 $\mu g/L$ and going as high as 9410 $\mu g/L$. Cobalt concentration oscillated between 20-89 $\mu g/L$. Lead concentration ranged from 5-0560 $\mu g/L$. Values were higher in the municipal drain (62-560 $\mu g/L$.)

Table 2

Heavy metal concentrations in the waste drains at Ichalkaranji meeting the River Panchaganga.

Drains, Month	Cooper Cu	Manganese Mn	Cadmium Cd	Nickel Ni	Zinc Zn	Iron Fe	Cobalt Co	Load Pb
KABNOOR DRAIN								
March 88	--	870	--	40	170	620	31	80
May 88	--	1120	10	30	210	780	40	50
July 88	--	100	10	30	530	420	30	--
September 88	20	520	10	10	1100	290	20	120
November 88	--	760	60	10	1320	680	40	90
January 89	--	930	20	30	1060	860	35	190
MUNICIPAL DRAIN								
March 88	150	3140	5	90	40	8320	70	330
May 88	20	4210	40	50	130	9410	80	560
July 88	10	140	20	60	51	7440	60	70
September 88	50	2430	10	30	70	5320	30	210
November 88	80	2360	40	56	98	4160	89	62
January 1989	70	5160	50	86	72	3130	65	130

All values are in µg/L. -- = absent.

DISCUSSION

The overall data of heavy metal suggest that the level of heavy metals was not very high in the river, Howeve, Ajmal and Razi-uddin (1988) record that the heavy metal concentration in water depends mainly on the pH of the system. Metzner (1977) studied the fate of copper and zinc at different pH values in waste waters and found that solubility of these were inversely proportional to the pH of the system and higher solubility was found at pH 7 or below. Polprasert (1982) has reported that the precipitation of heavy metals is enhanced at pH above 7.

In the present study copper contnet was 10 to 50 $\mu g/L$ which is in fact the highest desirable limit for water quality. World's mean stream concentration of copper has been reported by Goldberg et.al (1971) as 7 $\mu g/L$. Manganese concentration at site 1 to3 was in the range of 20 to 130 $\mu g/L$ but the same was usually above 1000 $\mu g/L$ and upto 2160 $\mu g/L$ at site 4. Thus the municipal drain is the main source of manganese as is evidenced from Table-3.

Table 3

Metal	Renge of concentration obtained in the river Panchaganga $\mu g/L$		Tolerance Limits			
			Who (1971) $\mu g/L$ Ministry of works and Housing			
	Lower Limit	Upper Limit	Highest desirable	Maximum Permissible	Acceptable	Cause for rejection
Cu	10	50	50	1000	50	1500
Mn	20	2160*	50	500	50	500
Cd	1	5	--	10	10	10
Ni	10	60*	--	--	--	3000* (BIS Std.)
Zn	30	900	6000	15000	5000	15000
Fe	30	7110*	100	1000	100	1000
CO	10	50	--	--	--	--
Pb	10*	180*	--	100	100	100

Site 1.	Fe	Pb	Mn	Zn	Ni	Co	Cu	Cd.
	3	2	1					
Site 2.	Zn	Fe	Mn	Pb	Cu	Ni	Co	Cd
	3	2	1					
Site 3.	Fe	Zn	Pb	Mn	Co	Ni	Cu	Cd.
	3	2	1					
Site 4.	Fe	Mn	Pb	Ni	Zn	Co	Cu	Cd.
	3	2	1					

Fe Zn Mn Pb at all study period.

BIS - Bureau of Indian Standards

Cadmium concentration was below 5 mg/L in the river water, while the prescribed standard for the same is 10 $\mu g/L$ in the river water.

Nickel concentration in the industrial drain (Kabnoor drain) was in the range of 10 to 40 $\mu g/L$, while in the municipal drain it was 30 to 90 $\mu g/L$. Ajmal et al. (1985) have reported 2.3 to 9.9 $\mu g/L$ nickel concentration in the river Yamuna.

The concentration of zinc was much higher in the Kabnoor drain (170 to 1320 $\mu g/L$), while it was only 51 to 130 $\mu g/L$ in the municipal drain. The higher concentration in the former drain is due to industrial discharge. The influence of this discharge is very clear as the concentration of zinc shows up at site 2 (110 to 370 $\mu g/L$) as compared to site 1 (3 to 40 $\mu g/L$). It continues to remain high at site 3 but declines at site 4. Since pH is also higher at site 4 precipitation of zinc can not be ruled out.

The concentration of cobalt was in the range of 10 to 50 $\mu g/L$ however, lead concentration at site 3 and 4 was around 100 $\mu g/L$ or above on few occasion which is above the prescribed limit.

The overall situation of heavy metals in the river is shown in table 3.

❑ REFERENCES

Ajmal, M. Khan, M.A. and Nomani, A.A. 1985. Distribution of heavy metals in water and sediments of selected sites of Yamuna River (India). *Environ -monit. Assess.* 5:205-124.

APHA - AWWA - WPCF 1986. *Standard methods for examination of water and waste-water* (16th Ed.) American Public Health Association, Washington D.C.

Metzner, A.V. 1977. Removing soluble metals from waste - waters, *Wat-Sewage - Works.* 124:98-101.

Polpraset, C. 1982. Heavy metals pollution in Chaophraya River Eustuary, Thailand *water Research.* 16:775 - 784.

Authors :

1. **R. K. Trivedy,** Deptt of Pollution Studies, Y.C. College of Science, Vidyanagar, Karad - 415 124. M.S. (India)
2. **S. S. Nakate,** Sabaero Organics Ltd., Sarigam, Gujarat

■■

32

Monitoring the Water Quality of River Ramganga At Bareilly

❑ INTRODUCTION

River Ram Ganga is a major tributary of River Ganga in Uttar Pradesh. It originates from Pauri Garhwal district and passes through Bijnor, Moradabad, Rampur, Bareilly, Badaun and Shajahanpur districts and finally merges with river Ganga at Farrukhabad. A number of drains loaded with sewage, domestic wastes and industrial effluents open directly into the River. Various anthropogenic activities and rapid pace of industrialization in the catchment area of river RamGanga has aggrvated the pollution problem. The river also receives silt fertilizers, insecticides, pesticides used for agricultural land through surface run off.

In order to check further deterioration in the quality of water of river RamGanga, it was thought desirable to assess the pollution load and determine the sources responsible for deterioration in water quality.

STUDY AREA AND METHODOLOGY

River RamGanga is located at a distance of 1.8 km towards north-east of Rampur road.

Six sites were selected for the study.

Site I	-	River RamGanga before its confluence with river Shankha.
Site II	-	River Shankha.
Site III	-	River RamGanga after joining river Shankha.
Site IV	-	River Deorania.
Site V	-	River RamGanga after joining river Deorania.
Site VI	-	3 km. from Site V in the upward flow.

Water samples were collected at monthly intervals (April 1994 to March 1995) from six sites. Standard methods APHA (1989) were employed during collection, preservation and analysis of water samples. Temperature, pH, DO, TDS and EC were determined by digital potable water analysis kit CT-105; Na, K, Mg, Ca, (By flame photometer) and other parameters viz., BOD, COD, total alkalinity, chloride, sulphate, total hardness, dissolved silica were ananlysed according to APHA, 1989 (Table - 1).

RESULTS AND DISCUSSION

Physico-chemical study of River RamGanga revealed unsuitability of its water for drinking, agriculture, fishery and other uses. The contamination was due to effluent discharge of industries as well as domestic and sewage disposal of Bareilly city.

Colour of water was yellow at site II and black at site IV. It is an important criterion which expresses the quality of water when it is put for potable use. Colour of water appears to be due to the presence of suspended matters, metallic ions, humic acids and other industrial wastes as has also been reported by Sharma and Habib (1994) for river Shankha. Due to these impurities at all the sites under study the water has not been found suitable for drinking.

Temperature of water varied at different sites during different months. Minimum values (18°C) were recorded in January 1995 while maximum (34.5°C) being at site I in September 1994. High water temperature was recorded sometimes due to various chemical reactions occurring among the pollutants. This is in agreement with Shrivastava et al. (1994) for river Tone.

Lower pH (6.7) was recorded at site IV in July 1994 while it was maximum 8.8 at site II in November 1994. The pH did not vary greatly with changing seasons. Increased pH of surface waters obviously is related to metabolic activities of autotrophs which by utilising CO_2 during photosynthesis reduces hydrogen ion concentration. The pH was low due to dilution of water during monsoon in the month of July. Similar response was recorded by Gupta and Sharma (1994).

Electrical conductivity values indicate the ionic load of water at different sites. The values ranged from 30-86.72 $mSem^{-1}$. Minimum EC was recorded in April 1994 at site VI while maximum and comparatively low in monsoon season.

Total suspended solids ranged from 68.0-270.0 mg/l. Maximum values were observed at site IV in September 1994 while minimum being obtained in April 1994 at site II. Higher concentration of TSS brought about increase in turbidity of water which was mainly due to clay, organic matter and other micro-organisms. Due to higher values of suspended solids water of site II and site IV was not fit for potable use.

TABLE : 1

Physicochemical characteristics of Water of River Ram Ganga at Bareilly (Studied at 6 sites) during April 1994 to March 1995.

Site	Temperature °C		pH		Colour		E.C.msm'		TDS		TSS		DO		BOD		COD		Chlorides	
	Max.	Min.	Max.	Min.	Max.	Min.	Max.	Min.	Max.	Min.	Max.	Min.	Max.	Min.	Max.	Min.	Max.	Min.	Max.	Min.
I	34.5 Sept.	16.0 Jan.	8.1 Nov.	7.33 July	Dark Brown Mar	Light Yellow April	72.62 Feb.	36.2 Nov.	760.0 Feb.	216.0 Nov.	170.0 Feb.	100.0 Sept.	8.6 Oct.	2.37 July	19.86 May	5.8 Sept.	157.0 Feb.	20.6 Nov.	66.8 Sept.	11.06 June
II	34.5 Sept.	16.0 Jan.	8.8 Nov.	7.19 Aug.	Muddish dirty Nov.	Clear March	83.33 June	40.0 March	860.0 Feb.	451.33 Oct.	148.0 Nov.	68.0 April	8.0 Jan.	1.1 July	36.0 Aug.	4.6 March	200.0 Aug.	39.0 Feb.	120.0 Feb.	28.2 Nov.
III	34.0 Sept.	16.0 Jan.	8.3 Nov.	7.23 July	Dark Brown Mar.	Light Yellow Sept.	68.72 Feb.	35.33 May	640.00 Feb.	278.00 Nov.	183.33 Oct.	118.00 Nov.	8.33 Oct.	2.60 July	15.00 Aug.	3.73 Sept.	96.80 Jun	26.50 Oct.	74.66 Dec.	17.13 July
IV	34.10 Sept.	16.50 Jan.	8.33 May	6.71 July	Dirty Blackish	Blackish	86.63 July	52.40 Aug.	830.00 July	460.00 March	270.00 Sept	156.66 Dec.	0 Mar.	0 Nov.	101.14 Jun.	68.00 Oct.	360.00 Aug.	148.00 Oct.	115.32 April	76.00 Jan.
V	34.00 Sept.	16.00 Jan.	7.98 Jan.	6.98 Aug.	Dark Brown	Light Yellow	69.45 Feb.	37.00 Oct.	722.00 Aug.	218.66 Oct.	200.00 Jan.	99.33 Oct.	6.03 April	1.20 Jan.	56.00 Aug.	58.00 Dec.	102.00 July	20.80 Oct.	74.00 Jan.	32.00 Oct.
VI	34.00 Sept.	16.00 Jan.	8.00 Dec.	7.12 July	March	Light Yellow April	65.56 July	30.00 April	680.00 July	271.66 Oct.	190.00 March	90.60 Sept.	6.23 April	2.00 July	36.00 Aug.	4.55 Jan.	108.66 Aug.	27.80 Nov.	46.66 July	10.13 May

Site	Sulphate (Mg/1)		phosphates (Mg/1)		Na (Mg/1)		K (Mg/1)		Ca (Mg/1)		Mg (Mg/1)		Dissolved Silica (Mg/1)		Total Hardness(Mg/1)		Total Alkalinity (Mg/1)	
	Max.	Min.	Max.	Min.	Max.	Min.	Max.	Min.	Max.	Min.	Max.	Min.	Max.	Min.	Max.	Min.	Max.	Min.
I	100.0 March	34.0 Nov.	2.50 March	0.43 Sept	43.0 June	14.0 Nov.	15.0 May.	5.66 Nov.	94.4 March	25.51 July	52.64 April	20.05 Aug.	7.0 Jan	2.0 July	296.0 March	110.0 Aug.	160.0 March	30.0 Aug.
II	642.0 May	44.0 Aug.	3.53 Sep.	1.17 July	156.0 May	18.0 July	33.0 June	7.0 July	55.33 Dec.	17.63 Aug.	40.0 Dec.	13.63 Aug.	6.0 Feb.	2.26 July	222.0 Dec.	74.66 Aug.	110.0 Feb.	31.0 Aug.
III	176.0 March	44.66 Aug.	2.80 March	1.11 July	52.0 Jun	11.0 Sept.	15.0 June	9.30 Sept.	64.33 Mar.95	13.8 Sept.	51.61 Jan	16.89 July	6.0 March	1.48 Aug.	272.0 Jan	127.33 Aug.	164.0 Feb	30.33 Aug.
IV	350.0 Dec.	56.0 May	3.93 Sep.	1.75 July	59.0 June	23.0 Aug.	25.0 June	9.0 Sept.	72.0 Oct.	38.4 Aug.	67.68 April	29.76 Oct	6.2 March	3.05 July	343.33 April	162.0 Aug	160.0 Feb.	50.0 Aug.
V	140.0 Jan.	30.0 Nov.	2.70 Feb.	1.4 Aug.	41.0 June	14.0 Oct.	16.0 May	7.0 Nov.	83.0 Dec.	39.26 Sep.	61.1 Feb95	30.96 Oct.	7.6 Feb	1.58 July	312.0 Feb.95	194.6 July	194.0 Jan.	52.2 July
VI	92.66 Sept	24.0 Oct.	2.60 March	1.10 June	42.0 Feb.	15.0 Nov.	16.0 May	8.0 Aug.	76.66 Dec.	41.0 April	65.15 March	29.28 Nov.	6.0 March	1.43 July	335.0 March	192.0 Nov.	164.0 Mar.95	42.0 April

Oil & Grease was obtained nil at all the sites during April'94 to Mar.'95.

Total dissolved solids ranged from 216.0 - 860.0 mg/l, being minimum at site I in November 1994 and maximum at site II in February 1995. Analysis of water revealed the presence of chlorides, sulphates and phosphates which collectively affected the salinity of soil and potability of water.

Dissolved oxygen was zero at site IV during April 1994 - March 1995 while the same exhibited maximum values (8.6 mg/l) at site I in Ocober 1994. Low values of dissolved oxygen in water which results in high BOD is the outcome of oxygen depletion brought about by oxygen demand of microbes and bichemical oxidation of putrecible dissolved and suspended organic matter in water, rendering the same unsuitable for aquatic life. At site IV low values of DO and poor biological system are due to higher concentration of toxic substances and oxidisable organic matter in the water. Similar findings have been reported by Roy et. al (1984), Verma et al. (1988) and Shrivastava et al. (1994).

Maximum BOD values were obtained at site VI (56.0 mg/l) in August 1994 while minimum (3.73 mg/l) being obtained at site III in September 1994. In summer, high BOD was due to prevailing high temperature and high concentration of oxidisable organic matter. Lowest BOD values at site III indicated poor biological activity. Since the BOD values exceeded tolerance limits, the water was not found suitable for drinking purpose.

COD ranged from 20.6-300.0 mg/l. Minimum values were recorded at site I in November 1994, maximum being at site IV in August 1994. COD values have been found to be much higher than BOD values which is due to the presence of chemically oxidisable carbonaceous matter such as suplhides and reduced metal ions.

Concentration of chlorides was maximum (120.0 mg/l) at site II in February 1995 while the same was minimum at site IV in May 1994. High chloride content indicates severity of pollution in the river. Important contributors of chlorides in water are effluents from local industrial units and decomposition of dead aquatic organisms. Present findings are substantiated by Munawar (1970) for fresh water ponds.

Sulphate contents ranged from 24.0 - 642.0 mg/l, minimum being at site VI in October 1994 and maximum at site II in May 1994. High sulphate content may be attributed to high input of organic matter. Sulphates are quickly reduced to hydrogen sulphide under high organic pollution and deplete oxygen availability as has also been shown by Steiver (1967).

Maximum concentration of sodium was obtained at site II in May 1994. High sodium content might be attributed to industrial discharge of Synthetics and Chemicals Ltd. This Industry utilises maximum NaCl for softening water for industrial use. Maximum sodium content was recorded at site II in July 1994 due to dilution of water during mansoon.

Concentration of potassium was maximum 33.0 mg/l at site II in June 1994 while minimum was at site I (5.7 mg/l) in July 1994. Excessive availability of potassium has been found to be associated with poor intake of Fe, Mn, Cu, Zn and Ca at higher pH levels.

Concentration of calcium varied from 13.8 to 94.4 mg/l. Maximum values were recorded at site I in March 1995 while minimum values were obtained at site III in September 1994. Calcium being an important contributor to hardness in water rejects its utility for domestic purpose. The rise in calcium content during winter months could be attributed to its higher solubility at low temperature. The decline in the values of calcium during summer might be on account of its dilution in rain water and its utilization by phyto-planktons in the river. Present findings are more akin to those of Swarup and Singh (1979) for lake waters.

Magnesium concentration was beyon the tolerance limits at all the sites and ranged from 13.68 to 67.68 mg/l. Though it is an important constituent of chlorophyll, its higher concentration increases the hardness in water and rejects its utility for drinking purpose.

Concentration of silica has ranged from 1.4 to 7.6 mg/l. Maximum concentration was observed at site V in February 1995 while minimum was recorded at site VI in July 1994. Silica is one of the major nutrients and is important for diatoms.

Total hardness ranged from 74.66 to 343.33 mg/l. Maximum values were recorded at site II in August 1994. This characteristic has been found to be associated with the presence of cations like calcium, magesium and iron. The anions responsible for hardness are mainly carbonates, sulphates, chlorides and silicates whoe concentration has been found to vary. Water of river RamGanga being hard is not suitable for potable use.

Total alkalinity exhibited wide fluctuation at different sites from April 1994 to March 1995. Maximum alkalinity was recorded at site V in January 1995, minimum beings at site I (30.0 mg/l) in August 1994. Most of the Indian rivers have been reported to exhibit slight to moderate alkalinity. (Bhargava 1985, and Venkateshwarlu 1986). In the present study also pH of water was high. The unpolluted sites with lower vlaues of hardness had slight alkalinity in the water However, water at polluted sites with higher values of total hardness was highly alkaline.

Oil and grease were to be found nil during April 1994 to March 1995.

❑ REFERENCES

APHA 1989. *Standard methods for examination of water and waste water,* 17th ed., APHA washington, DC.

Badge, U.S. and Verma A.K. 1985, Limnogical study of J.N.U. lake, New Delhi, *India. Proc. National Symposium Pure and Applied Limnology.,* 32 : 16-23

Bhargava, D.S. 1985. Water quality variation in Yamuna river. *Envtl. Pllut.,* 37(B) : 355-376.

Gupta, M.C. and Sharma L.L. 1994, Variation in selected limno chemical paramters of AmarChand reservoir, southern Rajashtna. *Poll Res.,* 13 (2) : 217 - 222.

Prasad, B.N. and Saxena M. 1980, Ecological study of blue green algae in river Gomati. *Indian. J. Environ. Hlth.,* 22 : 151 - 158.

Rao, P.S. 1994, *A study of primary productivity, plankton and some physico-chemical features of the lake Jaisamand in relation to fisheries.* Ph.D. thesis, Sukhadia University, Udiapur.

Roy Dutta., Maity S. Dutta B.R., and Banerjee R.D. 1984, *Environment and Ecology.,* 2 : 29

Shrivastava S.K., Kumar R. and Shrivastava A.K. 1994. Effect of textile industry effluent on the biology of river Tone at Mau (U.P.) Physico-chemical characteristics. *Poll Res., 13* (4) : 369 - 373.

Steiver M. 1967. The sulphur cycle of lake waters during thermal stratification. *Geochem. Cosmochim. Acta., 31* : 2151 - 2167.

Swarup K. and Sing S.R. 1979, Limnological study of Suraha lake (Ballia) J. *Inland Fish Soc. Indian.,* 11 : 22 - 23.

Venkateshwarlu V. 1986, Ecological studies on the rivers of Andhra Pradesh with special refernece to water quality and pollution. *Proc. Indian Sci. Acad., 96 (6)* : 495 - 508.

Authors :

1. **Rakesh Chandra** : Environmental officer, Tata chemicals Ltd., BABRALA - 202521. Distt. Badaun. U.P. India.
2. **Y. Bahadur** : Deptt. of Chemistry, Bareilly College, BAREILLY - 243005. U.P. India.
3. **B.K. Sharma** : Deptt. of Botany, Bareilly College, BAREILLY - 243005. U.P. India.

■

33

Studies on the water quality of Tapi River at Gujarat and Maharashtra with reference to phytoplankton

❑ INTRODUCTION

Majority of the earth's surface is covered by aquatic environment. The ecology of the estuarine entourage is subject to wide fluctuations by various factors. It may be either marine influenced or even fresh water dominated. This influence and dominance may change within a year from season to season or ever within a day with tidal rythm. Under the Aquatic habitat, estuaries are the transition zones between freshwater and marine habitats. In esturine environs, the physico-chemical features are unique, highly productive and fertile. Considerable effect on aquatic life has been established by salinity gradient (Burton and Liss 1976).

Horizontal turbulance and eddy formation have been noticed by knight (1965). Contributions on the estuarine hydrography have been made by so many workes in east and west coast. Church (1981) studied the Tapi salt marsh on both chemical and radiometrical aspects. Zingde et al, (1981) have studied the Quality of water in Damanganga River. Church (1981) studied the Tapi salt marshes on chemical aspects. Sarin and Krishna swami (1981) studied particulate and dissolved trace metal in Narmada and Tapi rivers. Physico and chemical parameters and productivity were estimated in Tapi estuary by Ragothaman and Ramachandra Reddy (1982). Ragothamam and Ramaiah (1986) have studied the physico-chemical parameters of Tapi estuary and noticed wide fluctutions. Ragothaman and Jaiswal (1995) studied the water Quality of Tapi river with reference to phytoplankton at Jalgaon of Maharashtra region.

There are many estuaries in Gujarat. Out of them Narmada and Tapi are the longer ones. Tapi river meets the sea near Surat city (21.12' N and 72.50'E). The Tapi river has its origin in Multai pradesh. It flows to a stretch of 720 km. with a dam over it at Ukai around 98 km. away from the mouth of the estuary.

It has a drainage area of about 64,000 square and an hourly flow of 120,00,000 cubic yards during the season of extreme floods. The estuary formed out of the river Tapi is still undergoing the process of evolution. Before 3-4 decades, it was a coastal plain type which had submerged extension of a river opening towards the sea. The geomorphology of this estuary has changed from Coastal plain type to bar buiet type with the deposition of siet.

The Course of the Tapi westwards from the highlands of central India to the sea, may be divided in to 4 sections the first, of 150 miles, from its source in the west of great Satpura plateau through parts of the central provinces of berar till, a few miles below the town of Barhanpur, Where it enters the plains of khandesh; the second of 180 miles of its passage across khandesh; the third, where the waters of river through 50 miles of hill rock force their way down to the low lanes of Gujrat and the fourth, ninety miles more, across the alluvial plains of Surat.

Though the main stream may be traced for some miles, further, a sacred reservior in the town of Multai, in the Betul District of Madhyapradesh is considered as the traditional source of the Tapi. After leaving Multai, the stream runs at first through open and partially cultivated lands, and then cuts its way through a gap in the Satpura mountains, the Chikhaldra hills of Berar on the left, and the wilder range of kalibhit on the right.

Falling rapidly from the Satpura uplands, through a deepcut channel from 100-150 yards wide, the flood waters of the monsoon drain away, leaving in the day season a stream which passing over a rocky bed can in many places be forded. The banks, though high, are not steep, and except where sharply cut by a turn in the river's Course, they shape gradually to the level of the stream; and, like the country round, are covered with forest trees, brused wood, and grass, a shelter to various wild animals.

The next stretch of 180 miles, the Tapi develops the wide valley plain of khandesh. The elevation of khandesh plain varies between 700-750 feet above the sea level.

On leaving the Dang forests the Tapi enters on its last stage, the passage of 50 miles in a direct line, or including wide meanders of nearly 90 miles across the Surat plain to the sea. These 90 miles of the course of the Tapi are naturally divided into 2 parts by the tidal limit. of these, the upper or freshwater section includes about 60 miles, and the lower or tidal section a little more than thirty.

The author has studied the Tapi estuary (1982, 1986 and 1995) for water Quality aspects and phytoplankton in different seasons. Fluctuations in physical, Chemical and plankton occurance are described in all the season.

METERIALS AND METHODS

The sampling is done bimonthly froms above sites. The suitable container, time and chemicals of preservation were selected as per APHA (1986). Well mixed samples of each site were taken and acidified with conc. HNO_3. Estimations of heavy metals were done with the help of Atomic absorption spectrophotometer.

RESULTS

Preliminary studies were carried out at the estuary at Magdalla part of Surat from July'79 to May'80 for a period of one year and the data were published by Ragothamam and Ramchandra Reddy (1982). Collections were made between 8.00-10.00 hrs every fortnight during July 1979 to May 1980. Temperature of the surface water was taken in the field. Water samples were analysed for various components using standard methods. Productivity was estimated by colorimetry and expressed as optical density of the pigment extract (blue filter used). Phytoplankton counts were taken using a haemocytometer. Plankton were collected in silk boting cloth with 20 mesh size and identified upto species level.

During the year August 1981 and July 1982 water samples were collected from the estuary from 3 different place namely Dumas, Magdalla and Nanpura. Dumas is the place where the river Tapi mixes with the sea. Here the waters are mostly found turbid. Magdalla spot is about 8 km away from the mouth of the river. The water is comparatively clean. Collections were made near the jetty of the magdalla port which also have a ship building yard very nearby.

Nanpura is about 13 km away from the mouth of the river and it is also the nearest spot from the Surat city. The domestic waste of the city is drained into the river just near the collection spot. From all the three sites the inogranic nutrients were mainly from American Public Health Association (1976).

The monthly water analyses for physical and chemical parameters were carried out from 2 sites namely Dumas and Nanpura from August 1981 to June 1983 by Ragothamam and Ramaiah (1986.) (Fig. 1)

The hydrobiological study (with reference to physical, chemical and phytoplankton) were carried out for a period of one year Viz, May 1989 to April 1990 at Jalgaon of Tapi estuary by Ragothamam and Jaiswal (1995). Two sites were selected for the present study. They are Savkheda 21.08' N and 75.14' E. (site I) and Kurwel 21.09' N and 75.19' E. (site II). (Fig. 2)

The distance between the 2 sites were about 5 km. Jalgaon district received rain during July to September. During October a slight rise in temperature followed by cold winter up to February. From March to May the season was hot and dry. Water samples were collected fortnightly in the morning between 8 a.m. to 9.30

a.m. Temperature of the atmosphere and surface water were taken in the field. The chemical analyses was done following APHA. (1976). Phytoplankton were estimated Qualitatively and Quantitatively.

Desikachary (1959), Fritsch (1945), Bold and Wyne (1978), Venkataraman (1939), Sarode and Kamat (1984) were used for phytoplanton identification.

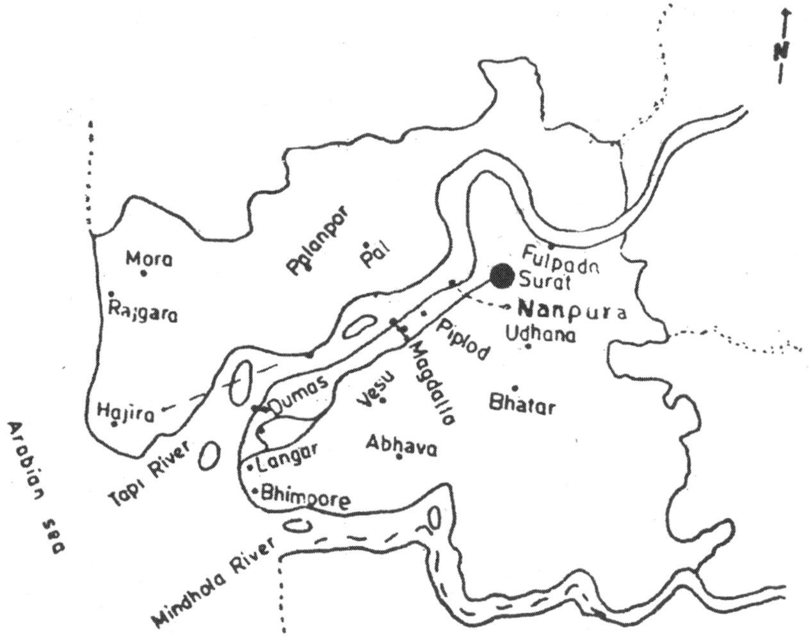

Figure 1. Map of Tapi estuary showing Sites of Study.

RESULTS

The data for water analyses and estimation of plankton are for the research work carried out during July 1979 to May 1980.

The minimum temperature was observed in January and maximum in May '80. The pH was low in August' 79 and high in January and May' 80. The salinity was maximum in January 1980 and minimum in July' 79. The dissolved oxygen was high in May' 80 and Low in July and August' 79. The minimum values for carbonate was noted in July, September, November, December 79 and Jan, 80. The maximum values for carbonate was noted in February to May 80. The bi-carbonate was less in July' 79 and more in May' 80. and minimum phosphates were observed in March and April' 80. and maximum in August' 79. The nitrates were more in September' 79 and less in May' 80 being maximum in July' 79. Calcium was high

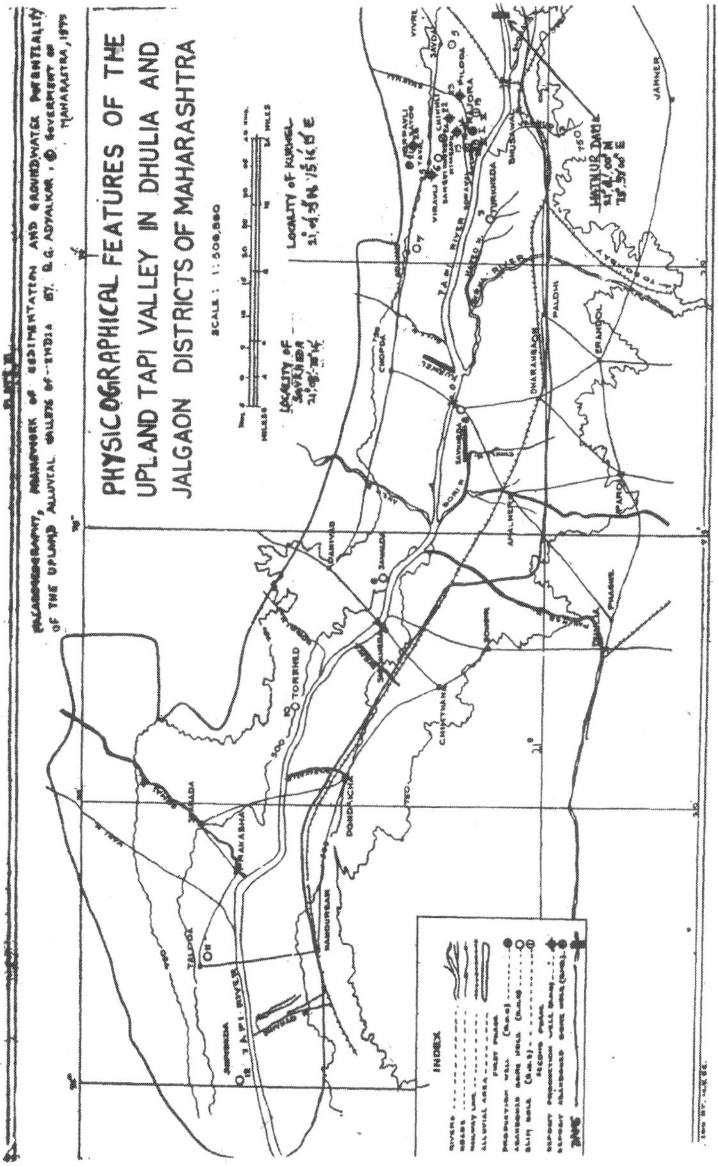
PHYSICOGRAPHICAL FEATURES OF THE UPLAND TAPI VALLEY IN DHULIA AND JALGAON DISTRICTS OF MAHARASHTRA
INDEX
TAPI RIVER

Figure 1.

in April' 80 and low in July' 79. The optical density of the pigment extract was more in December' 79 and less in July' 79. The plankton count was high in December' 79 and low in July' 79. The phytoplankton were represented with blue green algae, green algae and diatoms. The phytoplankton with reference to blue green algal were species of *Merismopedia, Rivularia, Nostoc, Oscillatoria and Lyngbya*. The green algae were represented with species of *Chlamydomonas, Gonium, Chlorella, Cladophora, Oedogonium, stigeoclonium, Mougeotia, Spirogyra, Closterium and cosmarium*. The diatoms were represented with species of *Cyclotella Chaetoceros, Fragilaria, Synedra, Eunotia, Gyrosigma, Pleurosigma, Navicula, Amphora, and Nitzschia.*

The organic nutrients were estimated at Tapi river from 3 different sites namely Dumas, Magdalla and Nanpura during August' 81 to July' 82. At Dumas site, Phosphates were found tobe maximum during April and May' 82 (31 mg 1) and minimum in November' 81 (9.0 mg 1). At Magdalla site maximum phosphate was observed during January' 82 (23.0 mg 1) and minimum during November' 81 (9.0 mg 1). At Nanpur site, During April' 82 (35.5 mg 1) was the maximum amount of the phosphate noticed and minimum in November' 81 (7.5 mg 1).

At Dumas site, silicate was maximum in July'82 (14.0 mg/1) and minimum in September' 81 (3.7 mg/1). At Magdalla site, during July' 82 (12.5 mg/1) silicate was found maximum and in September' 81 (4.0 mg/1) was the minimum. At Nanpura site, maximum silicate was noticed in January' 82 (16.1 mg/1) and minimum in September' 81 (3.8 mg/1).

At Dumas site, Nitrate was found maximum during August' 81 and Ocotober' 81 (145 mg/1) and minimum during April' 82 (22.5 mg/1). At Magdalla site, Nitrate was showing high value in August' 81 (260.0 mg/1) and less during April' 82 (16.0 mg/1). At Nanpura site, Nitrate was found maximum during August' 81 (231.5 mg/1) and minimum during April' 82 (25.5 mg/1).

At Dumas site, Nitrate was maximum during March' 82 (22.0 mg/1) and minimum during January' 82 (4.0 mg/1). At Magdalla site, the Nitrate showed maximum during July' 82 (84.0 mg/1) and minimum during February' 82 (4.0 mg/1). At Nanpura site, the maximum Nitrate was recorded during July' 82 (60.0 mg/1) and minimum during October' 81 (5.0 mg/1).

At Dumas site, Ammonia showed during October 81 (0.59 mg/1) and minimum in December 81 (0.14 mg/1). At Magdalla site, during October 81 and February 82 (0.38 mg/1) maximum ammonia was observed and minimum in May 82 (0.06 mg/1). At Nanpura site, maximum ammonia was noticed in February 82 (0.68 mg/1) and minimum in April 82 (0.12 mg/1).

The physico-chemical analyses were carried out in Tapi river and the 2 sites namely Dumas and Nanpura during August 81 to June 83.

At Dumas site the water temperature was high in April, May and June 82 and May and June 83. The low values were noticed in Jan 82.. At Nanpura site the temperature was high in May and June 82 and May and June 83 and the low values for temperature were observed in Dec 81.

The pH was found to be high in April 82 at Dumas site and less in January 82 and January 83. At Nanpura site high pH was recorded in January 83 and low in July 82 and March 83. Salinity showed high values on January 82, July 82 and April 83 at Dumas. The low values were observed in August and September 81. At Nanpura it was more during December 81 and less in February 82.

At Dumas site, dissolved oxygen was more in April 82 and low during March 82. At Nanpura high values were noticed in April 82 and low values in October 82. At Dumas site, the carbonates were high in May 82, June 82 and June 83 and low during November 82. At Nanpura it was high during May 82 and June 82 and low during October 81. At Dumas site, the bicarbonates were high in March 82 and low in April 83. At Nanpura the high values were recorded in February 82 and September 82 while low value was observed in July 82. At Dumas site, calcium was high in January 83. The less reading was recorded in July 82. At Nanpura it was high in January 83 and low in July 82. At Dumas site, phosphate was high in April 82 and May 82. It showed low values in July 82. At Nanpura site, it was high during April 82 and low in November 82. At Dumas site, silicate was high during December 82 and was low during July 82. At Nanpura site, it was high in December 81 and low during September 81. At Dumas high values of Nitrate were obtained in August 81 and October 81. and low values in January 83. At Nanpura site, it was high during August 81 and low during April 82. The Nitrates showed high values in Octomber 82 at Dumas site, and low values in February 83. At Nanpura it was high during July 82 and low during February 83. At Dumas site, ammonia showed high values in Octomber 81 and low values in May 83. At Nanpura site, it was high in February 82 and low in April 83.

The Physico-chemical analyses were carried out in Tapi river of Maharashtra of Jalgaon district at 2 sites, namely Savkheda and Kurwel from May 1989 to April 1990. The atmospheric temperature at site I was maximum showing 36^0 in May 1989 and the minimum recorded as 19^0 in January 90. At site II the maximum atmospheric temperature was 36.5^0 in May 1989 and the minimum as 18.5^0 in January 90. At site I the maximum water temperature was recorded as 30.5^0 during September 89 and the minimum as 17.5^0 in January 1990. At site II the maximum temperature was shown as 29.5^0 in September 1989 and the minimum as 17.0^0 C in January 90.

At site I, the maximum pH was 8.3 in June 1989 and the minimum was 6.5 in November 1989. At site II, the maximum pH was 8.25 in June 1989 and the minimum was 6.7 in November 1989.

At site I, the maximum dissolved oxygen was recorded was 16.5 mg/1 in January 1990 and the minimum was recorded as 5.40 mg/1 in August 89. At site II the maximum dissolved oxygen was recorded as 16.22 mg/1 in January 1990 and the minimum was recorded as 5.40 mg/1 in May 1989.

At site, I the maximum salinity was found as 563.19 mg/1 in June 1989 and the minimum was found as 89.63 mg/1 in December 89. At site II, the maximum salinity was recorded as 537.63 mg/1 in July 1989 and the minimum was recorded as 76.83 mg/1 in December 1989.

At site I, the maximum carbonate was shown as 84 mg/1 in June 1989 and the minimum was shown as 4.80 mg/1 in September 1989. At site II, the maximum carbonate was shown as 84 mg/1 in July 1989 and the minimum was shown as 10.0 mg/1 in September 1989.

At site I, the maximum bi-carbonate was recorded as 954.72 mg/1 in June 1989 and the minimum was recorded as 341.60 mg/1 in November 1989. At site II, the maximum Bi-carbonate was recorded as 907.68 mg/1 in June 1989 and the minimum was recorded as 363.56 mg/1 in November 1989.

At site I, maximum calcium was found as 92.52 mg/1 in August 1989 and the minimum was found as 3.56 mg/1 in September 89. At site II, the maximum calcium was found as 79.90 mg/1 in August 1989 and the minimum was found as 9.25 mg/1 in May 1989.

At site I, the maximum silicate was recorded as 4.30 mg/1 in April 90 and the minimum was recorded as 0.43 mg/1 in June 89. At site II, the maximum silicate was recorded as 4.15 mg/1 in April 90 and the minimum was recorded as 0.40 mg/1 in June 89.

At site I, the maximum concentration of phosphate was noted as 90.00 mg/1 in June 90 and the minimum was noted as 7.50 mg/1 in October 89. At site II, the maximum concentration of phosphate was noted as 85.50 mg/1 in July 89 and the minimum was noted 6.00 mg/1 in June 89.

At site I, the maximum nitrate was shown as 450 mg/1 in September 89 and the minimum was shown as 35 mg/1 in June 89. At site II, the maximum nitrate was shown as 380 mg/1 in September 89 and the minimum was shown as 55 mg/1 in June 89.

At site I, the maximum Nitrate was recorded as 67.50 mg/1 in September 89 and the minimum was recorded as 2.50 mg/1 in June 89. At site II, the maximum Nitrate was recorded as 51.25 mg/1 in September 89 and the minimum was recorded as 5.0 mg/1 in May 89.

At site I, the maximum Ammonia was found as 79.50 mg/1 and the minimum was found as 7.00 mg/1 in June 89. At site II, the maximum Ammonia was found as 71.50 mg/1 in July 89 and the minimum was found as 7.00 mg/1 in September 89.

Total standing crop of phytoplankton was estimated as optical density of pigment extract. At site I, the maximum optical density of pigment extract was noticed as 0.48 in January 90 and the minimum was noticed as 0.25 in August 89. At site II, the maximum optical density of pigment extract was noticed as 0.43 in December 89 and the minimum was noticed as 0.045 in August 89.

At site I, the phytoplankton was found maximum as 42,00,000/1 in November 89 and the minimum was found as 31,000/1 in September 89. At site II, the phytoplankton was noticed as 37,85,000/1 in December and the minimum was noticed as 27,000/1 in September 89.

The phytoplankton were represented by blue green algae, green algae and Diatoms. At site I and II the maximum blue greens were found in January 1990 and the minimum in September 89.

At site I and II the green algae were maximum in January 1990 and the minimum in September 1989.

At site I, the diatoms were found maximum in November 89 and the minimum were found in September 89. At site II, the diatoms were found maximum in December 89 and the minimum were found in September 89.

From both the sites 27 species of blue greens, 3 species or greens and 60 species of diatoms were recorded.

Following algae were recorded :
Cyanophyceae : *Microcystis, Chroococcus, Gloeocapsa, Gloeothece, Aphanocapsa, Aphanothece, Synecho-coccus, Synechocystis, Merismopedia, Dactylococcpsis, Sticosiihon, Trichodesmium, Phormidmum and Anabaena.*

Chlorophyceae :

Platydoriana, Chlorella and Spirogyra.

Bacillariophyceae :

Cyclotella, Fragilaria, Syhoaedra, Eunotia, Cocconeis, Achnanthes, Anomoeneis, Navicula, Pinnularia, Amphora, Cymbella, Gomphonema, Hantzschia and Nitzschia.

Discussion

The first part of the discussion deals with the work carried from July 79 to May 80 of Tapi river at Surat in Gujrat. Variations in different physico-chemical parameters and productivity in Tapi estuary during the study period and given in Table No.I.

Table 1

Distribution of Chlorinity (mg/litre), Salinity (mg/litre) and pH at two stations of Tapi River, Savkheda (Station I) and Kurwel (Station II)

S.No.	Date of Collection	pH		Chlorinity		Salinity	
		Savkheda	Kurwal	Savkeda	Kurwel	Savkheda	Kurwel
1	1.5.89	7.10	7.20	141.82	159.40	256.03	287.73
2	15.5.89	6.90	7.20	140.79	150.10	254.16	270.96
3	1.6.89	6.50	7.60	140.80	134.71	255.97	243.18
4	15.6.89	** *8.30*	** *8.25*	** *312.00*	290.74	** *563.19*	524.81
5	1.7.89	7.30	7.15	177.85	** 297.84	320.02	** *537.63*
6	15.7.89	7.25	7.30	163.10	141.83	294.43	256.03
7	1.8.89	7.20	7.30	212.74	198.56	384.03	358.43
8	15.8.89	7.30	7.20	106.37	148.92	192.03	268.83
9	1.9.89	7.40	7.25	056.73	078.00	102.43	140.83
10	15.9.89	7.20	7.20	078.00	085.10	143.83	153.63
11	1.10.89	7.30	7.25	092.19	099.26	166.43	179.23
12	15.10.89	7.00	7.15	063.82	056.73	115.23	102.43
13	1.11.89	* *6.50*	* *6.70*	070.91	081.55	128.03	147.23
14	15.11.89	7.10	7.35	092.19	078.00	166.43	140.83
15	1.12.89	7.15	7.10	106.37	113.46	192.03	204.83
16	17.12.89	7.10	7.15	* *49.64*	* *42.55*	* *89.63*	* *76.83*
17	1.1.90	7.15	7.20	060.28	063.82	108.83	115.23
18	16.1.90	7.20	7.40	085.10	078.00	153.63	140.83
19	1.2.90	7.90	7.60	078.00	092.19	140.83	166.43
20	18.2.90	7.70	7.40	113.46	099.28	204.83	179.23
21.	2.3.90	7.50	7.35	106.37	127.64	192.03	230.43
22	16.3.90	7.20	7.15	148.92	163.10	268.83	294.43
23	1.4.90	7.15	7.35	134.74	170.19	243.23	307.23
24	15.4.90	7.30	7.25	156.01	163.10	281.83	294.43

Maximum (**) Minimum (*)

Table 2

Range of Environmental & Surface Water temperature in ^0C and Dissolved oxygen in mg/litre. at two stations of Tapi river. (Savkheda Station I & Kurwel Station II).

S.No.	Date of Collection	Environmental Temp. Savkheda	Environmental Temp. Kurwal	Surface Water Temp. Savkeda	Surface Water Temp. Kurwel	Dissolved Oxygen Savkheda	Dissolved Oxygen Kurwel
1.	1.5.89	35	35	28	28	8.24	5.68
2.	15.5.89	** 36	** 36.5	28.5	28.5	7.77	* 5.40
3.	1.6.89	29	29	25	25	7.63	8.04
4.	15.6.89	27	28.5	25	25	9.39	13.38
5.	1.7.89	27.5	27.5	25	25	9.05	11.55
6.	15.7.89	28	28	26	26	5.95	5.74
7.	1.8.89	27	28	25	25	5.40	5.68
8.	15.8.89	27	28	23	23	9.12	10.34
9.	1.9.89	25	24	23	23	9.12	9.32
10.	15.9.89	28	28	** 30.5	** 29.5	11.82	11.49
11.	1.10.89	25	25	23	23	12.50	10.82
12.	15.10.89	24.5	24.5	23	23	12.03	12.16
13.	1.11.89	23	23	21.5	21.5	11.15	10.61
14.	15.11.89	22	22	20.5	20.5	11.08	10.74
15.	1.12.89	21	21	18.5	18	12.84	11.02
16.	17.12.89	20	21	18	18	15.61	15.14
17.	1.1.90	* 19	* 18.5	* 17.5	* 17	* 16.35	* 16.22
18.	16.1.90	22.5	23	21	21	16.01	15.54
19.	1.2.90	24	24	22	22	15.74	16.20
20.	18.2.90	25	26	22.5	22.5	12.16	11.96
21.	2.3.90	25.5	26	22	22	10.34	9.93
22.	16.3.90	27.5	27.5	24	24	10.00	9.66
23.	1.4.90	29	29.5	25	25	08.58	7.84
24.	15.4.90	32	33	26.5	26	6.96	7.30

Maximum (**) Minimum (*)

Table 3

Distribution of Carbonates (mg/litre) & Bicarbonates (mg/litre) alkalinity at two stations of Tapi river, Savkheda (Station I) and Kurwel (Station II).

S.No.	Date of Collection	Carbonates (CO_3) Savkheda	Carbonates (CO_3) Kurwal	Bicarbonates (HCO_3) Savkeda	Bicarbonates (HCO_3) Kurwel
1.	1.5.89	31.20	38.40	473.40	409.72
2.	15.5.89	35.40	42.37	507.63	438.73
3.	1.6.89	26.40	33.60	673.44	597.80
4.	15.6.89	** 84.00	72.00	** 954.72	** 907.68
5.	1.7.89	28.80	** 84.00	522.16	439.20
6.	15.7.89	12.00	24.00	475.80	500.20
7.	1.8.89	16.80	14.60	478.24	446.52
8.	15.8.89	14.40	31.20	458.72	522.16
9.	1.9.89	* 04.80	* 12.00	429.44	434.32
10.	15.9.89	16.80	18.00	448.96	458.72
11.	1.10.89	16.80	14.40	556.32	534.36
12.	15.10.89	55.20	54.40	370.88	390.40
13.	1.11.89	45.60	40.80	* 341.60	* 363.56
14	15.11.89	40.30	36.00	361.12	392.84
15.	1.12.89	33.60	38.40	432.36	417.24
16.	17.12.89	28.80	31.20	439.20	448.96
17.	1.1.90	43.20	38.40	431.88	461.16
18.	16.1.90	36.00	38.40	463.60	512.40
19.	1.2.90	45.60	50.40	575.84	512.40
20.	18.2.90	40.80	45.60	575.84	488.00
21.	2.3.90	36.00	21.60	492.80	500.20
22.	16.3.90	33.60	21.60	412.30	407.48
23.	1.4.90	31.20	33.60	466.04	427.00
24.	15.4.90	38.40	43.20	480.68	483.12

Maximum (**) Minimum (*)

Table 4

Distribution of Calcium ions (mg/litre) and Silicate (mg/litre) at two stations of Tapi river, Savkheda (Station I) and Kurwel (Station II).

S.No.	Date of Collection	Ca^{++} Savkheda	Ca^{++} Kurwal	Silicate Savkeda	Silicate Kurwel
1.	1.5.89	9.25	** 9.25	0.435	0.615
2.	15.5.89	9.80	9.80	0.605	0.650
3.	1.6.89	10.08	10.92	0.645	0.655
4.	15.6.89	59.64	50.40	* 0.430	* 0.400
5.	1.7.89	14.30	17.66	3.000	3.175
6.	15.7.89	67.28	71.49	2.375	2.975
7.	1.8.89	65.60	69.81	2.875	3.825
8.	15.8.89	** 92.52	** 79.90	1.550	2.000
9.	1.9.89	* 03.36	13.46	2.000	2.200
10.	15.9.89	45.42	37.01	2.200	2.400
11.	1.10.89	32.80	34.48	2.000	2.625
12.	15.10.89	33.64	36.16	2.975	3.750
13.	1.11.89	29.44	24.39	2.775	2.600
14.	15.11.89	31.12	25.23	1.725	2.000
15.	1.12.89	27.75	28.59	1.375	1.550
16.	17.12.89	23.50	22.71	1.250	1.550
17.	1.1.90	32.80	33.64	1.400	1.725
18.	16.1.90	36.16	37.85	1.550	1.250
19.	1.2.90	38.68	36.16	1.350	1.475
20.	18.2.90	34.48	31.98	2.000	1.850
21.	2.3.90	23.55	26.91	2.625	2.775
22.	16.3.90	21.87	24.39	3.375	3.225
23.	1.4.90	21.02	21.87	4.150	3.850
24.	15.4.90	15.14	13.46	** 4.300	** 4.150

Maximum (**) Minimum (*)

Table 5

Distribution of Inorganic phosphats and ammonia (mg/*l*) at two stations in Tapi River.

Sl. No.	Date of Collection	Phosphate Savkheda	Phosphate Kurwal	Ammonia Savkheda	Ammonia Kurwal
1.	1.5.89	44.10	29.40	12.50	9.00
2.	15.5.89	30.00	32.60	14.00	10.50
3.	1.6.89	24.50	23.40	16.00	12.50
4.	15.6.89	12.40	* 06.00	* 07.00	35.00
5.	1.7.89	** 93.00	44.10	14.00	15.00
6.	15.7.89	79.00	** 85.50	** 79.50	** 71.50
7.	1.8.89	85.50	67.50	61.50	53.00
8.	15.8.89	73.50	79.00	51.00	61.50
9.	1.9.89	16.00	17.25	12.50	9.00
10.	15.9.89	08.50	15.00	11.00	* 7.00
11.	1.10.89	11.00	11.75	21.50	23.00
12.	15.10.89	* 07.50	08.00	15.50	14.00
13.	1.11.89	12.50	10.00	20.50	28.50
14.	15.11.89	18.50	09.50	26.50	30.50
15.	1.12.89	13.50	11.00	35.50	37.00
16.	17.12.89	08.50	06.25	25.00	19.50
17.	1.1.90	10.50	09.25	23.50	22.50
18.	16.1.90	08.50	07.50	16:00	24.00
19.	1.2.90	11.00	08.50	21.50	21.50
20.	18.2.90	17.50	16.00	23.00	26.50
21.	2.3.90	19.50	23.50	32.00	34.00
22.	16.3.90	27.50	29.50	30.00	28.00
23.	1.4.90	23.00	23.50	12.50	11.00
24.	15.4.90	21.00	23.50	09.00	14.50

Maximum (**) Minimum (*)

Table 6
Distribution of Inorganic nitrate and nitrite (mg/*l*)
at two stations in Tapi River.

Sl. No.	Date of Collection	Nitrate (NO_3)		Nitrite (NO_2)	
		Savkheda	Kurwal	Savkheda	Kurwal
1.	1.5.89	65	140	11.25	* 5.00
2.	15.5.89	135	145	07.50	08.75
3.	1.6.89	70	125	11.25	16.25
4.	15.6.89	* 35	* 55	* 02.50	06.25
5.	1.7.89	100	65	16.25	13.12
6.	15.7.89	145	115	18.12	13.12
7.	1.8.89	135	145	17.50	14.37
8.	15.8.89	120	95	16.25	11.25
9.	1.9.89	210	320	03.75	06.25
10.	15.9.89	** 450	** 380	** 67.50	** 51.25
11.	1.10.89	260	235	8.75	14.37
12.	15.10.89	310	340	17.50	11.25
13.	1.11.89	345	330	16.25	19.37
14.	15.11.89	310	305	08.75	11.25
15.	1.12.89	250	275	14.37	17.50
16.	17.12.89	260	270	11.87	19.37
17.	1.1.90	285	320	10.62	17.50
18.	16.1.90	315	330	11.25	16.25
19.	1.2.90	390	340	06.25	06.25
20.	18.2.90	275	285	11.25	14.37
21.	2.3.90	250	240	17.50	19.37
22.	16.3.90	220	205	13.12	17.50
23.	1.4.90	170	155	8.75	8.75
24.	15.4.90	150	135	8.75	9.37

Maximum (**) Minimum (*)

Table 7

Optical Density of Pigment Extract & Total Organisms (in Thousands) at two Stations in Tapi river, Savkheda (Station I) and Kurwel (Station II).

S.No.	Date of Collection	Optical Density		Total Organisms/l (in thousands)	
		Savkheda	Kurwal	Savkeda	Kurwel
1.	1.5.89	0.15	0.12	327	262
2.	15.5.89	7.07	0.05	154	110
3.	1.6.89	0.12	0.17	720	1020
4.	15.6.89	0.09	0.11	540	660
5.	1.7.89	0.20	0.23	114	131
6.	15.7.89	0.075	0.09	40	51
7.	1.8.89	0.09	0.08	135	120
8.	15.8.89	* 0.025	* 0.045	33	63
9.	1.9.89	0.12	0.09	60	45
10.	15.9.89	0.07	0.06	* 31	* 27
11.	1.10.89	0.13	0.16	62	64
12.	15.10.89	0.15	0.17	2040	2312
13.	1.11.89	0.35	0.30	** 4200	3600
14.	15.11.89	0.42	0.36	3780	3240
15.	1.12.89	0.30	0.37	3150	** 3785
16.	17.12.89	0.39	** 0.43	2935	2945
17.	1.1.90	** 0.48	0.41	3672	3609
18.	16.1.90	0.32	0.29	2718	2610
19.	1.2.90	0.14	0.15	1750	1860
20.	18.2.90	0.11	0.13	980	1130
21.	2.3.90	0.16	0.14	740	650
22.	16.3.90	0.12	0.09	620	520
23.	1.4.90	0.08	0.09	530	640
24.	15.4.90	0.13	0.16	410	513

Maximum (**) Minimum (*)

Table 8
Distribution of Diatoms (Bicilliariophyceae) and
Bluegreen algae (Cyanophyceae) during the stuty.

S.No.	Date of Collection	No. of Diatoms/lit. (in thousands)		No. of Bluegreens/lit. (in thousands)	
		Savkheda	Kurwal	Savkeda	Kurwel
1.	1.5.89	277.9	222.7	29.4	23.6
2.	15.5.89	130.9	93.5	15.4	11.0
3.	1.6.89	633.6	897.0	64.8	91.8
4.	15.6.89	486.0	594.0	43.2	52.8
5.	1.7.89	95.0	110.0	12.5	14.4
6.	15.7.89	33.6	42.8	04.4	05.6
7.	1.8.89	117.5	104.4	12.2	10.8
8.	15.8.89	29.0	55.4	03.3	06.3
9.	1.9.89	52.8	39.6	04.8	03.6
10.	15.9.89	* 27.9	* 24.4	* 02.2	* 01.9
11.	1.10.89	50.2	51.5	07.1	07.0
12.	15.10.89	1805.4	2069.2	173.4	185.0
13.	1.11.89	** 3696.0	3240.0	399.0	288.0
14.	15.11.89	3269.00	2851.0	396.0	291.6
15.	1.12.89	2278.3	** 3349.7	293.0	331.2
16.	17.12.89	2538.8	2512.1	308.2	329.8
17.	1.1.90	3066.1	3092.9	** 459.0	** 389.8
18.	16.1.90	2337.5	2270.7	285.4	261.0
19.	1.2.90	1515.5	1618.2	182.0	189.7
20.	18.2.90	843.8	972.9	103.9	118.6
21.	2.3.90	639.6	559.6	77.7	69.5
22.	16.3.90	545.6	462.3	52.7	43.2
23.	1.4.90	463.	565.8	48.8	60.2
24.	15.4.90	355.5	447.9	38.1	49.3

Maximum (**) Minimum (*)

Table 9

Distribution of Green algae (Chlorophyceae) during the study period.

S.No.	Date of Collection	No. of Greens/Litre (in thousands)	
		Savkheda	Kurwal
1.	1.5.89	19.6	15.7
2.	15.5.89	7.7	5.5
3.	1.6.89	21.6	30.6
4.	15.6.89	10.8	13.2
5.	1.7.89	5.7	6.5
6.	15.7.89	2.0	2.5
7.	1.8.89	5.4	4.8
8.	15.8.89	0.7	1.3
9.	1.9.89	2.4	1.8
10.	15.9.89	* 0.9	* 0.6
11.	1.10.89	4.6	5.4
12.	15.10.89	61.2	57.8
13.	1.11.89	105.0	72.0
14.	15.11.89	113.4	97.2
15.	1.12.89	78.7	104.1
16.	17.12.89	88.0	103.1
17.	1.1.90	** 146.9	** 126.3
18.	16.1.90	95.1	78.3
19.	1.2.90	52.5	52.1
20.	18.2.90	22.4	38.4
21.	2.3.90	22.9	20.8
22.	16.3.90	21.7	14.5
23.	1.4.90	14.8	17.9
24.	15.4.90	16.4	15.9

Maximum (**) Minimum (*)

Water temperature was high during May when other parameters like salinity, dissolved oxygen, pH and calcium were also high. During this period the nutrient concentration were low, but the plankton population was fairly normal. Among the plankton, diatom population was high, as evident from high silicate values.

Considerable variation was found in salinity in Tapi estuary. However with increasing salinity, there is an increase in plankton population. Decrease in salinity may be due to monsoon run off/ Ukai dam water. When the productivity was more, no significant changes were found in other parameters. When factors like salinity, pH, dissovled oxygen, nutrients were least, the productivity was also less and only blue green algae were present during the period. The present results with agree those of shetty et al. (1961). who observed that the production and composition of plankton are governed by several hydrobiological features.

The second part of the discussion deals with the research work carried out from August 81 to July 82, of Tapi river at three sites namely Dumas, Magdalla and Nanpura.

Phosphate showed wide fluctuations in all the 3 sites, 2 peaks were observed during January 82 and April 82 in all the three zones. Similar fluctuations were observed by Krishnamurthy (1970) and Bhattathiri (1976). According to the observations of Pai and Reddy (1981) the levels of phosphate decreased in September 81 and gradual rise noticed in our study also. The decrease may be due to the absorption by plankton. Jitts (1959) showed that majority of Phosphorus may be trapped by sediments. Low phosphate levels were associated with high silicate values. Similar observations were made by Jospech (1974). Assimilation of phosphate from the water by plankton would tend to realease more phosphate from the sediment. Butler and Liss (1972) state that any change in phosphate content may affect the plankton community.

The major source of silicate in natural waters is from the decomposition of Aluminosilicate mineral in the drainage basins. (Hutchinson 1975). During the mansoon the main source of silicate are land ran-off and precipitation. (Quasim and Sengupta 1981) De Souza et al, (1981); Joseph (1974) During estuarine mixing removal of silicate has been noticed (Burton and Liss 1976). High values observed during high tide might be due to increased pressure, which forces more silica to get dissolved. The presence of more plankton is observed with maximum silicate. Peterson et al; (1977) stated that since river discharge is the fundamental source of dissolvled silicon, when river discharge is high; utilisation of phytoplankton do not have much significant influence on its distribution.

Nitrogen is the main nutrient which limits the growth of phytoplankton.

Nitrogen discharge to a water rcourse will promote biological response and may even from plankton blooms. Most of the nitrogen in the water is involved in a continuous cycle in which planktonic organisms play an important role (Corner and Davies 1971). Most of inorganic nitrogen in the water will be in the form of nitrates. From the analysis it is clear that nitrotate-nitrogen is high in this river. Higher content of nitrate during mansoon may be due to heavy rainfall and land run-off. The main source which contrributes thoughout the year is the sewage disposal in the river. Phytoplankton growth and pigments had no linear relation with the increasing amount of nitrates. Nitrate is a limiting factor for the growth of phytoplankton.

Nitrite is an intermediate compund and source for the growth of phytoplankton. Fluctuations are wide. Leonard (1972) states that high nitrite in surface water indicate sewage pollution. The higher value of nitrite may be due to the reduction of nitrate by phytoplankton. Production of nitrite may be due to oxidation of ammonia. Oxidation of ammonia may be mediated bacteriologically, photochemically of even by phytoplankton (Hutchinson 1975).

Ammonia is the major nitrogenous end product of the bacerial decomposition of organic matter. Ammonia is also a source of nitrogen for grwoth of phytoplankton (Thomson 1970). It exists both in ionic and non-ionic forms. Ammonia concentration is very irregularly distributed. This is due to the adsorption of ammonia by particulate matter (Cooper 1948). Lower values of ammonia may be due to low activity of bacteria at low temperature and a lower ammonifiction. The presence of large quantities of ammonia in surface water indicate sewage pollution (Leonard 1971). From the observations of high ammonia content it is evident that this estuary is polluted largely by sewage.

The third part of the discussion deals with the research work carried out from August 81 to June 83 of Tapi river at two sites namely Dumas and Nanpura.

At Duma site, when temperature was high pH, dissolved oxygen and phophate were also high. When it was low, pH and nitrate were also low. At Nanpura site, temperature and carbonate pH exhibited more direct relationship.

In both the sites, high pH values concided with high temperature and phosphate. In Dumas site, low pH values conicided with low nitrate values. On the other hand, in Nanpura, low pH values conicided with low calcium and carbonate. The pH of the estuarine waters is more variable than that of the open sea (Harvey 1960). pH is very important in regulating respiration and enyme systems (Odum 1971). Welch (1953) considers pH as a limiting factor in estuary. High photosynthetic activity raises pH high (perking 1976). Seasonal fluctuation in water level is said to have a significant effect over pH and alkalinity (Holden and Green

1960, Talling and Rzoska 1967, and Adebexi 1981). However, in the present study, tidal effect over pH was not appreciable. Considerable variations in salinity were recorded in the present study confrming to earlier observations by Ragothaman and Ramachandra Reddy (1982). At Dumas site high salinity values coincided with high temperature, pH, Calcium and phosphate. On the other hand at Nanpura high salinity values coincided with high silicate and low values with low nitrate. Salinity varies in an estuary due to mixing of sea water or freshwater inflow. (Joseph 1974; Parulekar and Dwivedi 1974). From the observations made on salinity in Tapi estuary, it is clear that it is a stratified type. Freshwater flow is more dominant than tidal flow. It was also observed that there was a linear increase in salinity values from monsoon to postmonsoon period. Tidal influence was observed to be less in both the sites. The longitudinal distribution of salinity showed a decreasing trend in the upstream direction, High values were obtained at Dumas and low at Nanpura. The statement receives support from the work done by Parulekar and Dwivedi (1974) and Postma (1967). Reddy, Hariharan and Kurian (1979) stated that salinity values increased with rising tides and decreased with falling tides. Dilution by rainwater has resulted in low salinity. Similar observations were made by Sewell(1929) and Jayaraman. (1951). It has been reported by Asim et al.,.(1972) that enrichment of nutrients associated with sudden fall in salinity results in higher phytoplankton counts. Here in Tapi estuary also when salinity was low nitrate and nitrite were high.

At Dumas and Nanpurn site, when dissolved oxygen was higher, temperature, pH, and phsophates were higher. Dissolved oxygen occured in lower concentrations than the expected solubility almost throughout the period of study. This indicated a higher nutrient regeneration than utilisation. (Qasim and Segupta 1981).

Constant oxygen values were noticed during the mansoon period increasing trend is closely related to decreasing water temperature. Qasim and Sengupta (1981) state that during the cycle oxygen concentration decreases with the rising tide and increases with falling tide. In the present study also random relationship between dissolved oxygen and nutrients were noticed.

When Carbonate was more at both the sites, temperature was also more. When bi-carbonate was less at Nanpura, pH was also less. Carbonate plays a small role in contribution of alkalinity (Round 1973). Carbonate ions are used as carbon sources when pH of the water is above 9.5. When bi-carbonate was more at Nanpura, ammonia was also higher. When it was less pH was also less, wide fluctuations were observed in bi-carbonate at both the sites.

When calcium was higher, salinity was also higher at Dumas. When it was less, phsphate and silicate were also low. When calcium was low at Nanpura, pH was also low. Wide fluctuations were observed for calcium in both the sites. Concentrations decrease with increasing distance from the month. (Sengupta and Sugandi Naik 1981; Narvekar et al., 1981).

When phosphate was high in Dumas, Temperature, pH and carbonates were also higher. When it was low, calcium and silicate were also low. At Nanpura when dissolved oxygen was higher. Similar relationships were observed by Krishnamurthy (1970) in Portonovo.

When silicate was less, calcium and phosphate were also low at Dumas. The major source of silicate in natural waters is from the decomposition of aluminosilicate mineral in the drainage basins (Hutchinson 1975). During the mansoon the main source of silicate are land run-off and precipition. (Qasim and Sengupta 1981; Desouza et al., 1981). During estuarine mixing removal of silicate has been noticed Burton and Liss 1976). This is due to absorption of ammonia by particulate matter (Copper 1948). Lower values of ammonia may be due to low activity of bacteria at low temperature and slower ammonification. The presence of large quantities of ammonia in surface water, indicate sewage pollution (Leonard 1971). From the observations of high ammonia content it is evident that this estuary is polluted largely by sewage.

The fourth part of the discussion deals with the research work carried out from May 89 to April 90 at Jalgaon of Tapi Estuary (Two sites namely Sevakheda and Kurwel).

The physical and chemical parameters were correlated with phytoplankton. Seasonal changes observed in water temperature was always correlated with similar behaviour of atmospheric temperature. (Singarajah 1981). Qasim and Sengupta (1981) and Zingade et al. (1981) observed similar phenomenon. In the present study the atmospheric temperature in both the site are more than water temperature. This may be due to the cold wave present in the environment. Increase in temperature not only reduces the dissolution rate of oxygen but increase the rate of oxidation of organic matter also, hence the rate of oxygen consumption. (Zingde et al. 1981). The rate of production of planktons are closely dependent upon temperature conditions of the water body (Mc Camby 1953, Zutshi and Khan 1978). According to Roy (1955) low temperature in Hooghly was favourable for the growth of diatoms resulting in highest plankton peak in the year. The preset studies show highest diatoms in Winter. When the water temperature was low at both the sites.

When pH was higher, Carbonates, and bicarbonates were also higher but silicate, Nitrite, and Nitrate were less. When the pH was minimum, the planktons were abundant. When pH was high salinity was also maximum. High amount of pH is correlated with minimum amount of ammonia, nitrite and nitrate. The minimum pH is correlated with minimum bicarbonate.

In an esturine environment distribution of salinity is dependent on the influence of fresh water influx and the penetration of the sea water into the estuary and the mixing process. (Joseph 1974; Pritchand and Dwivedi 1974). Pritchand (1967) states that an estuary may vary from a well mixed to a stratified type. The Tapi river at Jalgoan district shows lot of variation in salinity at both the stations. When the salinity was maximum bicarbonate and phosphate were also found to be maximum. When salinity was low the optical density to the pigment extract was more. A linear decrease in salinity value was noted at both the site and also fluctuations were seen throughout the study. Similar observations were made by Goswami and Singhbal (1974), Joseph (1974),. Cherian et al., (1974) and Verma et al. (1975). Salinity is an important ecological factor controlling phytoplankton growth and distribution in the estuary and shore waters. (Rajagopal 1981). Devassay and Bhattathri (1974) observed rich plankton when salinity was very low during mansoon. In the present work it was observed that salinity was inversely related to plankton growth.

Subrahmanyan and Sen Gupta (1965) opined that high dissolved oxygen values will indicate increased photosynthetic activity. Peelan (1974) stated that rich phytoplankton population will result in an increased dissolved oxygen. In such a condition the assimilation of CO_2 and evolution of O_2 will be high. (Corner and Devis 1971). The decreased dissolved oxygen observed at both the sites, can be attributed to the large scale depositon of human, animal and sewage to the river periodicity.

When the bicarbonate are higher, the pH and salinity were also higher, but silicate, phosphate, ammonia, nitrite and nitrate were less. When bicarbonates were less, pH was also less. The plankton were more abundant when bicarbonates are low. The higher values of carbonates and bicarbonates might be due to absence of free carbondi-oxide, which might have been converted into carbonates and bicarbonate ions, as stated earlier. (Saxena 1987).

Calcium is an important micronutrient in an aquatic environment. It contributes to hardness in water therby reducing the utility of water for domestic use (Saxena 1987). When calcium was low, the carbonates were also low. The calcium is less in Summer, increases in monsoon, reduces in Winter and increase in spring. Sengupta et al. (1980) reported that the possibility of loss of calcium is due to precipitations as their insoluble phosphates. The present study did not show any relationship between calcium and orthophosphate, when these factors were inversely proportional to each other. This was in agreement with the observations of Sharma et al. (1987) and Wrobel (1967). The distributions of silicates at both the sites has shown that it is very less in late Summer, increases in Mansoon, decreases in Winter and gradually improved in Summer. From this, it is clear that the silicates are present throughout the study. Whenever silicate increase, diatoms also increase, so there is a correlation of silicates with diatoms. Hutchinson (1975) states that the diatoms consume silicate and bring considerable variation in silicate concentration. In the presence of water, CO_2 reacts with silicate minerals to form Carbonate and Silicate (Uray 1952). In the present study, carbonate data co-incides with silicate. Paterson et al (1977) states that since the river discharge is the fundamental source of dissolved silicon, when river discharge is high utilization of phytoplankton does not have much significant influence on its distribution.

When the phosphates were found to be less, Nitrates and Ammonia were also less but pH, Salinity, Carbonates and bicarbonates were more. Braylinsky and Mann (1986) reported that growing concentration of nitrate-nitrogen and phosphate phosphorous and declining dissolved oxygen in water body denotes its eutrophic conditon. Pair and Reddy (1981) have observed the peaks of phosphate during August followed by a gradual decrease in September and October. The present study has showed the similar peak in July, August and the depletion of phosphorous was correlated with gradual increase in phytoplankton count.

Estuaries are rich in nutrients because of the nutrient trap Zone formation. (Knight 1965). When the nitrates were maximum, the nitrites were also maximum. Water temperature was also maximum, but plankton were less. When Nitrates were less, Nitrite, Silicate and ammonia were also less, but pH, Salinity carbonate and bicarbonate were more. The higher concentration of Nitrate in water, correlates with higher dissolved oxygen, higher optical density of pigment extract and higher phytoplankton count.

When the Nitrites were higher, water temperature was also higher, but planktons were low. When also Nitrites were low, so also nitrates, silicate, phosphate, ammonia and calcium were less but pH, Salinity, carbonate and bicarbonate were higher. When Ammonia was low, plankton were also low. The peak for Ammonia was recorded in July andAugust. This is due to the domestic sewage surface run-off in Monsoon (Saxena 1987). Copper (1948) had accounted the irregular distribution of Ammonia due to the adsorption of ammonia by particulate matter. Bela (1965) reported that Ammonia exsists in both ionic and non ionic forms. Rajendran and Venugoplan (1977) showed that nitrate reduction and ammonia oxidation were significant process in estuarine water.

When the optical density of the pigment extract was maximum the organisms were also present in maximum number so also the dissolved oxygen but temperature was low, During the minimum values of optical density of the pigment extract, the calcium was found to be maximum. The present study shows that the productivity is high in December and January at both the sites. During the post Monsoon period the productivity was low.

When the plankton population was higher the diatoms were also present in larger number. The pH and bicarbonates were low when plankton were high. When planktons were minimum, temperature, nitrate and nitrites were found to be maximum.

From the planktonic studies diatoms were found to be dominant. Nearly 27 species of blue greens,3 species of Greens and 60 species of diatoms were recorded from both the sites. The plankton population is more in November, December 1989 and January 1990. Less planktons were seen in July, August and September 1989. The productivity of the plankton is high in November, December and January.

❑ REFERENCES

Adibisi A.A. 1981. Plankton composition in to estuaries of the konkan coast during the premonsoon season. *Mahasagar* 14 : 55 - 60

APHA. 1976. *Standard Methods for the examination of water and waste water. ed. 14th, New York* Bela. 1965. *Environmental engineering Hand book Vol. I. Water pollution*. Chitten Book Company, U.S.A.

Bhattathiri, P.M.A., Devassy V.P., and Bhargava, R.M.S. 1976. Production at different trophic levels in the Estuarine systems of Goa, *Ind. J. Mar. Sci.* 38 : 25

Bold H.C. and Wyne M.J. 1978. *Introduction to the algae. Structure* and *reproduction*. Prentice Hall of India Pvt. Ltd., New Delhi.

Bose B. 1956 *Ind. J. Fish* 1956.

Brylinsky M. and Mann. K.H. 1969. An analyses foı factors governing productivity in lakes and Reservoirs. *Limnol Oceanogr* 18 : 1 - 14

Burton J.D. and Liss P.S. 1976. *Estuarine chemistry*, Academic Press, New York. PP - 1 - 31.

Buter E.T. and Tibbitts S. 1972. Chemical Survey of Tamar estuary. *J. Mar. Biol. Ass.* U.K. 52 : 681 - 699.

Chandramohan, P. and Satyanarayana Rao, T.S. Proc. *Ind. Acad of Sc* 75:23

Cherian T. et al., 1974. Hydrography and suspended sediments load of oceanic and esturine water adjoining Mormugoa, Harbour during early summer. *Ind. J. Mar. Sci.* 3 : 99 - 101.

Cooper, L.H.N. 1948. Particulate ammonia in sea water. *J. Mar. Biol. Ass. U.K.* 27 : 322 - 325.

Coner, E.D.S. and Davies., A.G. 1971. *Advances in Marine Biology-Plankton as a factor in the Nitrogen and Phosphorous cycle in the sea.* Acadmic Press New York 9 : 102 - 192.

Charch, T.M. 1981. Chemical and Radiometric studies of Tapi salt marshes. Seminar on Estuaries Dec. 7 - 11. N I O. Goa. India.

Dehadrai, P.V. and Bhargava R.M.S. 1972. *Ind. J. Mar. Sci. 1* : 52

Desikachary T.V. 1959. *Cyanophyta,* I.C.A.R. New Delhi.

Desouza S.N., Sengupta R., Sujatha Sanzgin and Rajagopal, M.D. 1981. Studies on Nutrients of Mandovi and Zuari river system. *Ind. J. Mar. Sci. 10* : 314 - 322.

Devassy V.P. and Bhattathiri P.M.A. 1974. phytoplankton ecology of the Cochin back water. *Ind. J. Mar. Sci. 3* : 46 - 50.

Fritsch F.E. 1945. *Structure and reproduction of algae* Vol. II university press cambridge.

Ganapati S.V. 1964. *Arch. Hydrobiol.* 60 : 200

Goswami S.C. and Singbal S.Y.S. 1974. Ecology of Mondovi and Zuari estuaries I plankton community in relation to hydrographic conditions during monsoon months. *Ind. J. Mar. Sci.* 3 : 51 - 54.

Gopinathan C.P. 1975. *J. Mar. Biol. Ass. India* 17 : 223.

Harvey H.W. 1960. *The chemistry and fertility of sea water.* University press, Cambridge pp 224.

Holden J. M. and Green J. 1960. The Hydrology and plankton of river Skoto. *J. Anim. Ecol.* 29 : 65 - 84.

Hutchinson G.E. 1975. *A treatise on Limnology* Vol. I Part 2. John Wiley and sons, New York pp. 684.

Jayaraman R. 1951. Observations on the chemistry of the water of the Bay of Bengal off Madras city during 1948 - 1949. *Proc. Indian Acad. Sci.* 33B : 92 - 97.

Jitts H.R. 1959. The absorption of phosphate by estuarine bottom deposits. *Aust. J. Mar. & fresh water research.* 10 : 7 - 21.

Joseph P.S. 1974. Nutrient distribution in the Cochin harbour and its Vicinity. *Ind. J. Mar. Sci.* 3 : 28 - 31

Kwight C.B. 1965. *Basic concepts of Ecology.* Macmillan Co., New York pp 468.

Krishnamurthy K. 1967. *Int. Revue Ges. Hydrobiologia.* 52 : 427.

Krishnamurthy K. 1970. Contributions in Biological oceanography 1-8. The ocean world joint oceanography assembly, Tokyo.

Kurian C.V. and Sebastian, V.C. 1976. *Prawns and Prawn fisheries of India.* pp 280.

Leohard L., Ciaccioo 1971. *Water and Water Pollution Hand book.* Vol. I Marcel Dekker Inc. York. pp. 449.

MC cambie A.M. 1953. Factors influencing the growth of Phytoplankton. *J. Fish. Res. Board. can* 10 : 253 - 282.

Narvekar P.V. Zingde M.D. and Kamat Dalal 1981. Behaviour of Boron, calcium and magnesium in Mindhola river estuary. *Ind. J. Mar. Sci.* 10 : 90 - 92.

Odum E.P. 1977. Principles pertaining the limiting factors. In *Fundamentals of Ecology* W.B. Saunders Co. 1 Philadelphia. p. 106 - 138.

Pai R. and Reddy M.P.M. 1981. Distribution of nutrients off Naip, South Kanara Coast. *Ind. J. Mar. Sci.* 10 : 253 - 282.

Palria S. and Rana B.C. 1985. Ecological studies of a river polluted by dyeing and textile factories waste *comp. physiol. Ecol. Vol* 10 : No. 4 pp. 235 - 237.

Prkash C. 1980. *Limnological stury of River Jamuna with reference to pollution.* Ph. D. thesis, Agra University, Agra.

Parulrkar A.H. and Dwivedi S.N. 1975. Benthic studies in Goa estuaries *Ind. J. Mar. Sci.* 13 : 41 - 45.

Perkins, E.J. 1976 *The Biology of estuarines and waters.* Academic Press, New York, London p. 25 - 27.

Peterson D.H., Cononmos T.J., Broenkow W.W. & Serviami E.P. 1977. Process controlling the dissolved silica distribution in San Fransisco Bay. *Estuarine Research,* Academic Press Ed., London.

Postma, H. 1967. In *Estuaries* G.H. Lauff. Ed. American Association of Advancement of Sciences. 83 : 158 - 161.

Pritchard D.W. 1967 What is an Estuary ? Physical views point. *Amer. Ass. for Advan. of Sci.* 83 : 3 - 5.

Qasim S.Z., V.P. Devassy, P.M.A.W. Bhattathiri 1972. *Marine Biology* 12 : 200

Quasim S.Z. and Sengupta R. 1981. Environmental characteristics of the Mandovi and Zuari estuarine systems in Goa. Seminar on Estuaries Dec. 7 - 11. NIO, GOA, India.

Ragothaman G. and Ramaiah P.V. 1986. Studies on Physico-chemical parameters of Tapi estuary. *Marine Biology and Maiculture* 1986. pp. 1 - 14 Ed. V. Krishnamurthy.

Ragothaman G. and Ramchandra Reddy Y. - 1982. Hydrobiology of Tapi estuary, Surat. *Ind. J. Mar. Sci.* 11 : 256 - 258.

Reddy P.M. and Venkatewaralu V. 1987. Assessment of water Quality and pollution in the river Thungabadra near Kurnool, Andhra Pradesh. *J. Environ. Biol* : 8 (2) : 109 - 119.

Rajagopal M.D. 1981. Plankton studies in the estuarine and near shore regions of Mandovi and Zuari. *Ind. J. Mar. Sci.* 10 : 112 - 115.

Reid G.K. 1961 - *Ecology and Inland waters and Estuaries.*

Rajendra A. and Venugopalan V.K. 1977. Microbiology of Nutrients regeneration. Part II. Distribution of nitrifying bacteria in vellar estuary. *Ind. J. Mar. Sci.* 6 : 6 - 9

Round F.E. - *The Biology of the algae*, Edward Arnold Ltd., London. pp 148 - 150.

Roy H.K. - 1955. Plankton Ecology of river Hooghly, at Palta, West Bengal *Ecology* 36 : 169 - 175.

Sarin M.N. and Krishna swami. 1981. A study of dissolved and particulate heavy metals in the Narmada and Tapi rivers and their estuaries. Seminar on Estuaries. Dec. 7-11 N.I.O. Goa. India.

Sarode P.T. and Kamat N.D. 1984. *Fresh water diatoms of Maharashtra.* Saikripa Prakashan. Aurangabad 431003

Sewell R.B.S. 1929. Geographical research in Indian waters (part V) *Mem. Asiatic Soc.* 9 (5) : 207 - 356

Shetty H.P.C., Saha S.B. & Ghosh, B.B. 1961. *Ind. J. Fish* 2 : 326

Talling J.E. and K. Rzoska 1967. The development of plankton in relation to hydrobiological regime in the Blue Nile. *J. Ecol. 55* : 636 - 662

Thomson W.H. 1970. Effect of ammonium and Nitrate concentration in chlorophylle increase in natural tropical pacific phytoplankton population. *Limnol & Oceanogr* 15 : 386 - 394.

Uray, H.C. 1952. The Planets, their origin and Development. New Haven, Yale University Press. XVII : 245 p 221 - 790.

Venkataraman G. 1939. Systematic account of some South Indian Diatoms *Proc. Indian. Acad. Sci.* Vol 10 (6) : 293 - 368

Verma, K.K. et al. 1975. Temporal and spatical variations in Hydrographic conditions of Mandovi estuary. *Ind. J. Mar. Sci.* 4 : 11 - 15

Zingde M.D. 1979. Pollution in river Par and its adatement. *Ind. J. Mar. Sci.* 8 : 266 - 70

Zingde M. D., Sureshchander, Rokade M.A. & Desai B.N. 1981. Base-line water quality of river Narmada, Gujarat. *Ind. J. Mar. Sci.* 10 : 161 - 165.

Zutshi, D.P. and Khan M.A. 1978. On the typhology of Kashmir in D.N. sen & R.P. Bansal (eds). *Environmental Physiology and Ecology of plants* B. Singh and M.P. Singh. Dehradun P. 465 - 472.

Author :

G. Ragothaman : Department of Aquatic Biology, South Gujarat University, SURAT - 395 007. Gujarat, (India)

34

Changes in Physico-chemical aspects and Phytoplankton distribution in river Shivnath between Durg and Dhamdha receiving distillery and Rice mill effluents.

❑ INTRODUCTION

According to Guruswamy (1986) and Kaul et al. (1993) effluents from distilleries cause serious problems of water pollution causing bad odours and fish kills. However, there are few reports on the impact of distillery effluents on the physico-chemical and biological aspects of river systems. The characteristics of distillery effluents have been reported in the data collected from 15 factories surveyed by NEERI (Guruswamy 1986). The spentwash contain solids, Ca as CaO, potash as K_2O, sodium as Na_2O, phosphorus and nitrogen in large quantities. The pH generally ranges between 4.5 to 5.0 and BOD_5 (10000 to 70000 mg/L) is very high. Most by products are recovered and reused. For more details on distillery wastes see Kaul et al. (1993).

The impact of distillery effluents on the ecology of rivers has been investigated by a number of workers out of which the notable contributions have been made by Verma and Shukla (1969) and David and Ray (1966). Studies have been made by Paramasivam and Sreenivasan (1981) on the impact of distilleries on river Cauvery; Kudesia (1985) on Bengal distilleries and its impact on river Hooghly and Verma et al. (1984) on the Eastern Kalinadi. Unni (1989) made a detailed study of distillery effluents on the river Wainganga at Seoni, in M.P. The present studies have been made on river Shivnath since very few studies on the impact of distilleries on rivers exists in India.

The river Shivnath comprise the main tributery to Mahanadi river system. Shivnath river originates from the hills of Rajnandgaon district in the south eastern Madhya Pradesh, flows towards north and reach Durg township. It flows further towards north upto Dhamdha and then turns towards east. The river flows through the industrially developed districts Durg, Raipur and Bilaspur and receives effluents from a number of factories. The present studies were made between Durg and Dhamdha (35km.), the most polluted stretch of the river receiving effluents from a distillery and a number of rice mills around Durg.

Sampling Stations

A stop dam has been constructed at Durg over a distance of 4 km on the river for the water supply to Durg township. It is situated right before entering Durg town by road. The first sampling station S_1 was situated right below the bridge and station S_2 located at a distance of 10 km from station 1, Before this site the river receives effluents of Pulgaovnala (a drain) joining Shivnath at stopdam. Station S_3 is Samodanala an effluent drain carrying distillery effluents from a distillery situated right in the Bhilai township. S_4 is located in Shivnath at the confluence of Samodanala into Shivnath river at a distance of 22 km. by river from Durg. Station S_5 was located at 35 km. from Durg town below the bridge towards Dhamdha a small township (Fig. 1)

METERIALS AND METHODS

Water samples were collected in acid washed polythene bottles from December 1993 to June 1995 every month except during the rains. In all 15 monthly samples from 5 stations were collected over two years of the present studies.

The temperature was noted by immersing a mercury thermometer of 0.1°C graduation in water taken in a plastic bucket immediately after collection. DO was determined in the field by modified Winkler mehtod. Free CO_2, carbonate and bicarbonate alkalinity were determined by titrimetry in the field. Total hardness, calcium and magnesium hardness were determined by EDTA titrimetric method (APHA 1980). Phosphate and silicate were determined colorimetrically in the laboratory. Molybdate - antimony method for total phosphate and stannous chloride were used for silicates as recommended by Golterman (1971). Ammonical nitrogen and nitrate & nitrite nitrogen were determined by distillation and nesslerisation. Sulphate was analysed by turbidometric method and Chloride by titrimetric method using silver nitrate. BOD_5 was determined by measuring DO concentration in the sample before and after incubation in the dark at 20°C in BOD incubator for 5 days. COD was determined by open Reflux method. pH and conductivity measured electrometrically in the laboratary. Total solids, TDS and TSS were determined as recommended by Trivedy and Goel (1984). Sodium and Potassium were analysed using a digital flame photometer (systronic 121).

The samples for phytoplankton counts were collected in 100 ml. polyethyene narrow mouthed bottles and preserved in Lugol's iodine. The density of phytoplankton was calculated by drop count method. Colonial and filamentous forms were counted as single units.

RESULTS

Physico-chemical aspects -

Table 1. gives the range, mean and standard deviation of physico-chemical characteristics of the river Shivnath analysed over two years and 15 samples collected from different stations and seasons.

Table 1

Physico-chemical Characteristics of the River Shivnath

S.No.	Parameters	S1	S2	S3	S4	S5
1.	Air Temp.	19.5-44.5	19.0-35.0	22.5-45.5	24.0-43.5	21.0-45.0
	^0C	29.6±7.4	28.2±5.4	32.2±7.3	34.3±7.6	32.6±6.9
2.	Water Temp.	17.0-34.0	22.0-31.0	22.5-34.5	22.0-24.5	21.5-37.0
	^0C	26.8±4.9	26.0±3.0	27.6±4.0	29.0±5.0	29.0±4.5
3.	T.S.	100-600	200+1000	330-800	400-1275	210-700
	*500 mg l^{-1}	238±132	375±224	573±162	654±337	386±117
4.	T.S.S.	40-200	20-600	100-700	19-1140	10-290
	100 mg l^{-1}	122±53	144±162	224±180	378±388	127±87
5.	T.D.S.	30-400	100-400	200-700	130-140	140-500
		116±92	234±99	349±141	276±87	261±92
6.	P.H.	7.3-8.8	7.7-9.0	6.7-8.2	6.9-8.3	7.2-9.2
	*7-8.5	8.2±0.4	8.2±0.4	7.0±0.4	7.3±0.5	7.9±0.5
7.	Conductivity	103-366	206-650	305-3669	232-701	238-568
	μmhos/cm	247±67	322±115	701±826	500±134	419±104
8.	Turbidity	5.0-26.0	4.5-24.0	9.0-82.0	8.0-60.0	4.0-28.0
	N.T.U.	10.4±7.0	12.5±7.5	39.9±25.3	28.3±22.7	3.4±6.0
9.	Sulphate	1.0-19.8	3.2-28.8	14.2-104	12.0-104	2.5-87
		12.5±7.8	16.3±2.2	51.0±34.0	44.0±42.4	21.2±7.8
10.	Chloride	9.2-17.2	12.8-142.0	32.7-95.1	28.4-72.5	9.9-78.1
		13.4±3.0	32.6±36.7	56.8±16.3	50.4±16.0	32.9±198
11.	Free CO_2	0.0-8.0	0.0-12.0	1.2-50.0	0.6-35.0	0.0-23.6
		1.8±2.4	3.4±5.7	24.5±16.5	16.8±11.0	3.4±6.0
12.	D.O.	4.8-10.0	4.8-13.6	0-6.0	0-6.0	0-9.0
		8.1±1.5	8.4±2.2	0.6±1.5	1.2±1.84	5.9±2.8
13.	Percentage O_2	64.5-135.0	65.1-166	0.0-72.7	0.0-75.3	0.0-130

	Saturation	105.6±21	106.0±26	7.1±19	16.0±23	79.5±39.0
14.	B.O.D.$_5$ #300 mg 1^{-1}	0.8-2.8 1.9±0.7	1.2-4.0 2.4±0.9	9.6-138.0 52.4±42.0	7.0-69.0 22.6±18.0	1.0-17.6 5.1±4.2
15.	C.O.D. #250 mg 1^{-1}	3.0-85.2 16.0±20.0	7.0-37.1 16.8±10.5	70.5-514.3 199.0±147	50.0-240 106.5±63	7.2-148 40.3±38.5
16.	Acidity	0.0-6.8 3.1±2.5	1.9-8.5 4.6±2.8	0.0-50.0 16.0±12.6	4.4-30.0 16.6±8.0	2.0-30 8.2±8.4
17.	Phe. Alkalinity	0.0-15.0 5.0±5.3	0-14.0 4.9±5.3	0.0-0.0 0.0	0.0-0.0 0.0	0.0-14.0 2.6±4.3
18.	T. Alkalinity	60-170 125±29.0	105-150 130±13.0	110-202 163±30.0	132-204 170±25.0	106-240 183±38.8
19.	T. Hardness	90.0-272 127.0±44.4	82.0-376 142.4±82.5	136-400 201.2±67	20.0-240 182±36.6	110-392 81.5±70.6
20.	Ca^{++} Hardness	46.0-112 81.4±17.8	29.4-117.6 75.5±23	94.5-294 146.0±48	109-168 135±17.6	84.0-193 112±29.6
21.	Mg^{++} Hardness	3.4-43.0 1.0±9.6	2.0-75.0 16.2±20.8	3.2-32.9 13.5±9.2	3.4-17.5 11.9±5.0	3.8-48.3 14.5±11.0
22.	NH_4 - N	0.5-2.6 1.3±0.7	0.5-2.6 1.2±0.5	2.5-9.8 5.5±2.3	2.4-10.6 5.9±2.7	0.5-6.6 2.8±1.7
23.	NO_3 & NO_2 - N	0.8-4.5 2.0±1.2	1.3-3.4 2.2±0.8	3.1-10.8 6.4±2.2	2.4-11.8 7.5±3.1	1.0-6.5 4.0±2.0
24.	Phosphate µg 1^{-1}	15-400 52.7±96.0	20-740 105±213	100-1440 598.7±446	120-1480 614±447	30-1020 208±224
25.	Silicates	0.1-4.8 2.9±1.5	0.5-5.0 4.7±3.8	0.6-9.2 4.2±2.3	0.5-6.6 3.6±1.8	0.4-5.4 3.5±1.6
26.	Sodium	4.6-28.2 16.4±5.8	11.0-36.5 23.1±6.9	11.9-43.9 30.7±11.4	22.7-48.4 33.5±7.3	12.6-43.2 26.6±9
27.	Potassium	1.1-5.6 3.0±1.2	2.4-13.7 4.9±3.3	2.6-18.7 12.9±5.2	7.8-19.7 14.3±4.0	2.2-17.8 9.3±49

Note : * - WHO highest desirable limits. # - MINAS.

1. All the parameters in mg. 1^{-1} unless stated above.
2. Range above, mean & S. D. below.

Air temperature

The air temperature of site - 2 was found to have lower values than the down-stream due to local microclimatic effects. It varied between 19.5 to 44.5 with a mean around 29.2 -29.6 °C. However, at the downstream the river water recorded greater mean values ranging between 32.2 to 34.3 (Table 1). It has been found that the S_3 (distillery effluent) increased air temperature by 4°C. This has been observed from the upstream and downstream air temperature differences.

Water Temperature

The mean water temperature in the river shows wide differences between the upstream and downstream stretches. In the upstream, values fluctuated between 17.0 to 34°C (between stations 1 & 2).The water temperature undergoes wide differences between 21.5 to 37.0° C. Similarly the mean values fluctuated widely from 26.8 to 29.0° C between the stations. This wide difference is due to effluent impact and time lapse during sampling.

Total solids, T.S.S. and T.D.S.

The total solids ranged between 238 to 375 mg 1^{-1} between the stations S_1 and S_2 and increased two times at S_3 in the drain and the mean further increased to nearly 3 times (654 mg 1^{-1} at S_4) at the confluence and then gradually declined at S_5.

The T.S.S. recorded a two fold increase at S_3 in its mean values in comparison to station S_1, further increased to 3 times at S_4 and declined to 127 mg 1^{-1} at S_3.

Higher concentration in solids was due to effluents discharged from different polluting industries.

pH

From 7.3 pH at the upstream (Station 1) the minimum values declined to 6.7 and 6.9 at stations S_3 and S_4. This declining trend was seen at station S_5 also. The mean pH values also showed a similar trend between the upstream to downstream.

The distillery effluents contain organic matter which get decomposed, producing humic acids and release of humic acids resulted in decline in pH at the downstream station.

Specific Conductivity

The specific conductivity was the least at station S_1 (103 - 366) and increased 3 times at S_3 in the minimum values. While the maximum values recorded a 10 fold increase to 3669 µ mhos cm^{-1}. The mean conductivity increased from 247 at S_1 to maximum of 701 µ mhos cm^{-1}. At S_5 the values exceeded in comparison to upstream S_1 clearly indicating only a partial recovery from the pollution load. From effluent source and discharge station S_3 is situated at a distance of 15 km. and hence the settling of solids in between reducing the impact partially.

Turbidity

The turbidity increased nearly 4 fold between station S_1 to S_3 and declined later as evident from the mean values. The ranges of turbidity also recorded a similar trend.

DO, free CO_2 and acidity

The DO ranged between 4.8 to 13.6 at stations S_1 and S_2 and the mean around 8.0 mg 1^{-1} showed nearly saturation values. The addition of effluents in the river reduced the DO to 6.0 and DO was absent at all down stream stations on a number of occassions. Presence of DO was generally recorded in October during the post monsoon or during the nearly winter month in November.

The least percentage saturation of 7.1 ± 19 mg 1^{-1} was observed at S_3 and the values gradually increased to 79.5 ± 39.0 at S_3.

Free CO_2 showed an increase from 1.8 at S_1 to maximum of 24.5 ± 16.5 at S_3 and further declined towards the downstream.

The acidity ranged between 0 to 50 mg 1^{-1} maximum recorded at S_3. The mean values also enhanced at S_3 and S_4. five times in comparison to S_1 and declined to two times than S_1 at S_5.

Carbonate and bicarbonates

The phenolpthalein alkalinity giving the carbonate content of the water showed a decline from 5.0 at S_1 to 0.0 at S_3 and S_4. However the total alkanility showing both carbonate and bicarbonate content registered an increase. From the minimum value of 60 mg 1^{-1}, it increased 4 times to maximum value of 240 mg 1^{-1} at staion 5. The mean values of total alkalinity also showed an increase from 125 (S_1) to 183 ± 38.8 at S_5.

Total Hardness, Calcium and magnesium

Increase in Ca, mg and total hardness is evident between the stations S_1 to S_5. The total hardness increased from 90 mg 1^{-1} to 400 mg 1^{-1} at S_3 and slighly lower values of 392 mg 1^{-1} (maximum) was recorded at S_5. The mean value however increased from 127 to 201.2 ± 67 at S_3 and declined to 81.5±70.6 at S_3. The calcium hardness increased from 81.4 to 146 at S_3 and declined to 112 mg1^{-1} at S_5. The addition of magnesium was much greater than calcium from 1.0 mg 1^{-1} (mean) at S_1. It increased to 16.2 at S_2, about 16 times and declined to 14.5 at S_5. The input of Mg^{++} was maximum at S_2 and values declined towards downstream.

BOD_5 and COD

The BOD_5 values increased from 0.8 at S_1 to 138 mg 1^{-1} at S_3 and declined further to 17.6 mg 1^{-1} at S_5. The mean values also increased from 1.9 to 52.4 mg 1^{-1} and later declined to 5.1. The increase in BOD_5 was 25 times at S_3 and S_5 nearly 2 times greater than the upstream.

The COD increased from minimum values of 3.0 to maximum of 514.3 mg l^{-1} at S_3 declined to 240 at S_4 and to 148 mg l^{-1} at S_5. The mean values also showed a decline from the mean maximum of 199 mg l^{-1} to 40.3 mg l^{-1}.

Both BOD_5 and COD increased several times than the WHO standard of effluent discharge into open waters evidently showing pollution at stations S_3 and S_4.

Sulphates and Chlorides

Both sulphates and Chlorides increased from S_1 to S_3. Sulphate composition in the effluent was greater than chloride. The increase of sulphate from 1.0 to 14 mg l^{-1} and 19.8 to 104 at the lower and upper end of the scale. The mean values were four times greater at S_3 in comparison to S_1. The station 5 mean values of sulphate was 1.5 times and chloride 2.5 times compared to the upstream at S_1.

Ammonical N and NO_2 and NO_3-N

NH_4-N values and NO_2 & NO_3-N values registered a wide fluctuation range between S_1 and S_3. The maximum of 9.8 and 10.6 mg 1-1 of NH_4-N at stations S_3 and S_4 respectively showed greater input of nitrogen. The NO_2 & NO_3-N also registered maximum at S_4 (11.8 mg l^{-1}). The mean values also recorded maximum (7.5 mg l^{-1}) at S_4. Compared to upstream S_1, these values were nearly two times, still showing the impact of effluents after a long distance of effluent flow from the source.

Phosphate and Silicate

The mean values of phosphate increased two times at station S_2, 11 times at S_3 and 12 times at S_4. At S_5 the mean phosphate content was 4 times greater than at S_1. The maximum phosphate content was 1.44 mg l^{-1} and 1.48 mg l^{-1} at S_3 and S_4 respectively. At S_5 also the phosphate mean values were almost same, evidently showing much organic pollution similar to S_3 and S_4.

The silicate content did not enhance greatly towards the downstream since the organic matter do not contribute towards its increase at different stations.

Sodium and Potassium

Both sodium and potassium enhanced towards downstream evidently contributing through effluent input. The mean values of sodium showed 2 times greater concentration. The maximum value of 48.4 mg l^{-1} was observed at S_4 as compared to S_1.

Potassium increased 4 times towards the downstream. From 3.0 mg l^{-1} it enhanced to 14.3 mg l^{-1} at S_4. The increase in potassium is evident at S_5 since it did not show much reduction at this station.

DISTRIBUTION OF PHYTOPLANKTON

General trend in the groups

The phytoplankton composition in Shivnath comprised of six classes Bacillariophyceae, Cyanophyceae, Chlorophyceae, Euglenophyceae, Dinophyceae and Chrysophyceae. Fig. 1 shows average density and Fig. 2 the percentage distribution of different classes. The figure of average density evidently show reduction in phytoplankton as pollution increases. The reduction was nearly 33% in comparison to S_1 at station S_3 (148 ind. 1^{-1}) in the effluent and the dilution of effluents enhanced the density at S_4 and increased further to nearly 4 times at S_5 (610 ind. 1^{-1}) as a result of effluent impact.

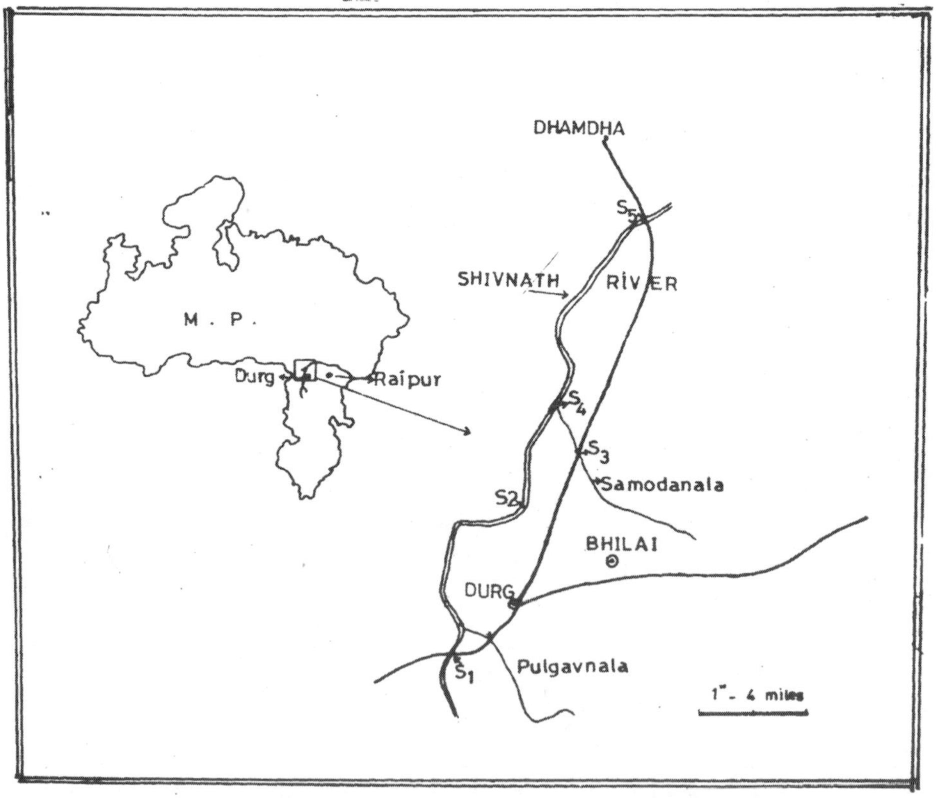

Figure 1. Shivnath river; its location and sampling stations.

It is interesting to note (Fig. 2) dominance of Bacillariophyceae at S_1 and least density of Cyanophyceae and Chlorophyceae. The effluent impact enhanced the Chloropyceae at S_2 and reduced Bacillariophycean density. Cyanophyceae dominated over all other groups at S_3, reducing Bacillariophyceae to the least density. Cyanophycean density continued at S_4 and S_5 but evidently shows that Chlorophycean groups are able to tolerate and thrive well under highly polluted conditions of organic decomposition.

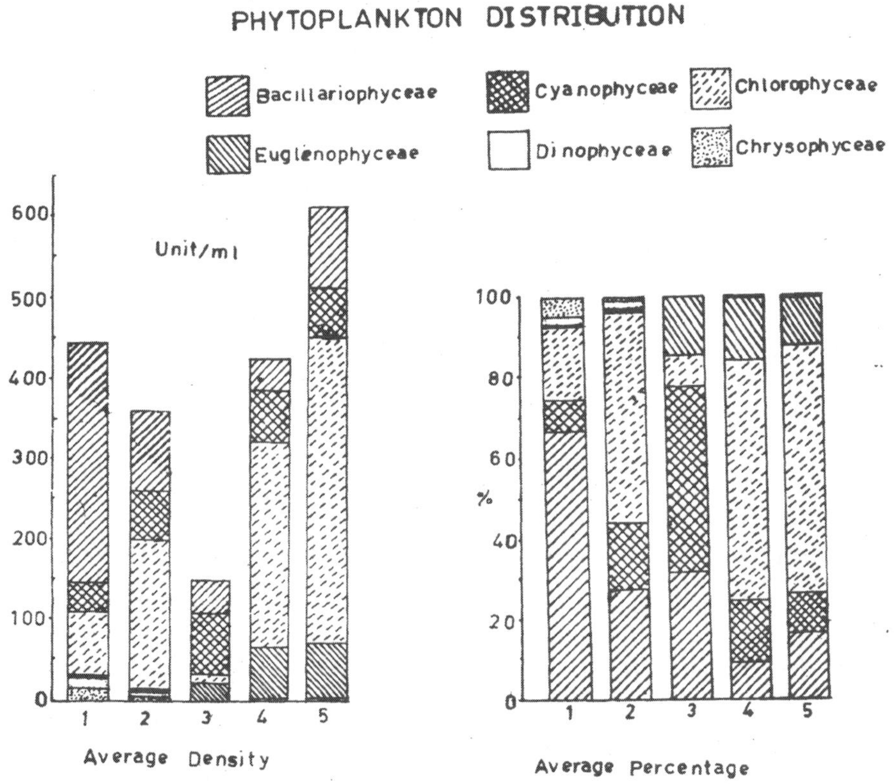

Figure 2. Phytoplankton density in River Shivnath receiving effluents.
S1 - Upstream Control Station.
S2 - Upstream of Samodanala.
S3 - Distillery effluent at Samodanala.
S4 - Downstream at confluence into River.
S5 - At Dhamdha bridge.

Figure 3. Average percentage contribution by different phytoplankton groups.

Fig. 3 shows deminance of Bacillariophyceae (65%) over other groups. It gets replaced by chlorophyceae at station 2 and by Cyanophyceae at S_3. Further downstream at S_4 and S_5 were Chlorophyceae dominated.

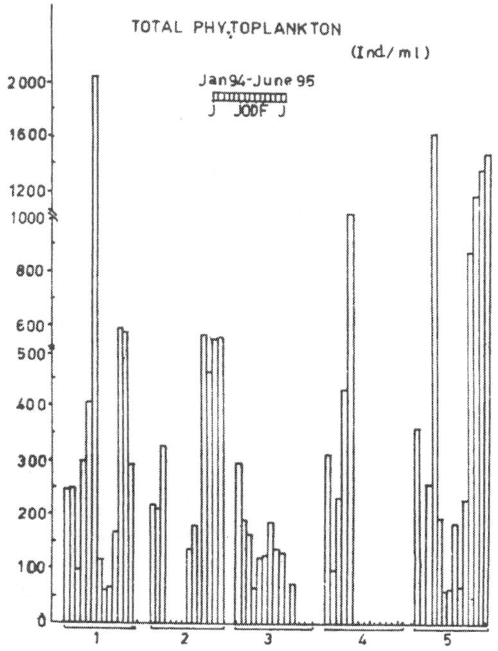

Figure 4. Monthly distribution of total phytoplankton at different stations.

An important aspect of the percentage composition is the clear enhancement of Euglenophyceae which reached maximum development as evident from the percentage distribution between S_3 to S_5. Euglenophyceae exceeded Chlorophyceae at station 3 and Cyanophyceae at station 4 and 5. Cyanophyceae is the most successful algal group which thrives well under distillery effluents.

Chrysophyceae and Dinophyceae do not tolerate distillery pollution and were totally absent between S_3 to S_5.

(ii) Periodicity

The monthly distribution of total phytoplankton is given in Fig. 4. The following trend was observed due to effluent impact.

Bacillariophyceae

Fig. 5 gives the monthly distribution of Bacillariophyceae between stations 1 to 5.

The least number of Bacillariophyceae was observed at S_4. at the confluence of distillery waste in to Shivnath showing the impact of pollution on the mainstream. The dilution and flow downstream enabled settling of organic matter promoting the growth of diatoms again as observed at S_5. Bacillariophyceae dominated the upstream waters which was nearly free from organic pollution. The maximum of 2514 ml 1^{-1} was observed in June 1994.

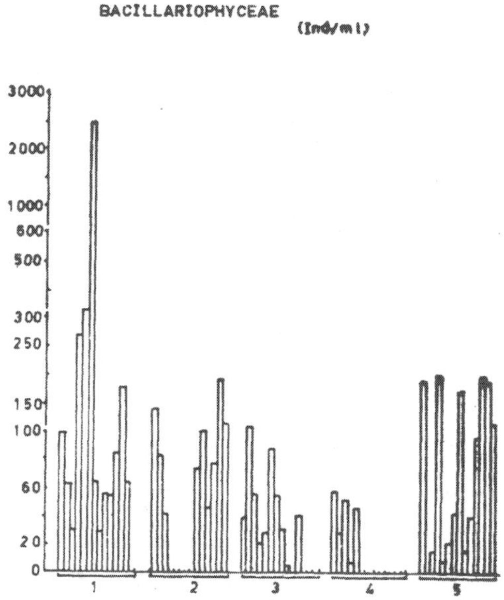

Figure 5. Monthly distribution of Bacillariophyceae.

Cyanophyceae

Fig. 6 shows distribution of the group Cyanophyceae. In general it dominated in summer at S_1 & S_5. The least total numbers were recorded at S_1 and maximum at S_3 in the distillery effluents. As the impact declined along with the course of the river dilution and setting seems to have little effect and the numbers of this group continued to show a decline to the least except in April and May during peak summer time. It dominated at station 3 in percentage (Fig. 3).

Chlorophyceae

The distribution of Chlorophyceae in River Shivnath shows its dominance (Fig. 7) at stations 4 and 5 replacing other algal groups in percentage composition and in density every month. The least number were observed in the drain itself in all months.

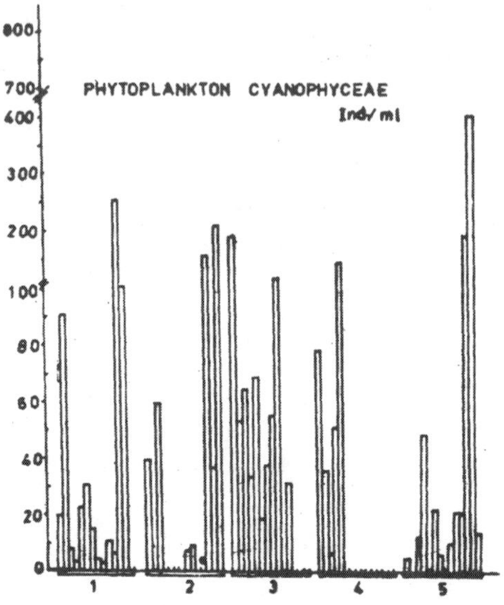

Figure 6. Monthly distribution of Cyanophyceae.

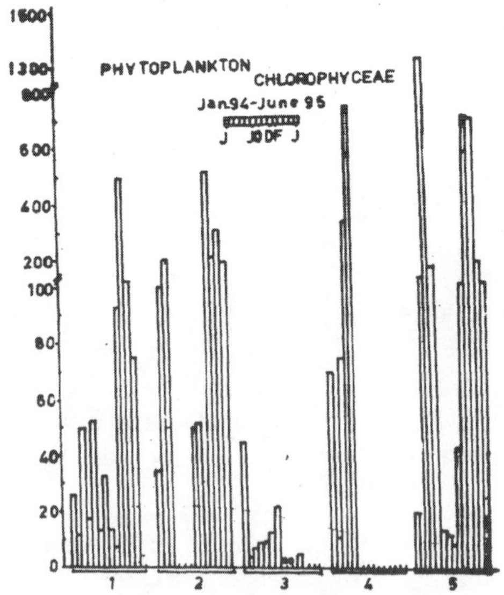

Figure 7. Monthly distribution of Chlorophyceae.

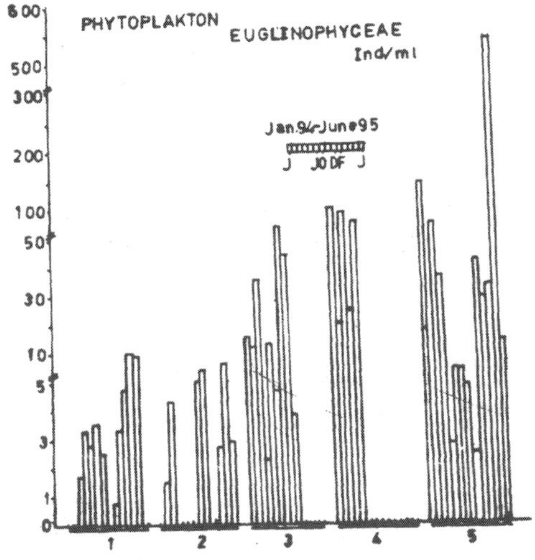

Figure 8. Distribution of Euglenophyceae in Shivnath.

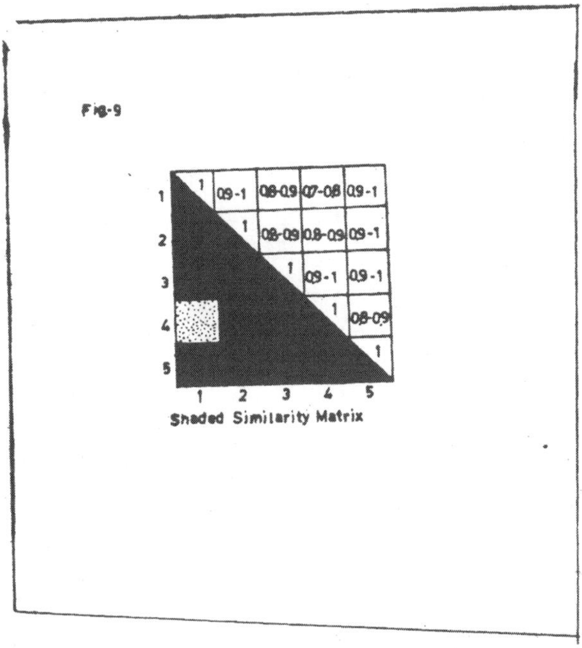

Figure 9. Similarity matrix between the different stations.

Periodicity of Chlorophyceae indicated a summer dominance at stations S_1 and S_2 between March to May and also at S_5. At S_5 the numbers exceeded, compared to all other stations and its dominance in month indicated that organic waste decomposition and relese of nutrient ions support prolific growth of this group.

Euglenophyceae

From the least numbers of Euglenophyceae observed at S_1 the number declined at S_2, further enhanced at S_3 in the drain. Dominance of Euglenophycean members in organic rich waters is evident from its maximum numbers out of all stations at S_4 followed by S_3.

Chrysophyceae and Dinophyceae appears to be intolerent and found restricted to stations 1 and 2 only. Dinophyceae were however found at station 5 in winter only.

Discussion

Effluents from distilleries are highly organic and the decomposing solid leads to oxygen depletion very quickly through microbial activities which create hazardous conditions in the receiving systems.

The study point of distillery effluent at Samodanala, the drain S_3 is located at least 8 to 10 km. from the distillery plant and the impact of heat is reduced since it gets cooled down during its flow. The mean and range of water temperature (Table 1) at S_3 do not differ much from the values recorded from the upstream and downstream sites.

The total solids ranged between 682 to 13.258 mg 1^{-1} in the river Reh (Shankar et al. 1986) and 110-5822 in Eastern Kalinadi (Verma et al. 1984). In river Wainganga Unni (1989) observed a range of 28 to 1345 mg 1^{-1}. In the present studies the mean value of 654 mg 1^{-1} and range between 400-1275 mg 1^{-1} indicated lower discharge of solids into the river, through distillery effluent.

The TSS values in the present studies show 19 to 1140 mg 1^{-1} at S_4 and 378±388. Verma et al. (1984) reported 50-2560 mg 1^{-1}. In Wainganga it ranged between 28 to 1345 and the mean was 617±1012 mg 1^{-1}. The sampling point in the river wainganga was located at only 2 km. from the discharge while in the present studies the distance of 8-10 km. enabled part of the total suspended solids to settle and the values showed a decline. The TDS also showed a similar trend in the present studies of 200-700 mg 1^{-1} and 334±141 mg 1^{-1}. in comparison to the mean of 586±553 in the Wainganga. Much greator values ranging between 28-2396 mg 1^{-1} in winter and 30-4680 mg 1^{-1} were reported by Verma et al. (1984) from eastern Kalinadi receiving sugar mill effluents.

The pH is neutral to alkaline and ranges between 7.7 to 9.0 at S_2 and 7.2 to 9.2 at S_5. However at S_3 and S_4 it was acidic and the values ranged between 6.7 and 6.9 at the lower end of the range. Much lower values ranging between 6.1 to 6.9 have been reported by Verma et al. (1984) due to acidic wastes. (Guruwswamy 1986) reported from his studies on 15 distilleries that the water are generally acidic in nature. In the river Reh the pH declined further than the above reported values to 5.5 to 6.4 clearly indicating organic acids produced during the microbial activities.

Paramasivam and Sreenivasan (1981) reported a conductivity range of 285 μmhos from the distillery outfall in to river Cauvery : 700μ mho cm^{-1} at the outfall in the river Wainganga (Unni 1989); 1028 to 14.399 μ mho cm^{-1} (Kudesia 1985) from river Hooghly and 1680 x 10^{-6} in summer from Eastern Kalinadi (Verma et al (1984). The present result show much higher conductivity range in River Shivnath (305-3669 mean 701±826) in comparison to river Cauvery and Wainganga as reported above.

The turbidity values in Easter Kalinadi range between 55-1520 during summer (Verma et al. 1984). In Wainganga the maximum ranged between 59.4 to 65.8 mg 1^{-1} and almost similar values were recored ranging between 9.4-82 mg1^{-1} (mean 39.9±25.3) in the present studies.

The sulphate values in the present studies (14.2-104) were much greater than the values reported by Verma et al. (1984) of 7.6-86 mg 1^{-1}. In Wainganga the values reported by Unni (1989) ranged between 93-416 mg 1^{-1} due to the effect of refinary effluent get mixed into distillery effluent and its combined effect enhancing the sulphate.

The chloride ranged between 12.8-142 mg 1^{-1} in the present analysis compared to very high values between 13.5 to 510 mg *l*-2 (Verma et al. 1984). Much lower ranges (13.5 - 79.5) were reported from Wainganga at Seoni.

Compared to Wainganga river (0-18 mg 1^{-1}) much grater values ranging between 0.6-50 mg 1^{-1} was recored in the present studies.

The dissolved oxygen values in the river is an indicator to organic pollution. In Eastern Kalinadi the values ranged between 0.2 - 8.1 mg 1^{-1} with mostly anaerobic conditions. David and Ray (1966) reported absence of DO from the distillery wastes discharged into river Daha (N. Bihar.) DO ranged between 0.0 to 12 mg 1^{-1} in the river Wainganga (Unni 1989). In the present studies the DO exceeded saturation level in S_1 and S_2 and anoxia was recorded a number of times at stations S_3, S_4 and S_5. The mean values between S^3 and S^4 were 0.6 and 1.2 mg 1^{-1} respectively producing a stretch of anoxic water of 2 to 10 km. downstream between S_3 and S_5 unable to sustain any fish population.

The BOD^5 of the water of Shivnath ranges between 0.8 to 4.0 in S^2 and increased to 52.4±46 mg l^{-1}, at S_3 receiving effluents. It ranged between 9.6-138 at S_3. 7.0 to 69.0 at S_4 and declined along with distance to 1.0-17.6 mg l^{-1}. Much lower values in the mean and range (18.2±3.5 and 14-24 mg l^{-1}) were recorded by Unni (1989)from the river Wainganga. David and Ray (1966) reported 8090-8750 mg l^{-1} from the river Daha. Verma et al. (1984) reported 9.4 and 4030 mg l^{-1} in Winter and Summer. Both the above values are much greater than that of the present studies.

COD ranged between 6.9-3522 in Winter and 10.8-5680 mg l^{-1} in Summer in the Eastern Kalinadi receiving distillery effluents. Much lower ranges were recorded in the present observation of 3.0-85.2 at S_1 and 70.5-514.3 at S_3 (Effluents). In Wainganga the values ranged between 28-224 mg l^{-1} (mean 106±66); Which were much lower than Shivnath (199±147).

Sodium ranged between 25-448 (mean 58±39.6) mg l^{-1} and potassium 0.15-1.7 (mean 1.15) mg l^{-1} in Wainganga.In Shivnath much lower sodium content was observed. However the potassium content was much greater and ranged between 7.8-19.7 mg l^{-1}.

Phenolpthalein alkalinity was absent in the effluent and at the downstream up to S_4 and ranged between 0.0-14.0 at S_5. While total alkalinity enhanced several times in Shivnath (110-240). A similar range of 60-292mg l^{-1} was observed by Unni (1989) in Wainganga at Seoni.

The calcium hardness content of the effluent in Shivnath ranged between 94.5-294 mg l^{-1} (mean 146±48). In Wainganga almost similar mean value of 149±77 mg l^{-1} weas observed.The total hardness in Shivnath ranged between 136-400 (mean 201.2±6) while in Wainganga river the range and mean values were 148-420 and 268±113 respectively indicating a similarity in hardness content of both river systems. The magnesium content was however much greater in (36-232). Wainganga (mean 119±63) than in Shivnath (3.2-32.9 mean 13.5±9.2) due to the difference in the effluent composition.

Some of the above physico-chemical aspects depict the difference and higher rate of pollution in Shivnath through increased BOD_5. COD and conductivity than other rivers.

Phytoplankton

David and Ray (1966) in their work on the river Daha (N. Bihar) receiving distillery wastes reported maximum density of Cyanophyceae of 351.28 per litre. and the dominant species were *Oscillatoria* and *Anabaena* and the maximum of 1900 l^{-1} shows density exceeding the present observation. Cyanophyceae dominated in the present studies at S_3 receiving distillery wastes. The reduction of

Bacillariophyceae in present observation at stations S_4 to S_5 is also in accordance with the reports of the above authors. *Oscillatoria* was the most common organism in the distillery waste at Seoni in the Wainganga river (Unni 1989) and Cyanophycea dominated in it in the downstream stations after receiving the waste. These observations are in agreement with the present studies on river Shivnath. The replacement of Bacillarophyceae by the more pollution tolerent group Cyanophyceae is clear by the dominance of the later at station S_3. After Cyanophyceae Chlorophyceae formed the next dominant form. This is the support of the findings by Verma and Shukla (1969) from the impact studies of sugar factory effluents on Khala Stream in U.P.

The replacement of Bacillariophyceae by Cyanophyceae and the latter by Chlorophyceae after receiving distillery effluents is the most distinct aspect in our studies not reported by any other workers on distillery impacts on phytoplankton. Mohan et al. (1989) also reported 57% of algae growing in distilleries belong to Cyanophyceae.

Reduction in phytoplankton density indicated the severe pollution strees. Verma and Shukla (1969) also reported a decline in phytoplankton alongwith an increase in pollution load and vice versa which corroborates with our findings in Shivnath.

An important aspect in the present observation in River Shivnath is the increase in Euglenophycean density. Euglenophycean density at a station can thus be taken as an indicator to classify and differentiate between the least polluted S_1, moderately polluted S_2, and highly polluted S_3, S_4 and S_5. Station S_3 and S_4 is most polluted among them, because very high organic load and absence of DO supresses the growth of Eugenoids at this points. Statistical analysis of the log transformed values of physico-chemical factors (Fig. 9) show similarity between the different stations. Sampling station S_3 (the effluent) and (down stream) S_4 stands apart and are distinet from others. However station S_1, S_2 and S_5 are relatively similar.

Acknowledgement
Dr. K.S. Unni is greatful to the Department of Environment and Forests for granting him the project 19/127/90-RE under which the present work has been done. He is also thankful to late Principal Dr. F.A. Jellani and Dr. T.K. Sudarsan for granting him Laboratory facilities and encouraging him to undertake research.

❑ REFERENCES

APHA (1980). *Standard methods for examinationof water and waste water* 15th Ed. American Public Health Association New York.

David A. & P. Ray. 1966. Studies on Pollution Status of river Daha in Bihar by sugar and distillery wastes *J. Environ. Hlth.* 8(1) : 6-35.

Golterman H.L. 1971. *Methods for chemical analysis of freshwaters* IBP handbook, Blackwell Sci. Publ. oxford 166 pp.

Guruswamy. 1986. Sugar and distillery industry, *Pollution control handbook,* Utility Publ. Ltd., Hyderbad. 203-212 pp.

Kaul S.N., T. Nandy and R.K. Trivedy 1993 : *Pollution control in Distilleris.* Enviro Media, Karad, India.

Kudesia V.P. 1985. *Water Pollution.* Pragati Prakashan, Meerut.

Mohan M.N. Kumar & J. Mohan (1986) : Algal flora of Alcohol distillery effluent a sample survey In *Advances in applied phycology* II pp. 207-211. Today & Tomorrow Pub., New Delhi.

Paramshivam M. & A. Shrinivasan. 1981. Changes in algal flora due to pollution of cauvery river. *J. env. Hlth* 23-3 : 222-238.

Shankar V. P.R.S. Sangu and G.C. Joshi (1986) : Impact of distillery effluents on the water quality and ecosystem of river Reb in Doon Vally. *Poll. Res.* 5.3:4,137-142.

Trivedy R.K. & P.K. Goel. 1984. : *Chemical and biological methods for water pollution studies* Env. Publication Karad, India.

Unni K.S. 1989. *Impact of a distillery and vegetable oil waste on the river Wainganga at Seoni. M.P. - a case study* report submitted to C.S.I.R. pp. 79

Verma S.R.P. Sharma, A. Tiyagi, S. Rani. A.K. Gupta and R. C. Dalela. 1984.

Pollution and Saprobic status of Eastern Kalinadi. *Limnologica* (Berlin) 15(1), 69-133.

Verma S.R. and G.R. Shukla. 1969. Pollution in a parennial stream Khala by Sugar factory effluent near Laksar (Dist. Saharnpur), U.P. India, *J.Environ. Health* 11. 145-162.

Authors :

1. **K.Sankarn Unni :** Hydrobiology Laboratory, Botany Deptt. Govt. P.G. College, CHHINDWARA - 480001. (M.P.), India.
2. **Sanjay Pawar :** Hydrobiology Laboratory, Botany Deptt. Govt. P.G. College, CHHINDWARA - 480001., India.
3. **Surendra Bhade:** Hyrobiology Laboratory, Botany Deptt. Govt. P.G. College, CHHINDWARA - 480001 (M.P.)

■

35

Heavy Metal Pollution studies in the River Krishna and Koyana In Maharashtra, India

❑ INTRODUCTION

Heavy metals today have a great ecological significane due to their toxicity and accumulative behaviour (Purves 1985). These elements contrary to most pollutants, are not biodergradable and undergo a global eco-biological cycle (Nurnberg 1984) in which natural waters are the main pathways., Pardo etal (1990) state that the determination of the concentration levels of heavy metals in these waters as well as the elucidation of the chemical forms in which they appear is a prime target in enviromental research today.

Disposal of domestic and industrial waste in the river leads to increase in heavy metal content of the river water.

River Krishna is fast getting polluted due to agricultureal, industrial, and municiple wastes and its water quality at many places is a matter of great concern., River Koyana after its origin in Koyana dam passes through highly cultivated agricultureal area.

After a course of 64 km., it receives smaller quantity of demestic waste. When the river enters Karad town it receives wastewater from Hospitals, & Karad city., Finally Koyana meets the Krishna river at Karad. Although large number of studies are available on river pollution in India, there are very few in which heavy metals have been studied. Some studies on this aspect are of Ajmal et al. (1985), Mathur (1987), Paul & Pillai (1983)., Wahid (1990) & Wahid & Khurshid (1991)., However very few studies are available from Western India. (Trivedy and Khatavkar 1990). The present monitoring of heavy metals was undertaken in continuation of our previous studies (Trivedy & Khatavkar 1990).

Study Area and Methodology

Krishna river rises in Western Ghats of India at an altitude of 1337 m. near Mahabaleshwar, about 64 km. from the Arabian sea. A tributory of Krishna-river Koyana also originates in the Western Ghats & finally meets the river Krishna at Karad in frontal confluence. The stretch of the river Krishna & Koyana in Maharashtra is about 260 km. & 128 km. respectively.

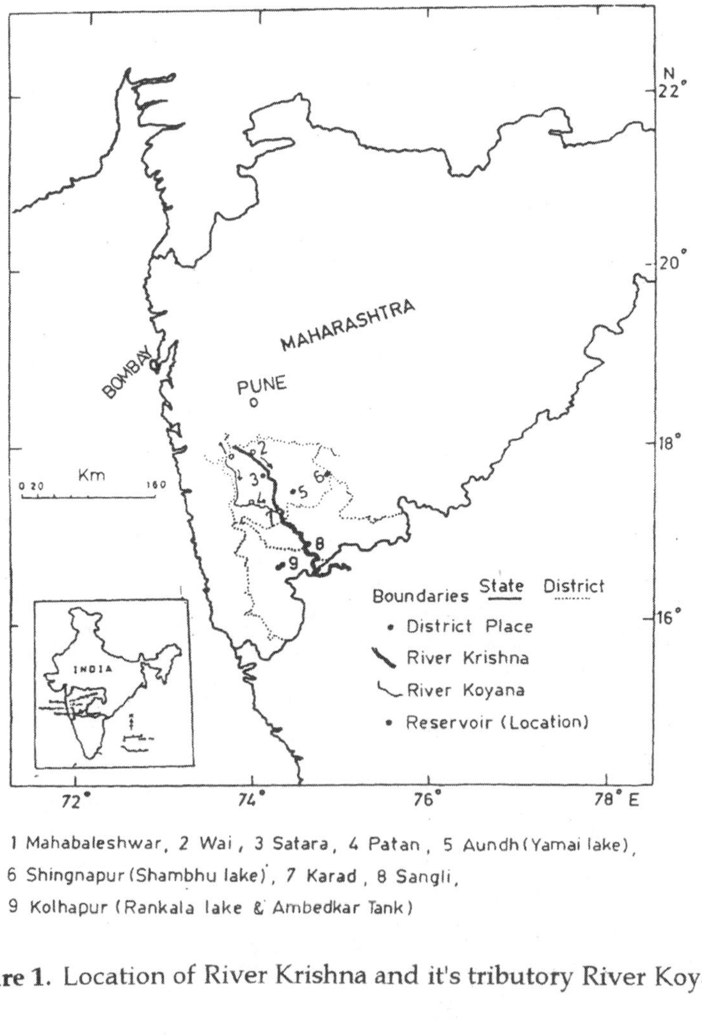

Figure 1. Location of River Krishna and it's tributory River Koyana.

The study area has been divided in to 11 sampling sites. Six (6) sites of River Krishna and Five (5) sites of River Koyana. Table 1. and Fig. 2 & 3 show the selected sampling sites with basin & drainage of the river Krishna & its tributory river Koyana.

The surface water samples were collected biomonthly by gently wading in to the water. Well mixed samples of each sites were taken & acidified with conc. HNO_3., Metal were determined on Atomic Absorption Spectrophotometer. *The location of river Krishna and its tributary river Koyana in Southern-Western Maharashtra is shown in fig. 1.

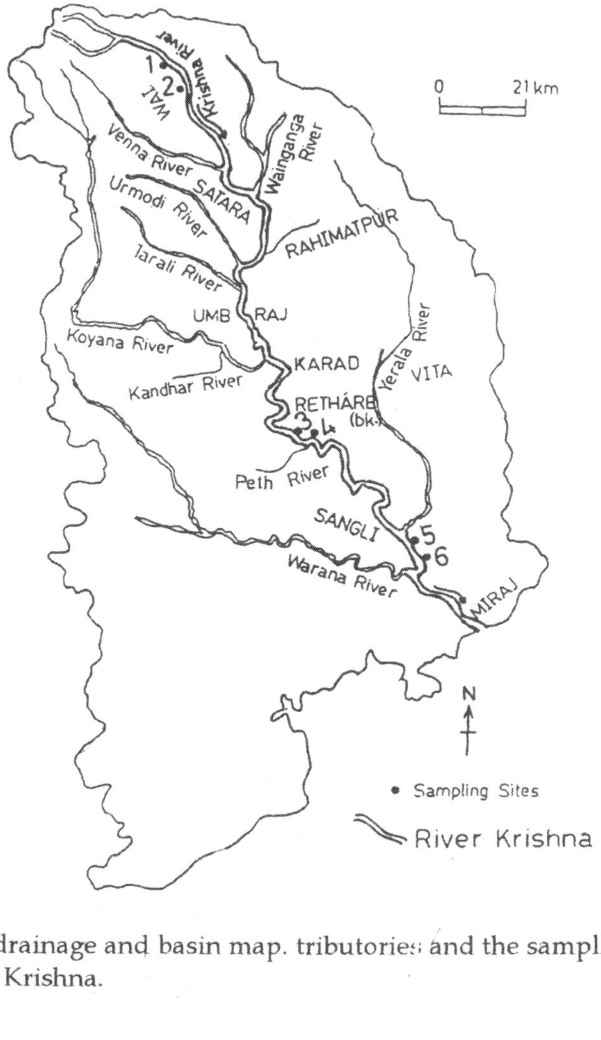

Figure 2. The drainage and basin map. tributories and the sampling locations of the river Krishna.

Table 1
Salient Features of the selected sites in the River Koyana
(Locations are shown in Fig. 3.)

Site No.	Site	Distance from origin (Km)	Characterstics
1.	Patan	20.0	Clothe washing, Cattle wading, bathing, excessive water drawing for drinking purposes for Patan township. Surrounded by agricultural fields.
2.	Karad (Upstream)	23.0	The site is before the river enters city.Water pumped for irrigation, No human activity; area surrounded by agricultural fields. A major nullah bringing hospital waste and sewage from Krishna hospital and Research Institute meets the river. Extensive sugarcane cultivation.
3.	Karad (Old bridge)	25.0	Some sewage drains meet the river before the site. Intense human activity including open defecation.
4.	Karad (Water Works)	26.5	The site is after water is pumped for Karad city. A major sewage drain bringing untreated sewage of Karad city meets the river. Intense human activity, extensive brickmaking on the banks is observed.
5.	Priti-Sangam (Downstream)	28.5	The site is just before the confluence of the River Krishna & Koyana. The place is a famous pilgrimage. Moderate human activity of bathing, cattle wading and clothe washing is observed.

Table 2

General Characteristics of various sites studied in the River Krishna.

Site No.	Site	Distance from source (Km)	Characterstics
1.	Wai-1 (Upstream)	20	The site is before the river enters in the holy township of Wai. Intensive cultivation of sugarcane, rocky bed.
2.	Wai-2 (Downstream)	23	The site is after the river has left Wai town In between the river receives untreated domestic sewage from the town. Mass bathing, Clothe washing, Cattle wading is rampant. River highly turbid, overgrown with aquatic macrophytes. The site is near cremation centre.
4.	Kole-1 (Upstream)	120	Extensive agriculture activity in the vicinity. The site is before a major sugar and distillery industry.
5.	Kole-2 (Downstream)	122	The site is intended to show the effect of discharge of sugar and distillery industry. Though the treatment plants in there units exist, the functioning is poor.
5.	Sangli-1	214	The site where the river enters Sangli City. it Is after the river has recived considerable quantity of dark and malodorous effluents from sugar factories, distillries, textile and engineering industries through a nullah - Sheri. The river appears to be extremely polluted here. Extensive sugar cane cultivation.
6.	Sangli-2	220	The site is after the river has received a major drain bringing the sewage of Sangli City (Sangli Nullah). Putrifying condition prevailing here.

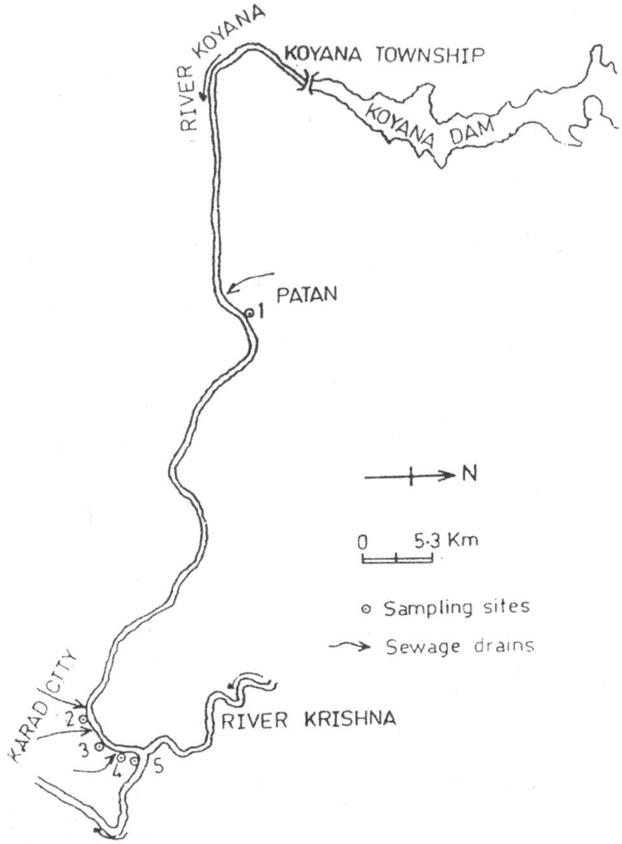

Figure 3.: Outline of the River Koyana and sampling sites.

RESULTS

The sampling of both Rivers & Nallas meeting the rivers was done for six months (Nov. 1987, Jan. 1988. Mar. 1988, May 1988, & Jul 1988) & the observed values of heavy metal content are given in fig. 4 & 5 and Table 1 & 2. Table 3 shows the average values of heavy metal concentrations along with the permisible limits of drinking water quality by W.H.O. (1984) & Recommended Maximum concentration of heavy metals in irrigation Water by F.A.O. 1985.

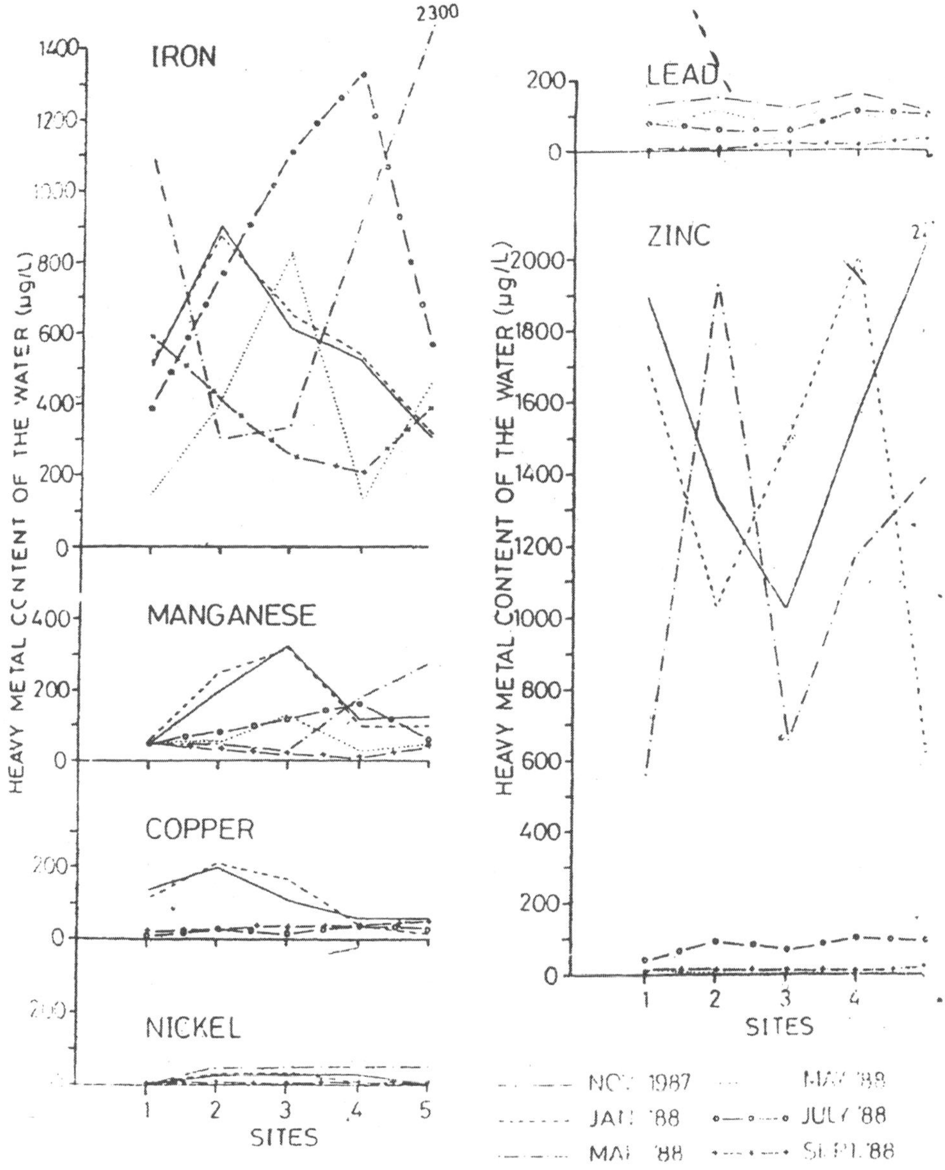

Figure 4. Heavy metal concentration in water in the River Krishna.

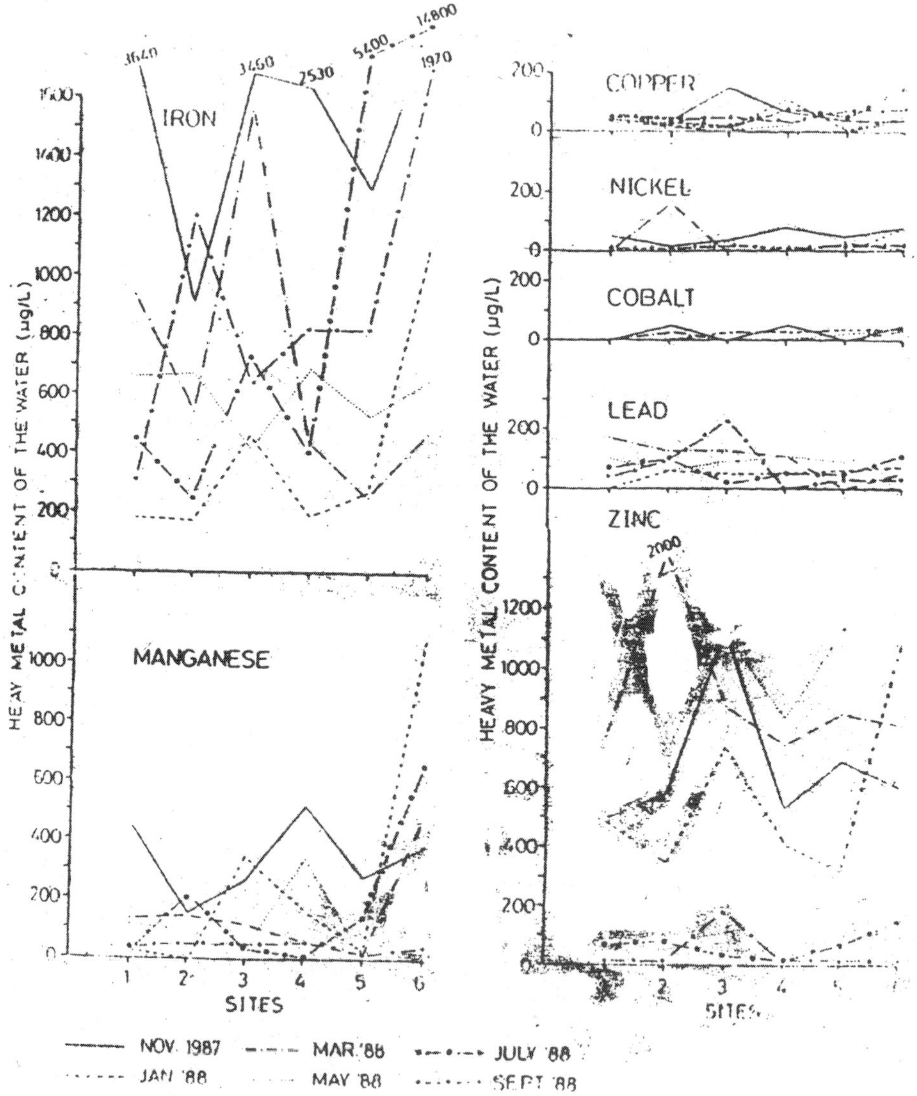

Figure 5. Heavy metal concentration in water in the River Koyana.

Following are the important observations

Heavy Metal Content of Drains

Heavy metal concentration in the 3 Nallas viz., Kole Nalla, Sangli Nalla & Sheri Nalla meeting the river Krishna is given in Table 1. The data on heavy metal concentration in water of the two major nallas viz. Patan & Karad Nalla meeting the river Koyana is presented in Table 2.

Table 3

Heavy Metal Concentration ($\mu g/L$) observed in the Nullahs meeting the River Krishna.

	Kole Nullah						Sheri Nullah						Sangli Nullah					
Heavy Metals	Nov. 1987	Jan. 1988	Mar	May	July	Sept.	Nov. 1987	Jan. 1988	Mar.	May	July	Sept.	Nov. 1987	Jan. 1988	Mar.	May	July	Sept.
Fe	3800	370	1170	1130	380	1970	1130	640	400	850	300	440	10000	1530	1170	580	710	420
Mn	460	260	420	260	10	50	290	300	40	60	200	190	1190	370	440	230	410	140
Zn	970	360	810	1320	30	30	660	1100	770	1280	50	60	1730	1080	1950	1050	90	30
Cu	100	40	-	20	20	90	40	280	-	30	30	30	210	90	40	30	40	80
Ni	60	20	-	-	-	30	40	30	-	-	T	30	80	20	-	20	-	40
Co	20	-	-	-	-	T	30	40	-	-	-	T	50	30	40	-	T	T
Pb -	50	50	130	20	40	-	80	40	120	70	60	-	90	80	130	60	10	
Cd	T	-	-	T	-	T	-	-	-	T	-	-	-	-	-	-	-	

\- - absent
T - Traces

Extremly higher concentration of iron, manganese, zinc & lead was found in Karad nalla. The values were higher compared to Patan Nalla & River Sites.

Heavy Metal Content of Rivers

Iron

Heavy metal concentration in the 3 Nallas viz., Kole Nalla, Sangli Nalla & Sheri Nalla meeting the river Krishna is given in Table 1. The data on heavy metal concentration in water of the two major nallas viz. Patan & Karad Nalla meeting the river Koyana is presented in Table 2.

Table 4

Heavy Metal Concentration (μg/L) observed in the nullahs meeting the River Koyana.

	Kole Nullah						Sheri Nullah					
Heavy Metals	Nov.	Jan. 1987	Mar 1988	May	July	Sept.	Nov.	Jan. 1987	Mar. 1988	May	July	Sept.
Fe	490	510	1120	330	910	1380	1500	1680	2030	1730	790	580
Mn	420	430	130	50	920	80	1720	1840	990	1310	730	730
Zn	750	1320	1090	40	80	30	2900	2430	940	70	100	30
Cu	350	400	20	-	910	50	90	70	30	20	790	50
Ni	40	30	-	-	-	-	50	20	40	20	-	20
Co	-	20	-	-	10	-	-	40	-	10	10	-
Pb	-	T	190	80	70	50	200	110	100	60	2130	40
Cd	-	-	-	-	-	-	-	-	-	T	-	-

Absent T : Traces

Iron

Iron was present in considerable quantity in the river Krishna and its concentration varied from 170 μg/Lit. to 14800 μg/Lit. In the river Koyana it was present in a concentration varying from 130 μg/Lit. to 1320 μg/Lit. Values were considerably higher at down stream sites in both the Rivers.

Zinc

Considerable quantity of Zinc was also recorded from both the rivers. The fluctuation in concentration were of enoromus magnitude ranging from a minimum values 20 μg/Lit. to the maximum values 2000 μg/Lit. The average value of zinc in various months has been between 500 μg/Lit. to 1000 μg/Lit. While these values were brought down during the rainy season.

Table 5
Heavy Metal concentration (μg/L) observed in the nullahs meeting the River Krishna.

Waterbodies & Nullahs	Fe	Zn	Mn	Cu	Pb	Ni	Co	Cd	AveragepH
River Krishna Nullahs									
Kole Nullah	1470	586	243	45	48.3	18.3	3.3	T	7.26
Sheri Nullah	626	653	180	68.3	61.1	16.6	11.6	T	7.64
Sangli Nullah	2401	838.3	463	81.6	61.6	26.6	20.0	-	6.91
River Krishna									
1	1028	516	135	40	63	10	-	-	7.72
2	623	626	111	26	75	31	13	-	7.56
3	1210	681	143	41	85	15	5	-	7.79
4	845	428	193	51	53	30	13	-	7.00
5	1428	515	85	31	38	21	11	-	7.55
6	3425	585	498	58	68	46	26	-	7.08
River Koyana Nullahs									
Patan Nullah	790	551	205	288	65	11.66	5.0	-	7.27
Karad Nullah	1385	1078	1220	175	440	25	10	-	7.33
River Koyana									
1	535	708	51	48	48	-	-	-	7.14
2	605	728	111	78	55	20	-	-	7.52
3	626	540	158	56	46	18	-	-	7.52
4	603	805	100	30	65	15	-	-	7.55
5	716	720	110	26	53	10	-	-	7.68

T - Traces - All the values are in μg/L - Not detectable

Manganese

The values at all the sites fluctuated between 10 to 520 μg/Lit. Few higher values were noted at last sites of both the rivers.

Copper

Copper concentration varied from 10 to 150 μg/Lit. in the rivers. The difference at various sites were not pronounced, except the last sites, where the values were usually maximum.

Table 6

A comparison of the Heavy Metal concentration of water bodies studied with prescribed limits (PL)

Sites	Iron			Zinc			Manganese			Copper			Lead			Nickel			Cobalt			Cadmium		
	PL	RC	Ave	PL	RC	Ave	PL	RC	Ave	PL	RC	Ave	PL	RC	Ave	PL	RC	Ave	PL	RC	Ave	PL	RC	Ave
River Krishna																								
1.	0.1	5.0	1.02	5.0	2.0	0.51	0.05	0.20	0.13	0.05	0.20	0.04	0.10		0.06	*	0.20	0.01	*	*	-	0.01	0.01	-
2.	0.1	5.0	0.62	5.0	2.0	0.62	0.05	0.20	0.11	0.05	0.20	0.02	0.10		0.07	*	0.20	0.03	*	*	0.01	0.01	0.01	T
3.	0.1	5.0	1.21	5.0	2.0	0.68	0.05	0.20	0.14	0.05	0.20	0.04	0.10		0.08	*	0.20	0.01	*	*	0.005	0.01	0.01	T
4.	0.1	5.0	0.82	5.0	2.0	0.43	0.05	0.20	0.19	0.05	0.20	0.05	0.10		0.05	*	0.20	0.03	*	*	0.01	0.01	0.01	T
5.	0.1	5.0	1.42	5.0	2.0	0.51	0.05	0.20	0.08	0,05	0.20	0.03	0.10		0.04	*	0.20	0.02	*	*	0.01	0.01	0.01	T
6.	0.1	5.0	3.42	5.0	2.0	0.58	0.05	0.20	0.49	0.05	0.20	0.06	0.10		0.07	*	0.20	0.04	*	*	0.03	0.01	0.01	T
River Koyana																								
1.	0.1	5.0	0.53	5.0	2.0	0.71	0.05	0.20	0.05	0.05	0.20	0.05	0.10		0.04	*	0.20	-	*	*	-	0.01	0.01	-
2.	0.1	5.0	0.60	5.0	2.0	0.73	0.05	0.20	0.11	0.05	0.20	0.07	0.10		0.05	*	0.20	0.02	*	*	-	0.01	0.01	-
3.	0.1	5.0	0.63	5.0	2.0	0.54	0.05	0.20	0.16	0.05	0.20	0.06	0.10		0.05	*	0.20	0.01	*	*	-	0.01	0.01	-
4.	0.1	5.0	0.60	5.0	2.0	0.80	0.05	0.20	0.10	0.05	0.20	0.03	0.10		0.06	*	0.20	0.01	*	*	-	0.01	0.01	-
5.	0.1	5.0	0.71	5.0	2.0	0.72	0.05	0.20	0.11	0.05	0.20	0.03	0.10		0.05	*	0.20	0.01	*	*	-	0.01	0.01	-

PL Prescribed limits of heavy metals by W.H.O. (1984) for drinking water quality.

RC Recommended maximum concentrations of heavy metals in irrigation water (adapted from F.A.O., 1985)

Lead

Lead concentration was usually higher than the concentration of copper in the river. At sites 5 & 6 in the Krishna, values usually fluctuated from 30 µg/ Lit. to 150 µg/ Lit. While in Koyana higher concentration at sites 4, averaged around 45 µg/ Lit to 50 µg/ Lit in the river Koyana.

Nickel

A higher variation was noted at all the sites from Nil to 200 µg/ Lit. in the river Krishna, Low concentration ranging from 10 µg/ Lit. to 50 µg/ Lit. was noted in the river Koyana.

Cobalt

Cobalt concentration at various sites ranged from traces to 50 µg/ Lit. in the Krishna, While it was very low in the river Koyana.

Cadmimum

Cadmium was practically absent in both the rivers.

DISCUSSION

Table 4 gives the comparison of heavy metal concentration in water with the prescribed standards. The overall data suggest that the level of heavy metals was not very high in the rivers. Maximum concentration of all the heavy metals was always found at sites 6 in the river Krishna, which is earily explainable by high concentration of metals brought about by Sheri Nala.

Ajmal & Razi-Uddin (1987); Metzner (1977), and Trivedy & Khatavkar (1990) report that pH plays an important role in the availability of heavy metals in water. However, in the present studies pH of most of the points was more or less in the same range & hence the concentration of metals in water does not appear to be pH dependent, as far as the difference between various sites are concerned. However the metals at high pH can sink down to sediments, (Dean et al - 1972).

Iron concentration in the river Krishna varied from 623 to 3425 µg/ Lit. against a world average of 670 µg/ Lit. (Turakin 1969). Ajmal (1985) and Wahid & Khurshid (1991) reported very high concentration of iron in the Yamuna River in U.P. & attribute it to sewage discharge. In present studies too, it appears that the high iron content at site 6 was only due to a heavy inflow of sewage. In the river Koyana iron concentration was much less than the river Krishna but the nullahs (especially Karad Nalla) had higher concentration which is again due to Sewage & Strom water.

Worlds mean concentration of Zinc is 20 µg/ Lit. Both the rivers had much higher concentration than this. Although the zinc appears to be naturally high in this area, waste discharge also contibutes significantly as zinc concentratron was quite high is nullahs meeting the River. Similar trend was found with ragard to manganese. Average Copper concentration in both the rivers varies from 26 to 58 µg/ Lit. against a world mean strem concentration of 7 µg/Lit. Considerably high concentration of lead was found in both the rivers (38 to 85 µg/Lit. against a world concentration of just 3 µg/Lit.) Ajmal et al. (1985) have reported concentration up to 510 µg/Lit in the Yamuna River. Such exceptionallly higher concentration of lead in the river is a cause of great concern. Nickel & Cobalt concentration in the river Krishna was much higher than Koyana and Cadmium was not detected from any of the rivers.

Various drains meeting the rivers constitute a major source of pollution with regard to heavy metals.

Following trend of occurence of concentration of various heavy metals in decreasing order was obtained.

RIVER KRISHNA :

Site No. 1	:	Fe>Zn>Mn>Cu>Pb>Ni>Co>Cd
Site No. 2	:	Fe>Zn>Mn>Pb>Cu>Ni>Co>Cd
Site No. 3	:	Fe>Zn>Mn>Pb>Cu>Ni>Co>Cd
Site No. 4	:	Fe>Zn>Mn>Cu>Pb>Ni>Co>Cd
Site No. 5	:	Fe>Zn>Mn>Pb>Cu>Ni>Co>Cd
Site No. 6	:	Fe>Zn>Mn>Cu>Pb>Ni>Co>Cd

RIVER KOYANA :

Site No. 1	:	Fe>Zn>Mn>Pb>Cu>Co>Ni
Site No. 2	:	Fe>Zn>Pb>Cu>Ni>Co
Site No. 3	:	Fe>Zn>Mn>Cu>Pb>Ni>Co
Site No. 4	:	Fe>Zn>Mn>Pb>Cu>Ni>Co
Site No. 5	:	Fe>Zn>Mn>Pb>Cu>Ni>Co

A comparision of the heavy metal concentration of these water bodies with the prescribed standards show that Iron, Manganese & Zinc concentration of all the water is much higher than the prescribed drinking water standards (W.H.O. 1984).

❑ REFERENCES

Ajmal, M., Khan M.A. And Nomani A.A. 1985. Distribution of heavy metals in water and sediments at selected sites of Yamuna River (India) *Environ Monit Assess* 5 : 205 - 214

Ajmal, M. Razziuddin & Khan M. A. 1987 Heavy metals in water sediments, fish and plants of River Hindon U.P. India, *Hydroibologia* 48 : 151-157

Dean, J. G. Bosqui, F.L. and Lononotlo, K.H. 1972. Removing heavy metals from wastewater. *Env. Sc. & Tech.* 6 : 518-522.

Mathur A., Sharma Y.C., Rupainwar D.C., Murthy R.C. & Chandra S. 1987. A study of River Ganga at Varanasi with special emphasis on heavy metal Pollution. *Poll. Res.* 6(1) :P37-44

Matzner A.V., 1977, Removing soluble metal from Waste. *Water & Sewage Works* 124:P. 98-101.

Nonberg H.W. 1984. The volumetric approach in trace metal chemistry of natural water and atmospheric precipitation. *Anayst. Chem. Acta.* 164, P.1-21

Pardo, R., Enriqual, B., Lourdes, P. and Marisol, V. 1990. Determination and specialization of heavy water metals in sediments of the Pisuerga river. *Water Res.* 24, (3): 373-379

Paul A. C. and Pillai K.C., 1983. Trace metals in a tropical river environment speciation & biological transfer. *Water, Air and Soil Pollution* P. 75-86, Reidal Publishing Co., U.S.A.

Purves D. 1985. *Trace element contamination of the environment.* Elsevier, Amsterdam.

Turekian K.K. 1969., The oceans, Streams & atmosphere In : *Hand Book of Chemistry* Vol. 1 Wedepoul K. H. (ed.)., Eerlin Weiddberg, New york, Springer, P.297-323.

Wahid, I. A. Nabi Ald., Hasan Q. and Naquis S. 1990. Heavy metal Pollution Studies in Western Uttar Pradesh, *Poll. Res.* 9, 57-61

Wahid, I. A. and Khurshid S. 1991, Distribution of heavy metals in Yamuna River Water & Sediments from Delhi to Allahabad. *Poll. Res.* 10(4), 215-222

W.H.O. 1971, *International Standards for drinking Water,* Geneva.

Trivedy R. K. 1990 and Khatavkar S. D. 1990. Heavy metals in the river Krishna, Maharashtra, India. *Asian Environment* 212 (2) : 67-75.

Trivedy R. K. 1990. *River Pollution in India,* Ashish Publishin House, New Delhi.

Authors :

1. **R. K. Trivedy :** Deptt. of Pollution Studies, Y.C. College of Science, KARAD-415 110, (India).
2. **S. D. Khatavkar :** Deptt. of Pollution Studies, Y.C. College of Science, KARAD-415 110, (India).
3. **B. L. Arjugade :** NFES Labs, Nehru Nagar, KARAD - 415 110. (India)

■

36

Impact of Oil-field effluents on The Dikhow River, Assam, India

❑ INTRODUCTION

Rudrasagar oil field (RDS-field) is confined within 26°57′N and 94°35′E in the South-western part of Sibsagar, Assam (India). Since its discovery in the early sixties exploration and exploitation of crude oil and gas is going on in full swing due to National endeavour to attain self sufficiency in oil production. The Dikhow river (a tributary of the mighty Brahmaputra) which rises from the Naga-Patkai hills ranges in the south, drains through this area and large number of oil installations (well sites and group gathering stations) of the RDS field is located on its either banks (fig.1). The river in this stretch, thus receives solid and liquid discharges drill cuttings, drilling mud and its additives, crude oil, oil field brines, untreated/paratially treated oil-contaminated waste water, either directly or through spillage, seepage or overflowing due to heavy rain besides aerosols generated from the non-stop flaring of oil and gas at the GGSs for the last 30-35 years. These solid and liquid discharges and gaseous emissions of the oil industry are reported to contain various toxic substances including heavy metal contaminants (Hofstadar et. al. 1976 Beckmann 1981 Ogden 1983, Callaghan 1191, Wahi et. al 1992). Entry of these objectionable substances into the river may degrade the water quality of the river which in turn may affect thousands of rural people living in the down stream side, which use this water for drinking as well as for agriculture. Although reports of degradation of the river water within RDS-field has figured in various local media, no attempt has yet been made to assess the impact of RDS-field activities on the river. Therefore, an invastigation of the water quality of the Dikhow river within RDS-field was called for. Thus the main objectives are:

1. To investigate the level of various water quality parameters in the river within RDS-field and see their variation with respect to time and space.
2. To investigate the heavy metal load of the river water and find the anomaly if any.

Their work will thus provide the most important background data base for future investigation.

Materials and Methods

Water samples in duplicate were collected fortnightly during the winter (Oct-Jan) and the spring (Feb-May) and monthly during the Monsoon (June-Sept) at six sampling points (s_1 to s_6) from the Dikhow river (fig.1) S_1 and S_2 are two points unlikely to be contaminated by the oil field effluents, S_3 is the point where the river recives oil contaminated waste water and aerosols from the GGSs located at Boliaghat, while S_4 to S_6 are three points down stream at about 0.5 km to 3 km gaps.

Water samples were taken in 1L polythene bottles and preserved according to standard methods (APHA standard methods 1989). For estimation of oil and grease and hydrocarbon contaminates water samples were taken in 56 glass bottles with ground glass stopper (Fresenius et al. 1988)

Analysis

Non conservative water quality parameters viz temp. Eh, pH, DO, COD, EC and TDS were measured at the collecting spot with the help of a portable kit. Other para meters were measured at the Laboratory within 48 hours using standard methodology (APHA standard methods 1989, Fresenius et. al. 1988). Metals (total) were analysed within a fortnight. Before analysis of metals, water samples were digested with HNO_3-$HCLO_4$(4:1). The digest after evaporating to dryness was taken in HCL. Metals in the solution were estimated by AAS using a parkin Elmer Model- 2380 instrument. Macro elements Na, K, Ca, Mg and micro elements Fe, Mn, Zn and Cr by flame technique, Cu, Cd, Co, Ni Pb, Graphite furnace technique after pre concentration by chelation-Extraction while As and Hg by arsine generation and cold vapour technique respectively. Correction for molar/ background absorbtion, ionisation and spectral interference were performed according to manufactures instructions.

RESULTS AND DISCUSSION

Results of analysis of water quality parameters and metallic contaminants are presented in tables 1 and 2 respectively. The value show sharp seasonal variation and water quality of the river is more or less safe expect for a small stretch near Boliaghat during the low flow period of winter (Oct-Jan) and spring (Feb-May).During these periods the river water near Boliaghat (S_3) shows slightly higher values of pH, EC, COD, alkalinity, hardness, Cl^-, S^{2-} oil grease and phenol along with Na, K, Ca, Mg, Fe, Mn and low value of Eh and DO. Than the natural background levels (S_1 & S_2). This indicates that water quality near Boliaghat is degraded. The cause of this degradation, might be, entry of untreated/partially treated waste water and aerosols generated from the nonstop flaring of oil and gas at the GGS located at Boliaghat. The presence of these contaminants in oil-field effluents have already been reported (Callaghan 1991, wahi el al.1992). Heavy metal load of the river is low and lies within their respective tolerance limits stipulated by ISI and WHO guidelines values,1984. At S_4 and subsequent points the river water is safe, indicating thereby sufficient self-purification

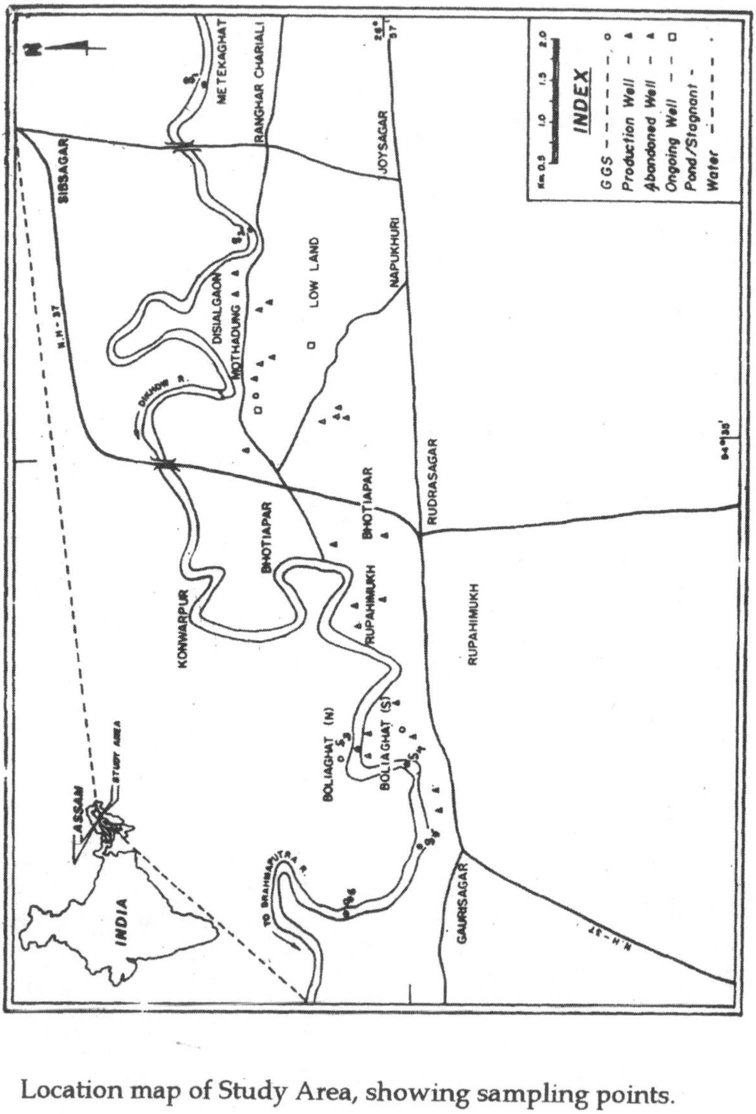

Figure 1. Location map of Study Area, showing sampling points.

capacity of the river.

During the Monsoon period (June-Sept.) values of various parameters do not show any change over the background values. This is due to large volume of water and highflow rate of the river.